T0312748

Telemedicine Technologies

Telemedicine Technologies

Information Technologies in Medicine and Digital Health

Bernard Fong
Providence University, Taichung City, Taiwan

A.C.M. Fong
Western Michigan University, USA

C.K. Li
Alpha Positive Clinic, Hong Kong

Second Edition

This edition first published 2020
© 2020 John Wiley & Sons Ltd

Edition History
John Wiley & Sons, Ltd (1e, 2010)

All rights reserved. No part of this publication may be reproduced, stored in a retrieval system, or transmitted, in any form or by any means, electronic, mechanical, photocopying, recording or otherwise, except as permitted by law. Advice on how to obtain permission to reuse material from this title is available at http://www.wiley.com/go/permissions.

The right of Bernard Fong, A.C.M. Fong and C.K. Li to be identified as the authors of this work has been asserted in accordance with law.

Registered Offices
John Wiley & Sons, Inc., 111 River Street, Hoboken, NJ 07030, USA
John Wiley & Sons Ltd, The Atrium, Southern Gate, Chichester, West Sussex, PO19 8SQ, UK

Editorial Office
The Atrium, Southern Gate, Chichester, West Sussex, PO19 8SQ, UK

For details of our global editorial offices, customer services, and more information about Wiley products visit us at www.wiley.com.

Wiley also publishes its books in a variety of electronic formats and by print-on-demand. Some content that appears in standard print versions of this book may not be available in other formats.

Limit of Liability/Disclaimer of Warranty
While the publisher and authors have used their best efforts in preparing this work, they make no representations or warranties with respect to the accuracy or completeness of the contents of this work and specifically disclaim all warranties, including without limitation any implied warranties of merchantability or fitness for a particular purpose. No warranty may be created or extended by sales representatives, written sales materials or promotional statements for this work. The fact that an organization, website, or product is referred to in this work as a citation and/or potential source of further information does not mean that the publisher and authors endorse the information or services the organization, website, or product may provide or recommendations it may make. This work is sold with the understanding that the publisher is not engaged in rendering professional services. The advice and strategies contained herein may not be suitable for your situation. You should consult with a specialist where appropriate. Further, readers should be aware that websites listed in this work may have changed or disappeared between when this work was written and when it is read. Neither the publisher nor authors shall be liable for any loss of profit or any other commercial damages, including but not limited to special, incidental, consequential, or other damages.

Library of Congress Cataloging-in-Publication Data

Names: Fong, Bernard, author. | Fong, A. C. M., author. | Li, C. K. (Chi Kwong), 1952- author.
Title: Telemedicine technologies : information technologies in medicine and digital health / Bernard Fong, A.C.M. Fong, C.K. Li.
Description: Second edition. | Hoboken : Wiley, 2020. | Includes bibliographical references and index.
Identifiers: LCCN 2020004340 (print) | LCCN 2020004341 (ebook) | ISBN 9781119575740 (hardback) | ISBN 9781119575757 (adobe pdf) | ISBN 9781119575771 (epub)
Subjects: MESH: Telemedicine–methods | Telemedicine–instrumentation | Medical Informatics Applications | Wearable Electronic Devices
Classification: LCC R858.A1 (print) | LCC R858.A1 (ebook) | NLM W 83.1 | DDC 610.285–dc23
LC record available at https://lccn.loc.gov/2020004340
LC ebook record available at https://lccn.loc.gov/2020004341

Cover Design: Wiley
Cover Images: Smart city and wireless © tcharts/Shutterstock, Telemedicine concept © Agenturfotografin/Shutterstock

Set in 9.5/12.5pt STIXTwoText by SPi Global, Chennai, India
Printed and bound in Singapore by Markono Print Media Pte Ltd

10 9 8 7 6 5 4 3 2 1

Contents

Foreword

Technology has come a long way since the publication of this book's first edition in 2011, such that capabilities of telemedicine systems have grown tremendously due to advances in information and communication technologies.

A wide range of healthcare services cannot be extended beyond hospitals and clinics without reliable telecommunication networks. Over the past few decades, advances in digital health and information technology have brought healthcare services to patients everywhere in a convenient and reliable manner. For example, a surgeon can now carry out a surgical operation away from the operating theater, while medical students can practice their surgical skills repeatedly without the risk of causing any harm to actual patients through augmented reality (AR) and virtual reality (VR), and a physiotherapist can monitor the progress of postsurgical rehabilitation anytime, anywhere. Technologies not only assist with medical practitioners and patients receiving treatment, perfectly healthy people can also benefit from a wide range of general health monitoring solutions that ensure optimal health can be maintained and any anomaly can be identified at the earliest opportunity through preventive care.

This second edition contains new information that offers readers a diverse range of possibilities through an introduction to telemedicine and its related technologies. Written by three experts in the areas of telemedicine, multimedia, and knowledge management, the book comprehensively covers many aspects of telemedicine applications that benefit both medical professionals and patients. It provides readers with fundamental knowledge in data communications without extensive mathematics, followed by a number of applicable areas, each with case studies in different areas of medical practice.

Nirwan Ansari
Distinguished Professor
New Jersey Institute of Technology

Preface

Telemedicine is the broad description of providing medical and healthcare services by means of telecommunications. Information and communications technology (ICT) in areas covering control, multimedia, pattern recognition, knowledge management, image, and signal processing has enabled a wide range of digital health and medical applications to be supported.

The combined effect of worldwide population growth and an aging population in most developed nations has given rise to a soaring demand on public health systems. The impact on national health systems in many countries has been further fueled by changes in lifestyle and environmental pollution. All these factors are stretching health systems to their limits, particularly under the constraints of limited healthcare resources. This is evident from the trend of chronic disease and obesity-related complications affecting younger people in recent years. The economic prosperity now enjoyed by many is a direct result of hard work by the previous generation and excessive consumption of natural resources, which may bring a range of problems for future generations. In response to all this, we should aim to take good care of the senior citizens who have devoted decades of work to today's prosperity. By the same token, we are working hard to enhance medical technologies to improve our health, and to provide a sustainable healthcare system for the next generation. Telemedicine forms the fundamental backbone in fulfilling our responsibilities for providing a diverse range of healthcare solutions to people of all ages.

There is an emergent interest among government authorities, healthcare service providers, academia, medical professionals, equipment manufacturers, and supplier industries to optimize the efficiency of providing a wide range of medical services in terms of both cost and time. The effective utilization of telemedicine and related technologies will be able to assist with, but is not limited to:

- Support more types of services.
- Bring services to more people in more regions.
- Make healthcare more affordable to the poor and older people.
- Optimize health for all ages.
- On-scene treatment for mobile medical professionals.
- Provide preventive care in addition to emergency treatment.
- Remote rehabilitation monitoring.
- Chronic disease relief and care.
- Ascertain service reliability and reduce human error.
- Safeguard patients' personal information and medical history.

To address the growing trend of telemedicine deployment in both urban and rural areas throughout the world, this book discusses different technologies and applications surrounding telemedicine and the challenges its implementation faces. This book also looks at how various signs of a human body are captured and subsequently processed so that they can be used for providing treatment and health monitoring. As conventional medical science tends to seek remedies according to symptoms, we explore how technologies in alternative medicine can go back to basics to address the root cause of many ailments by optimizing health in general.

Acknowledgments

First and foremost, the authors wish to thank all their readers for taking the time to learn more about telemedicine technologies. The authors are confident that this book will help readers to develop their expertise in further enhancing medical and healthcare technologies to benefit more people in the community. The main objective of telemedicine technologies has always been to extend medical services to more areas for more people so that people can live healthier for longer irrespective of where they are.

Over the years, the authors have seen numerous cases where people are unable to access healthcare either because they cannot afford it or because the service cannot be extended to their areas due to a number of reasons. The continuing advances of telemedicine technologies that break the geographical barrier of providing quality healthcare prompted us to write a book to share our insights together with the underlying technologies that can potentially benefit millions, if not billions, of people. Much of what we have learned over the years comes as a direct result of taking care of our retired parents as well as our delightful children, all of whom, in their unique ways, inspired us and, subconsciously, contributed a tremendous amount to the content of this book on promoting enhancement of telemedicine technologies to help people of all ages.

We also wish to thank the editorial team at John Wiley with their profusion of patience and talent, whose excellent work has led to a significant improvement on the presentation of this book.

Every effort has been made to trace rights holders. However, in case any have been inadvertently overlooked, the authors would be pleased to make the necessary arrangements at the very earliest opportunity.

About the Book

This book looks at the underlying information technologies providing telemedicine services. The text covers applications from traditional healthcare services to remote patient monitoring and recovery to alternative medicine and general health assessment for maintaining optimal health. It is primarily intended for readers ranging from medical professionals to final year undergraduate and first year graduate students in biomedical engineering or related disciplines. One of the book's main objectives is to help medical practitioners acquire fundamental knowledge in the technology behind the systems that help them with their work, and to serve as a reference for people who design and implement telemedicine systems. The text also provides detailed coverage of how technological advancements in high-speed wireless networking for the secure transmission of medical information may benefit both healthcare professionals and patients.

Book Overview

Chapter 1 is an introductory chapter that provides a general picture of what telemedicine entails and the importance of providing quality healthcare in various areas of medical practice with the aid of telemedicine technologies. The underlying concepts in various areas are summarized, and most of these will be elaborated on in more depth throughout the book. The technologies associated with individual applications would depend on their availability and specific regulatory limitations imposed by the respective authorities of a given country. Readers should be able to get a good understanding of how medical and information technology (IT) professionals are linked closely together through technological advancements. It is about how they help each other work better – and more importantly, how the general public improves the way they enjoy better health and medical services as a result of technology.

Chapter 2 provides technical coverage on what telecommunication technology is all about, and how it can be applied to better healthcare. Although this chapter primarily provides technical knowledge to readers, we do not go deeply into the engineering and mathematical aspects as the main scope of this book concerns technologies related to medical and healthcare applications. However, adequate knowledge will be provided to make use of underlying communications technology for healthcare. The chapter discusses what solutions are currently available and how to select the type of network most suitable for a given telemedicine application. Examples are given to demonstrate how each technology is applied. We also look at the harsh outdoor environment, where wireless communication systems are affected by various factors. The fundamental limitations of technology are dealt with, so what is and is not possible with technology is also discussed.

Chapter 3 looks at how life saving can be accomplished with technology developed for emergency rescue. It also looks at wireless communication systems used in remote patient monitoring. This is a particularly important application for servicing rural areas and older people. Such technology is also suitable for rehabilitation so that patients can recover at home with the assurance that they are being properly looked after even after they are discharged from the hospital. Various topics on body area networks (BANs) are also considered. These include different types of wearable monitoring devices, body sensors, data communication between devices, and practical difficulties faced.

Chapter 4 discusses the information theory behind the successful representation of various types of medical information with binary bits. It starts by looking at different ways of collecting data from patients; different applications require very different types of capturing devices. For example, measuring a person's heart rate and electrocardiograph (ECG) would require very different instruments. It also looks at what precautions are necessary for medical data transmission and storage, and covers the increasingly important area of artificial intelligence (AI) in various aspects of healthcare from medical training to preventive care.

Chapter 5 considers system deployment issues with examples of wireless telemedicine system development. It deals with a number of possible options and the importance of ensuring quality and reliability, something particularly important in critical life-saving missions. The concept of connectivity through the Internet of things (IoT) and cloud health management is also discussed.

Chapter 6 introduces the concept of information security and how to implement secure telemedicine systems for different applications. Patient privacy must be respected and any information collected needs to be safeguarded throughout the entire process from collection to analysis and subsequent storage. There have been reported cases of medical personnel losing removable storage devices, like USBs (universal serial buses) and memory cards, containing patient information. These irresponsible acts can be easily restrained by providing secure remote access to hospital staff. Any data collected for statistical analysis must ensure individual persons cannot be identified so that all such information remains anonymous. Since certain data needs to be shared between medical institutions and government agencies, a mechanism for maintaining data accuracy as well as anonymity is always crucially important. The chapter concludes with a look at the evolution of technologies related to biometric identification.

Chapter 7 introduces alternative medicine and may not be too relevant to certain medical professionals, although it is increasingly accepted as an effective way to treat prolonged illnesses such as colds and asthma. This chapter therefore aspires to give readers some background information on what alternative medicine entails and how information technology can be applied to serve the community better through practicing alternative medicine, often in conjunction with mainstream approaches. It also looks at an example of using biomedical databases for herbal medicine and acupressure aimed at treating patients who may require long-term treatment. The discussion then proceeds to using technology to optimize health, like progress monitoring in gyms or using an smartphone app when going on a short morning jog. Consumer healthcare products such as foot spas and massage chairs are becoming increasingly popular throughout the world. These products offer many novel features, including integration with existing audiovisual systems and other home appliances.

Chapter 8 addresses the issues of providing electronic healthcare from a user's point of view. This is considered important because as population aging becomes a more serious problem in most developed countries the demand for these services is expected to grow tremendously over the next few decades. Through the utilization of technology we will pay fewer visits to clinics and hospitals, and we will be better looked after. People living in rural areas will find this particularly helpful since not all remote small towns have medical facilities readily available at all times. Telecare becomes an important telemedicine application for providing easy access to healthcare for those with special needs. Although technology may not always reduce the risk of accidents occurring, we do have mechanisms for keeping an eye on people who need to be cared for so that necessary actions can be taken without delay should a mishap occur. In addition to providing special care for older people and those with special needs, we also look at how technology can help people recover from sports injuries. Some exercises may facilitate speedy recovery yet improper movement can worsen the affected area. So, technology that monitors the rehabilitation progress can therefore be very helpful for those struggling to recover from injuries.

Chapter 9 takes an in-depth look at spinoffs from the ever-growing list of wearable devices brought to us through the administration of devices, circuits, and systems that in turn provide numerous opportunities for digital health applications. The concept of consumer healthcare and medical devices is discussed in providing round-the-clock care anywhere in an affordable and efficient manner. The emerging topic of using transcutaneous electrical nerve stimulation (TENS) and electrical muscle stimulation (EMS) for rehabilitation is also introduced.

Chapter 10 presents an overview of smart and assistive technologies based on a telemedicine framework. It considers examples of deployment from a patient's point of view that include smart home healthcare and smart clothing. It also looks at improving treatment with an example on optimizing medication delivery through media such as hair treatment and smart pills.

The book concludes with Chapter 11 and a brief summary of the possibilities in telemedicine for the foreseeable future. The chapter explores cases like learning support for medical and nursing students as technology makes training healthcare professionals easier and more efficient. We also explore other emerging technologies for telemedicine advancements, such as haptic sensing by conducting various tasks through touch, and what future telemedicine and digital health technology have to offer, how the transition from 4G to 5G mobile communication technology changes the way patient-centered health services develop, and finally the benefits telemedicine brings to people of all ages.

1

Introduction

1.1 Information Technology and Healthcare Professionals

The history of modern telemedicine goes back to the traditional telephone about a century ago. Medical advice was given by physicians over the telephone. The term "telemedicine" is very simply a description of supporting medical services through the use of telecommunications. The prefix *tele* comes from the Greek for "distant." So, "telemedicine" literally means providing medical services over distance. Telecommunications used in medical applications can be categorized as sending medical information between a pair of transmitter and receiver. The "medical information" can be as simple as a doctor providing consultation from sophisticated data captured from a human body. "The Radio Doctor," which first appeared in the *Radio News* magazine (*c.*1924), is perhaps the earliest documented case of utilizing telecommunication technology for medical application. Although information technology (IT) has been used in healthcare since then, it was the first scientific literature formally addressing the application of technology in medicine that appeared. (Moore et al. 1975).

IT advances over the past decades mean a wider range of healthcare services can be supported. Indeed, the types of services that can be supported are so vast that any book which makes an attempt to provide comprehensive coverage of all areas will most likely contain thousands of pages over several volumes. This book aims to provide in-depth coverage of how wireless communications and related technologies are used in medical services. We will also look at the challenges and limitations of current technology associated with healthcare information systems.

We begin by taking a look at how simple wireless communication networks function and what a telemedicine system consists of. We look at a number of examples that describe how a primitive system supports healthcare services. Over the course of the book, more sophisticated systems will be described in more detail.

This chapter aims give readers an overview on how IT is widely used in assisting healthcare without going into any technical depth. To begin our discussion, we revisit the term "information technology," something often associated with computer science. Essentially, it is extensively interpreted as a blend of computing and telecommunications. This leads to the acronym ICT, which stands for information and communications technology, also known as "infocomm" for short. All these are merely descriptions of the use of technology to securely and reliably transmit information between two or more entities. IT is widely used in many areas that influence our daily life, for example banking, transportation, manufacturing, etc. This list is seemingly endless. When we see information technologies support so many things that we use on a daily basis, it will not be difficult to understand how widely it can be used in supporting healthcare and medical applications.

Telemedicine Technologies: Information Technologies in Medicine and Digital Health, Second Edition.
Bernard Fong, A.C.M. Fong, and C.K. Li.
© 2020 John Wiley & Sons Ltd. Published 2020 by John Wiley & Sons Ltd.

The IT industry has never quite recovered from the "dot-com bubble" bursting in 2000, which saw the industry lose much of its value for several years. And the IT sector was similarly hit by the financial crisis of 2008/09, triggered by the collapse of the subprime mortgage market. Put simply, although IT systems are widely used in many aspects of daily life, the industry remains tied to the global economic market, and so will always be affected by any changes to it. In contrast, health and medical services are two of the few domains that are in consistent high demand, simply because illness is not market led: it is experienced by everyone on the planet to a lesser or greater degree. For this simple reason, healthcare naturally becomes an essential part of daily life that will continue to be in high demand for many years to come.

Having realized the prime importance of healthcare, we go further into how IT is applied to healthcare and medical services. Long before the evolution of IT, herbal medicine practitioners millennia ago were already using the most primitive form of information exchange mechanism, namely communication system, to convey messages on medical services. Wang et al. (1999) documented a case where Shen Nong made use of information exchange for the treatment of respiratory syndrome as far back as 2735 BCE. This may not be the first case but it is certain that medicine and communications have been linked together for over 4000 years. As IT became more sophisticated over time, a more diverse range of medical services could be supported. To name a few, IT in medicine involves drug prescription, spread of pandemic modeling, patient monitoring, remote operation, medical database, and so on. This is by no means an exhaustive list and we cover these as well as many others throughout the book.

Obviously, healthcare professionals can make use of IT advancements in different areas. Advantages brought by IT include improvement in reliability, efficiency, precision, ease of information retrieval, accomplishing tasks remotely, and better organization. Healthcare therefore becomes more readily accessible and more efficient. We will look at how technology benefits healthcare professionals, with the assumption that readers have very little prior IT knowledge and know virtually nothing about the underlying technologies.

1.2 Providing Healthcare to Patients

In addition to facilitating medical practitioners perform their tasks, providing healthcare services to patients is also an important issue to address as they are the end users who must feel comfortable to receive the treatment given. Providing a technically feasible solution is not the only obstacle to deal with. Other important issues, including patients' acceptance and accessibility, must also be addressed. We strive to look at providing healthcare solutions to patients using IT from the perspectives of both providers and patients. End users, particularly children and older people, may not be too keen on accepting technology as a tool for healing. Convincing patients of the benefits of IT for healthcare may involve liability, security, and privacy issues. For example, in the case of monitoring or tracking a patient recovering at home, the patient must be assured that personal information is securely kept and no such information accessed in any way without consent.

With the number of senior citizens increasing steadily over the past decade, there has been a growth in demand for healthy aging care (Colby and Ortman 2014). Assistive care provides numerous opportunities for supporting independent living through smart home integration (Bonaccorsi et al. 2015). A comprehensive range of customized solutions for the care of older people has been made possible through advances in IT and digital health.

Before leaving the topic on providing care to older people, it is worth briefly noting the advantages brought to this group of users by telemedicine technology. As population aging is becoming more significant concern in many countries, it can widely be expected that more care and monitoring will be needed. A significant increase in the application of wireless communications in care for older people has been seen over the past few years as related technologies have matured. The cost of service becomes more affordable and portable devices become smaller and more user-friendly. As pervasive computing technology advances, more comprehensive and automated services will become available to the aging population in the years to come (Stanford 2002). The design of interconnected devices and sensors on the patients' side must ensure non-obtrusiveness and can be comfortably worn. Also, users' movement will not be restricted and reliability will not be affected irrespective of wearing condition. User-friendliness is another important design factor, as absolute minimal training should be necessary, especially for children and older people. These should be genuine plug-and-play devices. In this sense, the healthcare system in the patient's home can be installed by a technician during initial deployment. Thereafter, almost everything should be fully automatic, except for unavoidable scheduled maintenance such as battery replacement and calibration.

Let's elaborate more on a patient's point of view as an end user. The primary objective of telemedicine is to provide medical services remotely. Among numerous advantages brought to patients by telemedicine, an obvious convenience is reducing the need for clinical visits. Through the utilization of IT, a patient can rest at home while receiving full medical attention. Reviewing the level of medical support provided over the past two to three decades, it can be seen that IT has certainly provided tremendous benefits to the general public as a whole. The advancement of faster computers and more efficient bandwidth usage has allowed more types of services to be extended to more users. For example, a few decades ago a simple request for medical advice could be obtained by finding a fixed line telephone and dialing in to the clinic where a physician was stationed. With the availability of mobile voice over Internet protocol (VoIP) technology, one can now simply pick up a mobile phone and place a video-enabled call to a physician; the physician does not necessarily have to be situated inside the clinic in order to provide advice. This is just one among numerous examples where IT advancements have made healthcare more readily available. More such examples are presented throughout the book.

While the benefits to patients are obvious, there is a wide range of challenges that different parties face in order to serve patients. These concern people from developers, practitioners, and healthcare management and authorities. The following highlights challenges that different people face, starting from the initial planning stage to the final rollout and continuing maintenance.

From the IT perspective, the fundamental question is feasibility. The primary consideration is whether current technology is capable of doing something. After this comes practicality and cost effectiveness. We begin by considering an example where schoolchildren are to enroll into a program that ensures their school bags are ergonomically prepared so that there will be no issue with back pain. The advantage to participating children is very obvious because the program should ensure that they will not suffer from any back pain. However, how viable is the entire program? We need to understand more about the technology involved in order to answer this seemingly simple question.

In this case study, we have the following parties involved: engineers developing the monitoring system, clinical staff analyzing the captured data, funding bodies providing necessary resources, children participating in the study, and finally participants' parents giving consent to their children's involvement. Below, we look at the case with respect to benefits and concerns from each party's standpoint.

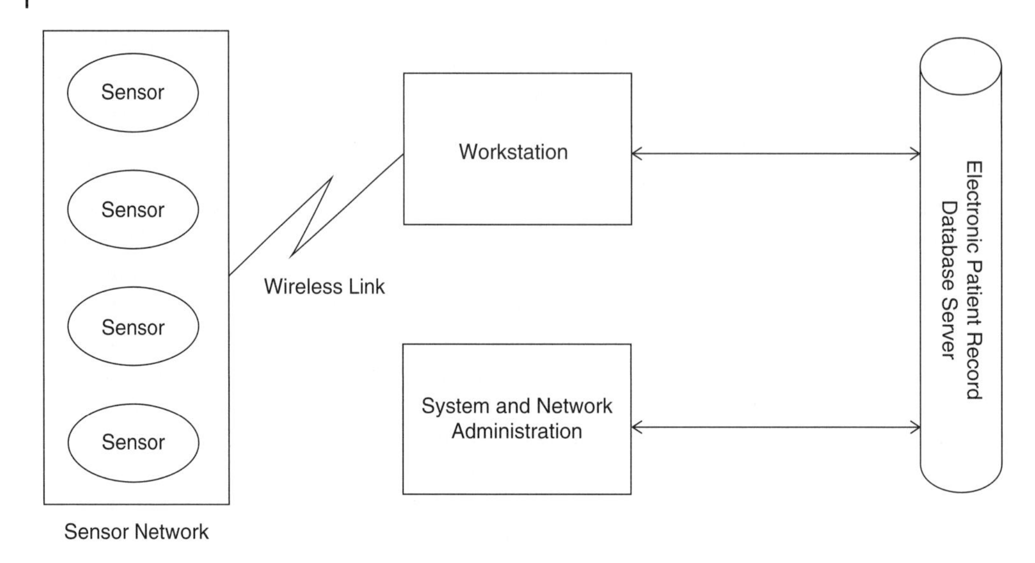

Figure 1.1 A simple telemedicine system.

1.2.1 Technical Perspectives

Biomedical engineers need to develop a system based on requirements specified by clinical staff, such as that illustrated in Figure 1.1, with the necessary sensors and data communication network. This simple system has a number of sensors forming a sensor network for capturing different types of information about a patient. It is linked to the system for analysis by a workstation and storage in an electronic patient record (EPR), and is monitored by the necessary system and network administration tools. In this discussion, we shall not go into the technical details while giving an insight into what is involved. Engineers analyze this by evaluating technical feasibility and practicality. Digging deeper into the technical challenges, one obvious issue to address is how to ensure whatever captured data is meaningful. There are several factors that influence the validity of data, most notably from what the sensors have picked up followed by what has been transmitted and subsequently received. In this respect, the sensors must be securely attached at the relevant points of the participant's body, and each sensor must be sensitive enough to pick up any subtle tilting of the body while not being too sensitive that it picks up any vibration from other sources. Having dealt with these problems, next we must ask whether the sensors are suitable for the specific application – the size may be too large for attaching onto a child, for example – and whether it will cause any discomfort. Are readings affected by any physical obstacles that may be separating the children from the backpack, such as clothing? How is captured data sent out for processing and analysis? Will sensors interfere with each other if placed too close together? These are just some of the questions related to sensors that need to be dealt with.

We shall proceed by assuming that the sensors are well designed and the problems listed above have been overcome. So, we are technically able to capture a set of valid data that tell us something about a child's behavior when carrying a backpack. We now look briefly at how telemedicine is utilized in a biosensor network; we will come back to this with more details in Section 3.5. In the previous paragraph we raise a question about how the captured data is sent out. Essentially, we have two choices: using wireless communications or connecting the sensors with wires. How they compare will depend on the system itself as there is no clear advantage with either option. This is one major topic that we cover throughout this book.

Briefly summarizing the discussion here, we have seen how many questions need to be addressed in relation to the deployment of such supposedly simple health monitoring systems. So, although the system may appear simple enough to patients, its design and implementation may not be as straightforward, and there are many limitations.

1.2.2 Healthcare Providers

Healthcare professionals should understand that technology is available for making their routine work easier and safer. Many may still prefer traditional methods, just like many people still prefer jotting down notes using pen and paper. Others may find technologies helpful when using a personal digital assistant (PDA) or tablet computer for the same purpose. There are, of course, many advantages with a tablet, although users may need to familiarize themselves with its user interface. Another concern to some is the risk of losing its stored data due to breakdown, or unauthorized access of data if backed up on a cloud server. We can see that people who are used to conventional ways of carrying out a task may need to be convinced of the associated benefits technologies bring, in order to impel them into learning to utilize these technologies. So, as a practitioner, a simple-to-use interface would be a fundamental design requirement. User experience (UX) developers should involve professionals throughout the design process. The entire process should be as automated as possible while maintaining a very high level of reliability. Different applications may have very different demands. For example, telesurgery requires ultra-high precision for control and crystal clear imaging details with no time delay, whereas teleconsultation may have much less stringent requirements.

Although technical advancements have made, and will continue to make, current technologies more efficient and fault-free, and hence do and will enable numerous tasks to be accomplished quicker and more reliably, the incentive to use them may not be compelling unless practitioners actually know how to use them. Getting used to something new, especially for critical tasks, can be a major challenge. A uniform change to new technology for all applications would be vitally important for a swift switch to make good use of the new technology.

1.2.3 End Users

The end users of the system are the patients. The term “patient” refers to someone who receives medical treatment or a medical service, which includes routine check-ups. We should clarify at this point that by definition a person described as a patient may not necessarily be unwell. A perfectly healthy person can be referred to as a patient in this regard. Here, in our case study we have a group of patients who participate in the study of schoolbags on children. They help with the study by having a set of sensors attached to their back while carrying a schoolbag of varying weights. An illustration on how the sensors are attached to the back of a patient is shown in Figure 1.2. We discuss the case study from a patient’s point of view by first looking at Figure 1.2. As shown, a number of sensors are attached to the back; each sensor is connected to a data capturing device by a wire. Movement is somewhat affected by the wires so we can readily see the advantages of using wireless sensors as far as the patient is concerned. So, why not wireless? This example exhibits three major technical challenges that make wires extremely difficult to eliminate. First, sensors attached to a child’s back must be very small. Powering the sensors can be an issue as installing an internal battery may be a problem. Also, wave propagation issues effectively rule out its use between the body and the bag, as absorption would be a very significant issue. Finally, measurement accuracy given the physical separation of individual sensors and the amount of movement would make a

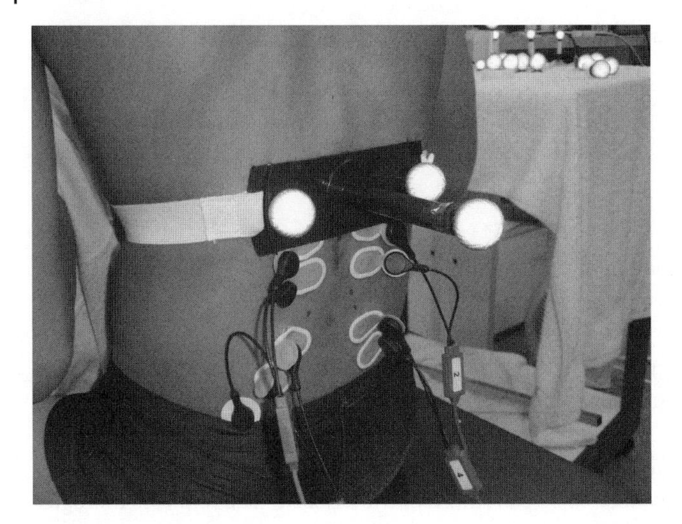

Figure 1.2 Sensors attached to the back of a patient using wires that severely restrict the patient's movement.

wireless solution impractical. For all these reasons, patients have to bear with the wires surrounding them while participating in the experiment.

1.2.4 Authorities

Funding agencies and authorities are most concerned about cost effectiveness. Long-term benefits to the community must be clear. In this particular case study, obtaining funding may be difficult despite all the benefits stated in the above subsections. This is primarily due to the projected time length of realizing the benefits; this will only be seen when a clear statistical trend of reduction in back pain is attained. The political details are far beyond the scope of the book so we will not discuss anything in detail here. As a general rule, those looking to acquire funding for projects on applying technology to healthcare services, by and large, need to prove that the benefits will be immediately realized. And this explains why there is generally a lack of financial support for healthcare solutions using innovative technology.

1.3 Healthcare Informatics Developments

In this section, we look briefly at how healthcare and bioinformatics have evolved over the past decades. Medical science has undergone consistent advancements for thousands of years and IT is certainly a much newer topic that has only really commenced from the first computer by Konrad Zuse (1936). Soon after the birth of computers, information storage devices were born. Health informatics is made possible only when computers are connected together to form a network, hence computer networking. The whole idea of health informatics kicked off after World War II as technology became more readily accessible. These networks provided a framework to link hospitals together in the cyberworld. More recently, computational intelligence has made a wide range of services available. In fact, by combining with multimedia and technological advancements, healthcare has been enhanced greatly. As illustrated in Figure 1.3, a diverse range of medical and healthcare services can be supported by technology.

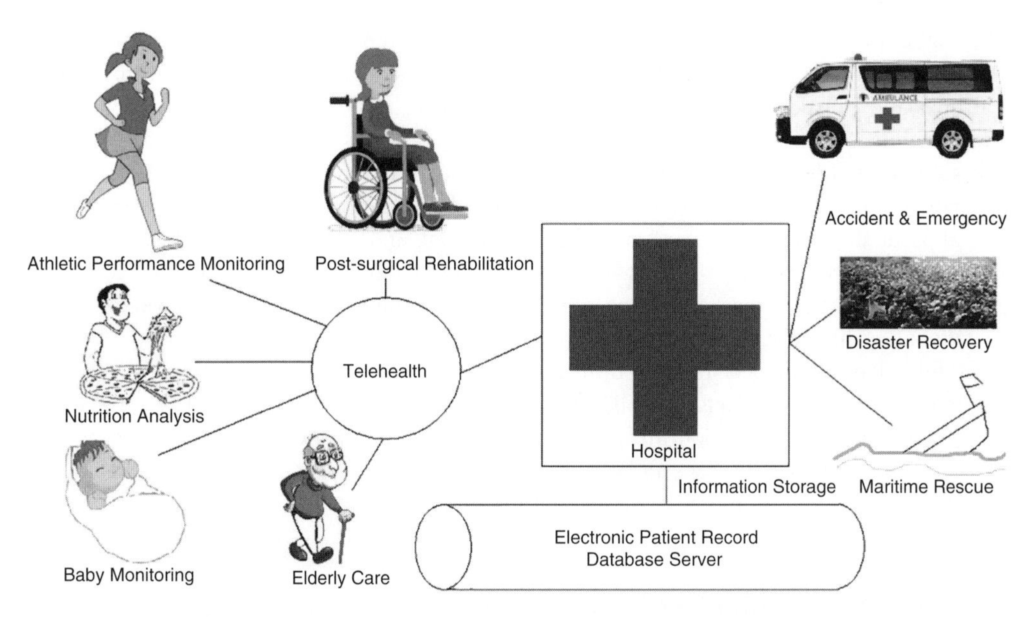

Figure 1.3 Some examples of digital health applications supported by telemedicine.

So, the eight decades since *The Radio Doctor* appeared have seen the blend of technology with medicine in just about all areas of practice. We have very briefly covered how health informatics has evolved from the first computer; we shall pay more attention to more recent developments that are directly related to possible future developments. The first challenge that many people will talk about are probably security and privacy issues. There are cases of patient information being leaked because of a wide range of reasons from breach of security to simply loss of storage devices. A significant part of health informatics involves ensuring the security of data keeping. This includes protection from theft or altering of information and policies in ensuring that data will not be misused by parties authorized to access patient records. A thorough discussion on security and privacy is presented in Chapter 6. In addition to assurance for safeguarding medical data and privacy, there are many other issues to address since health informatics entails a very wide range of topics in linking people, resources, and devices together, and many of these have developed independently over time. The first documented case of modern healthcare informatics deployment in the United States was around the 1950s in a dental project pioneered by Robert Ledley for the National Bureau of Standards (now the National Institute of Standards and Technology) (Ledley and Wilson 1965). More medical information systems were developed over the next few years across the USA and most projects were advanced independently of each other. It was therefore practically impossible to develop standards for health informatics systems. The International Medical Informatics Association (IMIA) was formed in 1967 with the main objective to coordinate the development of health informatics and related technological advancements. Soon after its formation came the programming language MUMPS (Massachusetts General Hospital Utility Multi-Programming System) for building healthcare applications which is still used today in electronic health record systems. There was soon a need for different variants of the programming languages to run on different computer platforms and a standard was inaugurated in 1974. It is now developed as "Caché" for medical application development on different computer platforms. It is worth noting that, although Caché is still currently used today, many present electronic health record systems are developed using relational databases.

The traffic flow in terms of passengers and cargo internationally has seen substantial growth. partly because of the synergistic effect between the lower cost of air travel with an influx of budget airlines and global e-commerce (Bigne et al. 2018). This has facilitated the flow of emerging pathogenic microbes across countries (To et al. 2015). The increasing international travel of passengers and animal products can rapidly import emerging or re-emerging infections into a region and export them to other parts of the world. The huge population and density of metropolitan cities, in particular, provide an ideal incubator for brewing and spreading new infectious agents and antimicrobial resistance. Thus this growing issue of infectious diseases and control of emerging and re-emerging microbes should be controlled and monitored through the development of international health informatics strategies.

So, a quick look at the development of healthcare informatics reveals that a vast collection of topics in IT are involved. It deals with all aspects of technologies related to preventive caring, consultation, treatment, rehabilitation, and monitoring. From this point onwards, we shall concentrate our discussion on communications and networking technologies for healthcare. Related technologies will also be covered from time to time as appropriate.

1.4 Different Definitions of Telemedicine

Telemedicine, the combination of ICT, multimedia, and computer networking technologies to deliver and support a wide range of medicine applications and services, has several widely accepted definitions. The definition given in *wiki* is: "Telemedicine is a rapidly developing application of clinical medicine where medical information is transferred through the phone or the Internet and sometimes other networks for the purpose of consulting, and sometimes remote medical procedures or examinations." This definition is simply a brief recapitulation of what is described in Section 1.5 below. Other definitions exist. For example, the Telemedicine Information Exchange (Brown 2005) gives its own definition as the "the use of electronic signals to transfer medical data from one site to another via the Internet, telephones, personal computers, satellites, or videoconferencing equipment in order to improve access to health care"; and (Reid 1996) defines telemedicine as "the use of advanced telecommunications technologies to exchange health information and provide health care services across geographic, time, social, and cultural barriers."

Variations of definitions do not stop here. The Telemedicine Report to Congress (Kantor and Irving 1997 gives "telemedicine can mean access to health care where little had been available before. In emergency cases, this access can mean the difference between life and death. In particular, in those cases where fast medical response time and specialty care are needed, telemedicine availability can be critical. For example, a specialist at a North Carolina University Hospital was able to diagnose a rural patient's hairline spinal fracture at a distance, using telemedicine video imaging. The patient's life was saved because treatment was done on-site without physically transporting the patient to the specialist who was located a great distance away."

Among these various definitions, there are several points in common, including that they were all given in the mid-1990s and are closely related to providing different kinds of medical services over distance by utilizing some form of telecommunication technology.

We mention what telemedicine is at the beginning of the book. Very briefly, it is about the use of telecommunications and networking technologies for the transmission of information related to medical and healthcare application. In modern telecommunications, information can be transmitted across many types of networks in many forms. By definition, telemedicine can be as simple as

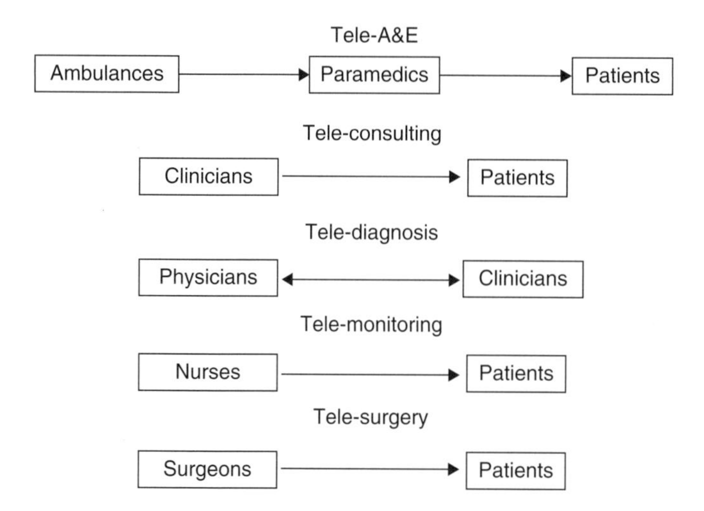

Figure 1.4 Subsets of telemedicine connecting patients to medical professionals.

two doctors talking about a patient over the telephone or as complex as a sophisticated global hospital enterprise network that supports real-time remote surgical operations with surgeons situated in different parts of the world controlling an operation that takes place in one hospital simultaneously. To elaborate on the vast coverage of telemedicine, Figure 1.4 summarizes a number of services that telemedicine is capable of supporting. It is not a complete list of all services that telemedicine is capable of supporting while showing all major services currently used worldwide. As we begin looking at these services, it is not difficult to see that there is one thing in common: conveying medical information from one entity to another. Before we proceed further, remember this is an introductory chapter so don't worry about the technical terms and details, as we shall cover them thoroughly throughout the book. Obviously, each application entails different types of information. We look at each of these examples and see what telemedicine does. A simple application like teleconsultation involves the delivery of advice, often verbally from an expert to people in need of medical information. In recent years this has extended to services using mobile devices. Telediagnosis lets experts carry out diagnostics with medical instruments from a remote location, quite simply by providing a communication link between the two locations. Telemedicine can be far more complex than this, like sophisticated tele-A&E (accident and emergency) services that can involve high resolution digital images along with vital signs of a patient collected in a remote location that must be transferred to the hospital with maximum reliability and minimum delay. Some systems may provide additional features such as video conferencing functions and the real-time retrieval of medical history records. Likewise, telemonitoring facilitates monitoring of patients recovering at home or moving around in locations away from the hospital by transmitting different types of data. Depending on the specific application, remote patient monitoring may involve the attachment of small wireless biosensors on a patient forming a body area network (BAN), where data captured by individual sensor is collected within the BAN before being sent collectively for subsequent processing. In this kind of situation, a telemedicine system may include different types of communication networks. While we cover networking in more depth in Chapter 2 with specific emphasis on the following telemedicine applications in Section 2.4, we refer to the example in Figure 1.5 to get some understanding of how three separate networks are interconnected to form a telemedicine system. Here, the patient under observation is surrounded by a BAN which the patient carries when moving around. Captured data captured is sent to a nearby local area network (LAN) that stores and

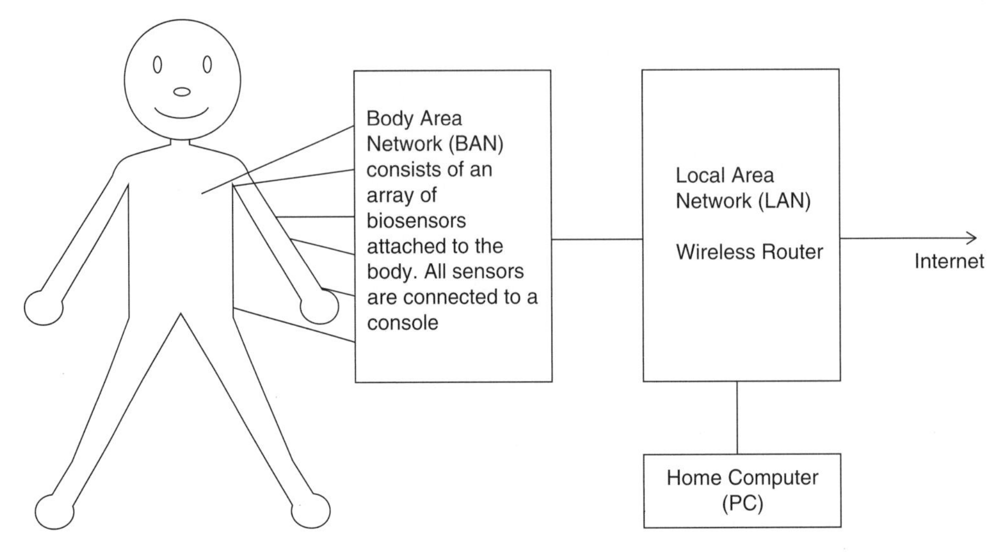

Figure 1.5 Simple network connection from the patient's body to the outside world for a wide range of digital health services from general health assessment to remote diagnosis and disease management.

processes the data. The LAN effectively serves as a bridge between the hospital that is served by the metropolitan area network (MAN) and the patient's home. The LAN is very simply an ordinary home network that is permanently installed at the patient's home. Through installation of appropriate equipment associated with the BAN and establishing a connection to the hospital via the MAN, a telemedicine system that performs telemonitoring can be set up.

Telesurgery is probably the most convoluted application, partially because of the precision involved. In order to perform a surgical operation from a remote location, apparatus must have a very high degree of movement in all directions and an unobstructed view must be delivered to the surgeon with good clarity. Therefore, the following basic requirements must be fulfilled to perform even a simple operation:

- Sensors capable of capturing the slightest movement of a surgeon's hand in real time with extreme precision.
- Cameras that can deliver incredibly sharp images of the patient without any obstruction. This is particularly challenging as movement of surgical tools must be taken into consideration. Maintaining a good view of the patient at all times is vitally important.
- Actuators that exactly replicate three-dimensional hand movements as interpreted by the sensors with no time delay.
- A communication network that is fast enough to deliver all types of data in both directions, and reliable enough to ensure that it is free of transmission errors throughout the entire operation.

By now we should be convinced that telemedicine entails technologies far more than simply POTS (plain old telephone system) that allow two medical professionals to share information verbally. In the later chapters we look at more telemedicine applications and the underlying technologies that make telemedicine possible. In particular, we will take a close look at how artificial intelligence (AI) and augmented reality (AR) change the way telerobotic surgery is changing how a surgeon practices.

Connecting people and resources together for better healthcare covers more than the examples given above. We have described ways that the general public can directly benefit from telemedicine;

there are other applications, such as connecting relevant authorities worldwide to track the spread of diseases in epidemiological surveillance, that had been found to be effective in limiting the crisis caused by severe acute respiratory syndrome (SARS) and avian influenza (bird flu) over the past decade and the Zika virus in more recent years (Chan et al. 2016). Another less obvious yet important application in safeguarding the community is telepsychiatry where psychiatrists are able to monitor acutely anxious patients so as to proactively prevent violent crimes using telemedicine.

Telemedicine covers almost all aspects of daily life. For example, we can easily access healthcare information by the touch of a 4G LTE cellular phone; getting nutrition information for a healthy diet while dining out has never be easier. Throughout the book we will see telemedicine virtually support all aspects of healthcare in daily life for consumers with a portable device such as a smartphone, tablet, or notebook computer.

1.5 The Growth of E-health to M-health

We all know what the Internet is, and almost certainly we access the Internet on a daily basis. It is widely understood that the Internet allows email access, video conferencing, information retrieval from websites, contents downloading of music, video clips, pictures, etc. The evolution of the Internet provides information sharing with worldwide coverage. In essence, long sequences of binary bits 1s and 0s are carried across the world, trillions of them a second. Although only two possible states are sent in the digital world, combinations of these can represent virtually anything that one can imagine. The Internet is about integrating devices and information together. In the cyberworld information can travel across any part of the world in a fraction of a second. To get a better understanding of how advances of Internet technology support telemedicine we must first look at the Internet's development from its birth and what it offers telemedicine.

1.5.1 Evolving from the Internet

The origin of the Internet was likely the Galactic Network documented by (Licklider and Clark 1962). We can see that telemedicine has a far longer history than the Internet yet the impact of the growth of the Internet on advances in telemedicine is very significant. This includes connecting computers and devices together, along with the development of packet switching (Kleinrock 1961), and the eventual evolution of networks capable of carrying different types of data to be delivered across a single transmission medium. With such a capability, communication networks can support telemedicine in many areas, such as:

- Reliability: quality of service (QoS) assurance
- Information sharing: medical web pages online
- Audio: teleconsultation, respiratory, cardiac, and pulmonary sounds
- Still images: X-ray, scans, medical images
- Video images: teleconferencing, telepsychiatry, medical education
- Databases: electronics patient records, e-pharmacy, alternative medicine
- Vital signs: electrocardiography (ECG), electroencephalography (EEG) analysis, and storage.

The Internet, in its early days, supported primitive services such as BBS (Bulletin Board System) and email. These were fairly adequate for teleconsultation services. It was not until 1984 that the Internet incorporated TCP/IP (transmission control protocol and the Internet protocol), which supports multimedia data traffic.

Today's "modern" Internet, which supports all types of telemedicine services described above, has not removed all of the various challenges that need to be considered (or dealt with) in the continued development of telemedicine. Interestingly, a computer virus that spreads across the Internet may in certain aspects replicate epidemiological control that we touch on in the previous section. A computer virus is defined as a program that interferes with a computer's normal operation. There are many ways viruses can spread across the Internet. Very commonly, they are transmitted as email attachments. Viruses can be disguised in various forms as embedded in other programs or files such as pictures and video clips. They can also be concealed in illicit software just like a human carrying hepatitis virus who looks healthy from the outside. It is well known that antivirus utilities can be installed on a computer to safeguard it from virus attacks. Telemedicine can actually do something similar in preventing bacterial and viral infections and spreading by proactively tracking down the pattern of the spread as well as the mutation of viruses using signal processing techniques.

1.5.2 Digital Health on the Move

The development of wireless communications technology allows more flexible deployment of telemedicine for supporting off-site applications. More recently, with the advancements of related technologies such as batteries and antennas, wearable devices have been made widely available for many medical and consumer healthcare applications. These open up numerous opportunities to new telemedicine services in providing treatment and preventive care as data can reach almost anywhere while the patient is on the move. Further, internal memory can store the collected data offline so that the monitoring device does not even have to stay connected at all times to support continuous health monitoring.

Now we have seen a few examples of different telemedicine application types that use the Internet, we need to ask what is really needed for supporting telemedicine. Unlocking the potentials of what telemedicine can offer entails a thorough understanding of various technologies, from ICT to medicine. The Internet does appear to support unlimited data flowing to virtually anywhere in the world. Of course, this perception is not exactly correct. The Internet, or storage facilities, can become saturated when too much data is dumped onto it. Further, processing requirement to extract relevant information from a large dataset also poses challenges to computational resources.

A good example of such limitation is revealed when we used to be pleased with the video quality that DVD brought to us before the turn of the millennium, this was soon followed by the HD (high definition, 720p where "p" stands for progressive scan) and FHD (full high definition at 1080p). The 4K ultra high-definition (UHD) format has been commercially available for several years and we are moving into the 8K era that is expected to be launched in time for the 2020 Tokyo Olympic Games (Miki et al. 2015). From the video's point of view the picture resolution gets better all the time while each iteration requires higher data throughput, more data storage, more efficient algorithm for data compression, and subsequent decompression. It is not difficult to understand why as of 2019 an affordable 4K UHD TV is readily available yet finding a good selection of 4K videos is not as easy. What we see in the video market here is that the amount of information that we can receive practically and economically can be limited by one or two of many factors that yield quality enhancement. In the same context, it is theoretically possible to monitor a wide range of health parameters of a patient continuously. What we need to consider is what relevant data is useful for helping the patient; how accurate is the useful data; how difficult it is to maintain the desired precision; how often is a reading taken; and how long does it take to capture, analyze, transmit, and store the useful data. So, the key point here is that we need to understand the limitation of what we collect and how we handle the data.

1.5.3 Data is Sent as a Sequence of "Packets"

For a start, telemedicine is about healthcare across the entire world. It does not necessarily mean all medical knowledge should be made available there. Flooding the networks with information will cause it to slow down and malfunction, eventually causing data loss. The Internet must be used in a responsible way since it is a shared medium. Minimizing overheads is therefore an important task for telemedicine system developers. Determining what kind of information should be sent requires an understanding of data composition. Data is sent across the Internet as "packets," a packet is a unit of binary bits sent from the source to the destination. Figure 1.6 illustrates a simplified structure of a typical data packet that is sent across the Internet (Mullins 2001). It shows that only a portion of the packet contains the actual information that needs to be delivered. The remaining bits are "overheads" that facilitate the transmission of information. Very similar to sending a letter through the postal system, we put the piece of paper that contains our actual message into an envelope, and the envelope contains basic information like the sender's address (source location), destination address (recipient location), airmail label (delivery method), and postage stamp (class of service). The protocol defines the delivery method, and the type of service marks the class of service. Finally, we also have the actual information. In addition, checksum is used for checking data integrity upon receipt, and additional services similar to registered or courier post are also available in the digital networking world. Certain communication protocols provide guarantees for successful delivery, and different QoS schemes can be set to prioritize data traffic across the network.

So we see that a data packet contains far more than the actual information. However, we need to bear in mind that we cannot change the way data is structured as it is necessary to comply with applicable standards for data transmitting across the Internet (both IPv4 and IPv6). What we need to do is to ensure that telemedicine services, especially when utilizing the Internet, should incur minimal overheads. We revisit the topic of transmission efficiency shortly in the next chapter. In summary, what we have seen in this section is that the growth of the Internet provides us with a platform for popularizing telemedicine services with more sophisticated applications possible. There is a need of ensuring that what is sent across the Internet is carefully chosen.

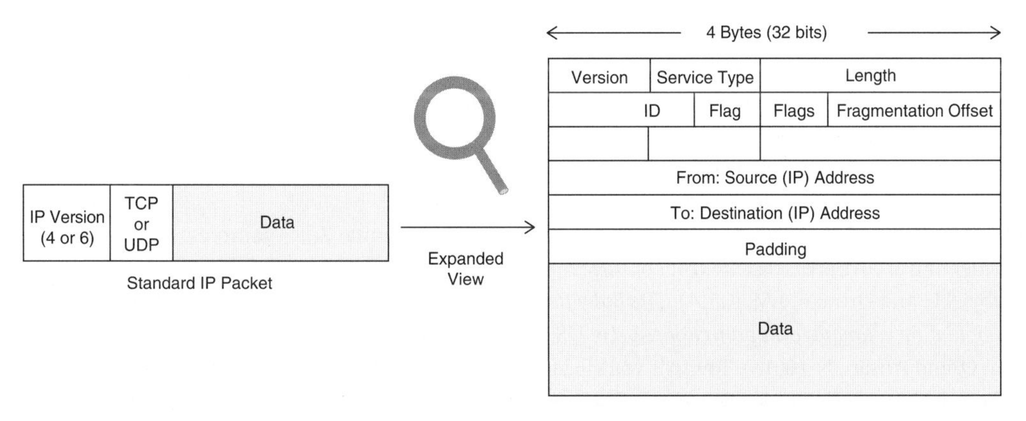

Figure 1.6 Simplified structure of a typical data packet that contains the actual health-related information along with redundancies such as origin/destination address, timing, and additional data checking codes.

1.6 The Connected World Between Human and Devices

Devices and machines are manufactured to work for people. The concept of the Internet of things (IoT) links "things" together via the Internet with a high degree of mobility (Sun and Ansari 2016). "Things cover basically everything connected together like sensors, devices, systems, consumerables, patients, medical, and technology professionals, etc. IoT is about connectivity and processing that provides a platform for telemedicine system to connect patients with their surrounding environment and digital health records, all readily available to their caregivers. Common consumer electronics devices such as smartphones and smartwatches can be wirelessly linked to a wide range of sensors and to an IoT network to acquire comprehensive health information about a patient, including physiological information, daily activities, environment, and potential hazards. Integration of patients and devices through the IoT can support decision making for treatment and recommendations for preventive care.

IoT is a key driver of digital health transformation across the medical industry. Low cost components such as sensors and radio frequency identification (RFID) tags are used in vast quantities in healthcare management of both patients and inventory (Manzoor 2018). Recent advances in IoT provide new opportunities in customized health management, assistive home care, smart pills, and medication management systems.

"Fog" computing is an emerging technology that distributes the load of processing toward to the edge of the network (Fan and Ansari 2018). The main advantages of using fog computing include minimizing latency (the delay of data transfer), network bandwidth conservation (efficiency), and optimal data processing with better analysis and insights patient state of health, all made possible through IoT integration.

Before leaving this introductory chapter, we should briefly return to data security. As the Internet is a shared medium, we should be reminded of the risk of security breach, as anyone can access the Internet. Telemedicine demands the highest standard of data security, both in terms of information accuracy and patient privacy. This is covered in Chapter 6.

References

Bigne, E., Andreu, L., Hernandez, B., and Ruiz, C. (2018). The impact of social media and offline influences on consumer behaviour. An analysis of the low-cost airline industry. *Current Issues in Tourism* 21 (9): 1014–1032.

Bonaccorsi, M., Fiorini, L., Cavallo, F. et al. (2015). Design of cloud robotic services for senior citizens to improve independent living and personal health management. In: *Ambient Assisted Living* (eds. B. Andò, P. Siciliano, V. Marletta and A. Monteriù), 465–475. Cham, Switzerland: Springer.

Brown, N.A. (2005). The telemedicine information exchange: 10 years' experience. *Journal of Telemedicine and Telecare* 11 (2 suppl): 7–11.

Chan, J.F., Choi, G.K., Yip, C.C. et al. (2016). Zika fever and congenital Zika syndrome: an unexpected emerging arboviral disease. *Journal of Infection* 72 (5): 507–524.

Colby, S.L. and Ortman, J.M. (2014). *The Baby Boom Cohort in the United States: 2012 to 2060: Population estimates and projections*, 1–16. US Census Bureau.

Fan, Q. and Ansari, N. (2018). Towards workload balancing in fog computing empowered IoT. IEEE Transactions on Network Science and Engineering. https://web.njit.edu/~ansari/papers/18TNSE.pdf (accessed 13 January 2020).

Kantor, M. and Irving, L. (1997). Telemedicine Report to Congress. http://www.ntia.doc.gov/reports/telemed/index.htm (accessed 13 January 2020).

Kleinrock, L. (1961). Information flow in large communication nets. PhD thesis. Massachusetts Institute of Technology. https://www.lk.cs.ucla.edu/data/files/Kleinrock/Information%20Flow%20in%20Large%20Communication%20Nets.pdf (accessed 13 January 2020).

Ledley, R.S. and Wilson, J.B. (1965). *Use of Computers in Biology and Medicine*. New York: McGraw-Hill.

Licklider, J.C.R. and Clark, W.E. (1962). On-line man-computer communication. In: *Proceedings of the Spring Joint Computer Conference*, 113–128. ACM.

Manzoor, A. (2018). RFID in Health Care-Building Smart Hospitals for Quality Healthcare. In: *Health Care Delivery and Clinical Science: Concepts, Methodologies, Tools, and Applications*, 839–867. IGI Global.

Miki, Y., Sakiyama, T., Ichikawa, K. et al. (2015). Ready for 8K UHDTV broadcasting in Japan. IBC 2015, Amsterdam (10–14 September 2015).

Moore, G.T., Willemain, T.R., Bonanno, R. et al. (1975). Comparison of television and telephone for remote medical consultation. *New England Journal of Medicine* 292 (14): 729–732.

Mullins, M. (2001). Exploring the anatomy of a data packet. *TechRepublic*. http://articles.techrepublic.com.com/5100-10878_11-1041907.html (accessed 13 January 2020).

Reid, J. (1996). *A Telemedicine Primer: Understanding the Issues*. Billings, MT: Innovative Medical Communications.

Stanford, V. (2002). Using pervasive computing to deliver elder care. *IEEE Pervasive Computing* 1 (1): 10–13.

Sun, X. and Ansari, N. (2016). EdgeIoT: Mobile edge computing for the Internet of things. *IEEE Communications Magazine* 54 (12): 22–29.

To, K.K.W., Zhou, J., Chan, J.F.W., and Yuen, K.Y. (2015). Host genes and influenza pathogenesis in humans: an emerging paradigm. *Current Opinion in Virology* 14: 7–15.

Wang, C.K., Wang, Z., Chen, P. et al. (1999). *History and Development of Traditional Chinese Medicine*. IOS Press.

Zuse, K. (1936). *Konrad Zuse's First Computer: The Z1*. Germany.

2

Communication Networks and Services

Communication networks provide support for a very wide range of healthcare and medical services. Telemedicine uses various types of networks so that physicians can share ideas, surgeons anywhere in the world can perform a single operation together irrespective of where the operating theater is, and nurses and paramedics can retrieve patient records anytime, anywhere. Hospitals and clinics use networks for everything from patient care to administrative work and inventory management. In this chapter, we learn about the fundamentals of telecommunication technology, with an emphasis on wireless networking since most telemedicine applications require the flexibility that wireless networking provides.

2.1 The Basics of Wireless Communications

To understand how telemedicine works, we must learn about fundamental telecommunications theory. Telecommunications is about the delivery to or exchange of information between different entities. The most primitive communication example is perhaps two people talking to each other, where the voice that conveys information is transmitted through the air and reaches the ears of the person who listens. Any communication system would consist of a transmitter (sender), receiver (recipient), and a channel (the path the information passes along), as illustrated in Figure 2.1. Here is how it works. The transmitter sends out information $s(t)$. The notation $s(t)$ is a function of time meaning the information content varies with time. For simplicity, we may interpret this as the "sent" information at a given "time." This passes through the communication channel and the receiver is presented via the channel with $r(t)$, the "received" information at a given time. This sounds simple enough. It is logical to think $s(t)$ and $r(t)$ are identical. However, in practice this is almost always not the case.

Unfortunately, the channel causes degradations such as additive noise, distortion, attenuation, etc. Before we proceed further, let's briefly explain what these terms mean. "Additive noise" is something that is induced to the information, and eventually becomes part of the information. In a way, additive noise is added to the original information sent as causing contamination. When two people talk, the person who listens may hear other background noise from different sources. Distortion is the warping of the information, causing the information to be altered. It should be noted that the effects of distortion are often considered as some form of noise. "Attenuation" is the weakening of a signal over the distance traveled; the intensity decreases as it propagates away from the sender and may eventually fade out completely. We shall discuss more degradation factors in this chapter. Having established the fact that information received is highly unlikely to be identical to what is sent, let us redraw the basic communication system to that of Figure 2.2. This block

Telemedicine Technologies: Information Technologies in Medicine and Digital Health, Second Edition.
Bernard Fong, A.C.M. Fong, and C.K. Li.
© 2020 John Wiley & Sons Ltd. Published 2020 by John Wiley & Sons Ltd.

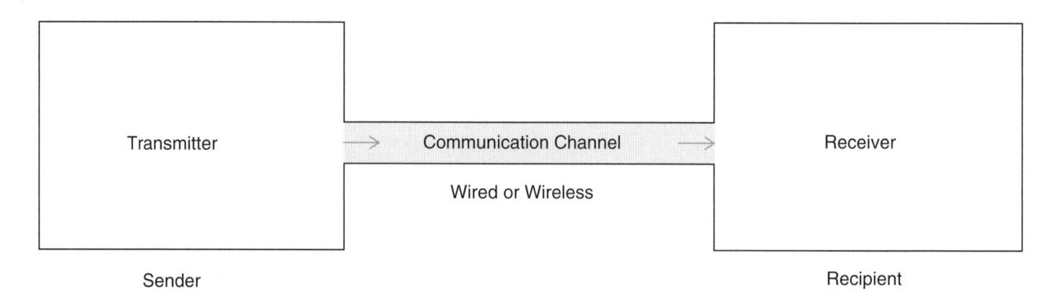

Figure 2.1 Block diagram of a basic communication system that consists of three key entitles: sender, receiver, and channel.

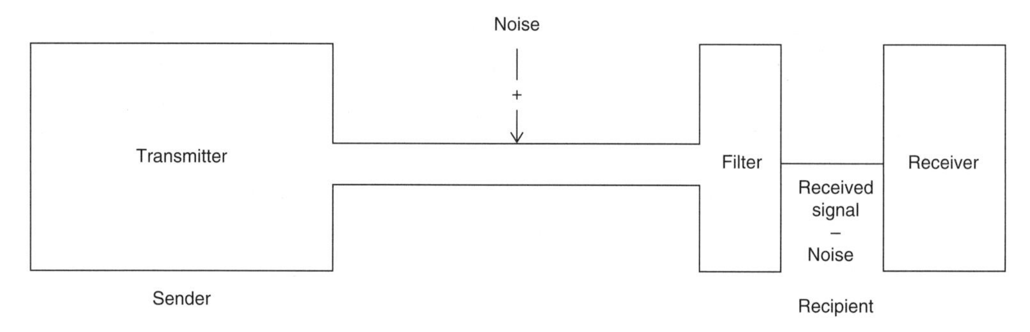

Figure 2.2 Communication system under the influence of noise.

diagram shows that noise is added along the channel. This does not necessarily mean noise cannot be induced at the transmitter or receiver. Here, we can write a simple expression to describe the process of communication:

$$r(t) = s(t) + n(t) \tag{2.1}$$

n(t) can take many different forms, the one thing in common is that it will degrade the received information quality. In severe cases, corruption will be so great that the information cannot be correctly interpreted by the receiver. For the sake of completeness a filter is added to remove the noise, but its effectiveness can vary significantly in different systems under different situations.

The distance over which information is transferred in a telemedicine system can be as short as a few micrometers within a device or even within an integrated circuit (IC) chip, or thousands of kilometers across continents. The channel can take the shape of copper conductors having physical connection between the transmitter and the receiver or can be "wireless" over the air. Regardless of what the channel is, maximizing the transmission speed is always of great concern, since more information can be conveyed within any given time period. This is similar to operating a bus where the bus company would like to maximize its utilization by having as many passengers as possible; there would be little difference to the operation between carrying 5 and 50 passengers. By the same token, a given communication channel should carry as much information as possible. Shannon (1948) describes how noise can affect the maximum transmission speed of a communication channel. We do not intend to go deep into Shannon's information theory, but an extract of his landmark work is worth mentioning to understand the effects on telemedicine performance.

Before leaving this overview of communication systems, it is timely to introduce the term "transceiver," as it will appear throughout the book. It describes a device that can simultaneously act as a TRANSmitter and a reCEIVER, hence combined together forms the word "transceiver." The words "transmitter" and "receiver" are frequently abbreviated as Tx and Rx, respectively.

2.1.1 Wired vs. Wireless

Wireless communication systems have been gaining popularity as a direct result of technological advancements that have effectively solved numerous reliability and security issues that have traditionally confined the use of wireless technology in low-cost critical applications. Mobility and convenience are undoubtedly driving factors for opting for wireless. Although both wired and wireless communications are widely used throughout the world, a comparison is given here.

Wired communications have been used for over a century. Following the invention of telegraphy in the mid-nineteenth century, the invention of telephony began when A.G. Bell and E. Gray worked on the first telephone that made use of a microphone to pick up a person's voice and a speaker that reproduced the voice. The audio signal picked up was transferred through wire that connected two telephones together. This formed the basis of using electric wire for telecommunications. Even before this, wired communication appeared as early as 1794 when C. Chappe started sending telegraphs visually through a line of sight (LOS) communication channel. The meaning of "LOS" should be quite self-explanatory. It simply means the receiver can "see" the transmitter, unobstructed. This means that, if you sit on the receiving antenna, the transmitter's antenna should be in sight either with the naked eye or through binoculars, depending on the distance separating the transmitter and the receiver. However, radio LOS is slightly broader than visual LOS because the radio horizon extends beyond the optical horizon as radio waves follow slightly curved paths in the atmosphere.

Combining the two ushered in the beginning of optical communications when J. Tindall discovered around the 1870s that light followed a curved water jet as it was poured from a small hole in a tank. This led to the idea of keeping traveling light within a curved glass strand (Hecht 2004). The works of these inventors formed the basis of wired communication technology that evolved over a century to support a wide range of services. Currently, wired technology is so reliable that it can easily provide at least 99.999% reliability, that is it has a failure rate of no more than 0.001% of the time or less than 5.5 minutes/year. We compare the two major types of wires for communication, namely electrical conductors and fiber optic cables in Section 2.1.2.

The commencement of wireless technology dates back to 1887, almost as early as the first telephone (see Garratt 1994). This was when D.E. Hughes and H. Hertz began generating radio waves using a spark gap transmitter. Such underlying technology formed the basis of radio broadcasting by pioneers M. Faraday and G. Marconi at the end of the nineteenth century. Over three decades after the first radio came television broadcasting in the 1930s followed shortly by commercially licensed television stations introduced in Pennsylvania and New York in 1941, long after the first electromechanical television appeared in Germany in 1929 (Sogo 1994). So far, radio and television broadcasts were both one-way communication systems, known as "simplex" communications.

Two-way radio communication was used during World War I but commercial use became popular only after World War II. Although N.B. Stubblefield held a US patent for his wireless telephone in 1908, cellular phones only became widely available from the early 1980s, when the FCC (Federal Communications Commission) approved the Advanced Mobile Phone System (AMPS). Up till now the perceived advancement of wireless communications may not be obvious to end users since these merely let users talk to each other verbally without any added features. "2G" GSM (Global System for Mobile Communications) was launched in Europe in 1991 and has supported text messaging since 1993. Shortly after, 2.5G and 3G came with an array of new features such as MMS

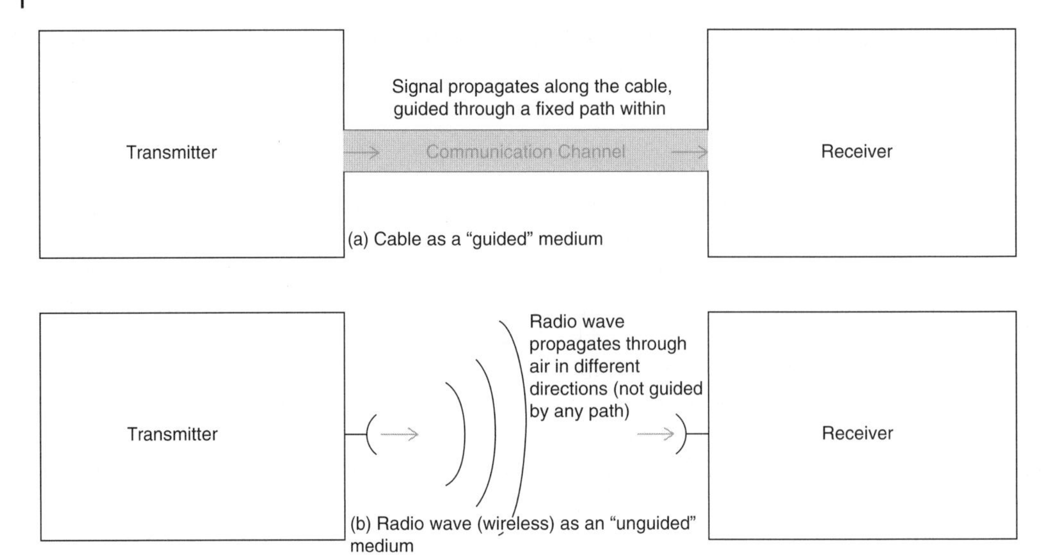

Figure 2.3 Guided versus unguided transmission medium, which describes whether the path which data travel across is bounded to a certain "track."

(Multimedia Messaging Service), video call, Internet surfing, just to name a few. We see how fast wireless technologies have advanced over the past decade or so. It is all about "speed." Section 2.1.3 discusses all we need to know about transmission speed.

Communication technologies have been evolving all the time with the introduction of Li-Fi (light fidelity, not to be confused with the popular Wi-Fi that we have all been using very widely over the past decade) and PLC (power-line communication). The former utilizes visible light communications (VLC) that literally travels at the speed of light and the latter sending data over existing power cables without additional wiring. Although they are fundamentally different in that Li-Fi is wireless whereas PLC is wired, one major advantage that they both share is the ability to operate through existing infrastructure since lighting equipment and power cables are things that already form an essential part of daily life.

So, wired and wireless technologies have evolved over a century and are now both well-established technologies. An interesting point to note is that wired and wireless are classified as "guided" and "unguided" media, respectively. Figure 2.3 explains the two types self-explanatorily. Information traveling across a cable is "guided" through a fixed path (namely the cable itself), whereas wireless communication does not have a fixed guidance for where the information can travel through, hence the description "unguided." Briefly summarizing their respective merits and drawbacks in telemedicine, wired communication is more reliable and cheaper for short length deployment, whereas wireless communication provides the convenience of high mobility and deployment flexibility. Wireless communication is a preferred option in most telemedicine applications because of the requirements for mobility: no one wants a clump of wires tangled all over the body!

2.1.2 Conducting vs. Optical Cables

While we have said mobility is an important decisive factor for the dominance of wireless communication in telemedicine applications, it is still important to learn the basic properties of metal conducting cables and fiber optic cables, because they are still needed in certain areas, such as

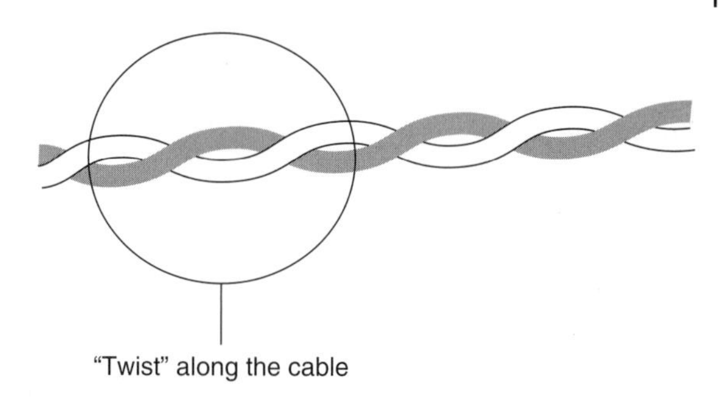

Figure 2.4 Within a twisted pair cable, a pair of wires are literally twisted together.

network backbone or connection between fixed devices. In this section, we look at how these cables convey information, and compare the properties that make them suitable for certain applications.

We briefly discuss the operation of metal conducting cable by looking at a "twisted pair" cable, which is illustrated in Figure 2.4. It shows two insulated wires twisted together in a helical structure. This is a type of copper cable commonly used in computer and telephone networks. The way they carry information is very simple: a certain voltage represents a logic "1" and another voltage level represents a logic "0." The exact representation depends on the specific encoding mechanism used, but for the sake of discussion, we may assume a positive voltage denotes a "1" while the lack of voltage (0 V) represents a "0." In this context, carrying information is simple: the cable simply carries a voltage that alternates between a positive voltage and a 0 V when transmitting a sequence of 1s and 0s.

Optical communications work in a very similar way. Looking at the illustration shown in Figure 2.5, a light beam travels through the center core when a "1" is transmitted. In contrast, the lack of light represents a "0." So, the light beam that comes out of the end of a fiber optic cable will be successions of on and off. Of course, the switching is far too rapid to be seen by the human eye, hence it may appear as always on. The cable can be bent, so there must be some kind of mechanism for retaining the light within the cable's core. Figure 2.5 shows a cladding that surrounds the center core. It is a highly reflective material that reflects the light back into the core and prevents it from escaping.

In both cases, 1s and 0s are transmitted across a cable by the presence or absence of a signal. It is important to mention at this point that what goes on behind the scenes may not be as simple, but the above discussion does illustrate how transmission is accomplished. Before we leave the discussion on cables, we should look at some major types of commonly used cables in wired telemedicine networks. Another type of metal conducting cable is the coaxial cable. It is no longer popular with telemedicine applications but warrants a brief mention as this type of cable appears in many places, most notably in TV antennas and decoding boxes. Its main feature is the center core conductor in much the same structure as the fiber optic cable, having another group of metal conducting strands surrounding the center conductor separated by an insulator. The main disadvantage of this type of cable is its bulkiness. There are also other types of wiring, like the very simple setup of a couple of

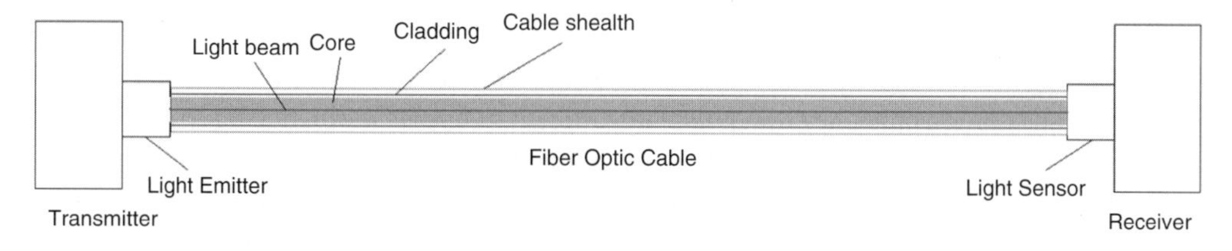

Figure 2.5 A simple fiber optic communication system.

wires running in parallel. For fiber optic cables, two major types are glass and plastic fiber, with the major difference being a tradeoff between performance and cost. In general, the former supports higher transmission rates and is more reliable, whereas the latter is usually cheaper per unit length.

2.1.3 Data Transmission Speed

"Bandwidth," which determines the amount of information a given channel conveys, is a vitally important term to understand regarding any aspect of communications. The bandwidth of any given channel is fixed. As a rule of thumb, higher bandwidth supports higher data rates. Since the bandwidth of a given transmission medium is fixed, it may be possible to increase the data transmission rate by stuffing more bits into one "baud." "Baud" is defined as a count of the number of changes of electronic states per second. For example, a copper cable of 1 K baud changes the voltage 1000 times a second. An important point to note is that it does not necessarily mean it only carries 1000 data bits per second. To illustrate this we will look at some of the mathematics that informs it, although we do not intend to go too deeply into proving the concepts.

In each baud, or a change of signaling state, there is a certain number of different signal levels L. An example would be voltage levels like 0.5–1.0 V. Combinations of binary bits can be assigned to these different levels, e.g. each level represents two bits such that 0.5 V represents "01" and 1.0 V represents "11." The number of bits n per baud has a simple relationship of:

$$n = \log_2 L \tag{2.2}$$

or:

$$L = 2^n \tag{2.3}$$

So, in this particular example we have two bits ($n = 2$) and use four different levels ($L = 4$) each representing: "00," "01," "10," and "11." By using more different signaling levels, more bits can be carried per baud, hence the data transmission rate (or bit rate), measured in number of bits per second or bps can be increased for a given fixed baud rate.

Bandwidth is a very important term used when describing the data transmission rate that a given channel supports. It refers to the band of frequencies an electronic signal occupies when transmitting data across the channel. Therefore, the bandwidth of a given channel is measured in hertz (Hz) as often the difference between the maximum frequency and the minimum frequency used. For example, a telephone channel that transmits voice data between 300 Hz (minimum frequency) and 3400 Hz (maximum frequency) has a bandwidth of 3.1 KHz. So, what is the relationship between the channel bandwidth and data transmission speed?

Nyquist theorem states that the bit rate R_b of a channel of bandwidth H is:

$$R_{b=2 \cdot H\ log2\ L} \tag{2.4}$$

This is, of course, the maximum data transmission rate that a channel can theoretically achieve. There are many factors that cause an actual communication channel to have a lower bit rate than this.

Remember, earlier we said that more bits can be carried by each change of signaling state to improve the transmission efficiency by using more different levels. However, having more different levels means signaling levels are squeezed closer together. For instance, in the example above each step is 0.5 V. Instead of representing two bits, we use eight levels to represent four bits per levels. We may reduce the separation between levels from 0.5 to only 0.25 V. The single most important problem here is noise, which may cause the signaling levels to overlap. The noise level N corresponds to the minimum separation between two levels before the noise causes error to cross the boundary of the adjacent level. The maximum number of levels L is given by:

$$L = \sqrt{\frac{S}{N} + 1} \tag{2.5}$$

where S is the maximum or peak signal power level. In general, the maximum data transmission rate R_b is directly proportional to peak signal power S, and inversely proportional to channel noise N. A communications system should provide the highest possible transmission rate at the lowest possible power under minimal noise. The above gives us some background theory about data transmission speeds.

2.1.4 Electromagnetic Interference

One major drawback of wireless communications is electromagnetic interference (EMI), since EMI effects are far more problematic than with conducting cables. This is particularly risky in healthcare applications because wireless transmitting devices can severely affect the operation of some delicate medical instruments. Tikkanen (2005) reports various ways of combating EMI effects in healthcare applications. Among various solutions available, providing proper shielding by the use of appropriate housing material for medical instruments can effectively safeguard the device from picking up unwanted interference. Many composite materials can be used for this purpose. Metallized plastic materials are suitable in housing many types of devices as they can be thermoformed into virtually any shape and are considerably lighter than most metal alloys while providing shielding effectiveness comparable to that of metals. There are three potential problems that cause EMI: source radiating noise, receiver picking up noise, and coupling channel between source and receiver.

All wireless transmitting devices are vulnerable to EMI from nearby radiating sources. These include laptop computers and cellular phones operating in the surrounding area. Such interference usually affects electronic circuits by causing capacitive coupling, meaning that energy is charged up inside the circuit. Here, a changing electric field is generated that can be capacitively coupled to nearby equipment. There are two major types of EMI, namely continuous and transient interference. The former is caused by the emission of radiation consistently by nearby sources such as other transmitting devices or medical instruments. The latter is intermittent such that sources radiate short duration energy. These can be caused by thunderstorms triggering lightning electromagnetic pulse (LEMP) or switching of high current circuits. Standards concerning the regulation of EMI are primarily handled by the International Electrotechnical Commission (IEC), while the Comité International Spécial des Perturbations Radioélectriques (CISPR, directly translates to "International Special Committee on Radio Interference") deals with radio interference related issues. It is, incidentally, worth noting the "CE" mark that commonly appears in electronic products, including healthcare and medical equipment. "CE" signifies Conformité Européenne (or "European Conformity" in English). Products bearing the CE mark indicate conformance of European Directives that require Electromagnetic Compatibility (EMC) tests to be conducted to ensure a given product complies with the European Union (EU) directive 2004/108/CE before it is permitted to be sold in any member states of the EU (93/42/EEC for Medical Devices).

2.1.5 Modulation

Before concluding our discussion on telecommunication fundamentals, we should look at the term "modulation." It refers to a process where a "carrier signal," which provides the necessary energy for the information to be delivered to the receiver, is altered in some way according to the information to be carried. This is essentially a procedure of stuffing data into a signal for transmission. The method of altering certain parameter(s) of the carrier signal is changed to represent the data. For example, in an FM (frequency modulation) radio broadcast the carrier signal's frequency is changed in relation to the voice information. Such frequency variation would be interpreted by the receiver (radio) as the voice carried over. In its primitive form, parameters that can be changed include the amplitude (the signal level), frequency (number of oscillations per second), or phase (the signal's relative position to time). In more complex modulation schemes, more than one parameter can be changed simultaneously so that more information can be represented per baud, thereby increasing the transmission efficiency. In general, the higher the spectral utilization efficiency (SUE), the more the receiver's electronic circuit structure complexity is required as resolving between different possible states of the signal becomes more difficult. SUE is a measure of how efficient a modulation scheme is to carry a certain amount of data for a fixed bandwidth.

2.2 Types of Wireless Networks

Wireless communications have been developed to such an extent that numerous options exist, and different network types are optimized for different applications, with coverage ranging from a few meters to thousands of kilometers. In this section, we introduce some commonly used networks in telemedicine applications and explain why they are suitable for specific situations. Key properties are summarized in Table 2.1.

2.2.1 Bluetooth

This technology provides short range coverage primarily for mobile devices connected in an ad hoc network called a "piconet" within a room. Key selling points are the low cost requiring simple circuitry and low power consumption. Its flexibility for connecting between devices in close proximity may pose a threat of spreading computer virus. Bluetooth uses adaptive frequency hopping (AFH) to reduce EMI by detecting other devices in the spectrum, and hops between 79 frequencies at 1 MHz intervals, so as to avoid the frequencies nearby devices are using. Bluetooth technology is overseen by the Bluetooth Special Interest Group (SIG). There are currently three classes covering around 3, 30, and 300 m.

Although it is widely seen in hands-free units of cellular phones, it is useful for small wearable biosensors, owing to low power (1 mW for 3 m or 10 ft, Class 3) and a simple, low-cost transceiver.

Table 2.1 Properties of some common wireless communication systems.

Network Type	Operating Frequency	Speed	Band	Maximum Range
Bluetooth	2.4–2.485 GHz	3 Mbps	Unlicensed ISM	300 m
IR	100–200 THz	16 Mbps	IR-B	5 m
Wi-Fi	2.4–5 GHz	108 Mbps	ISM to U-NII	100 m
ZigBee	900 MHz	256 Kbps		10 m
Cellular Networks	850–1900 MHz	20 Mbps		5 km
WiMAX	10–66 GHz	75 Mbps;		40 km
LMDS	10–40 GHz	512 Mbps		5 km

2.2.2 Infrared (IR)

The IR wave sits between the microwave and visible red light of the spectrum. A considerable amount of IR radiation is emitted from the sun and is usually associated with heat. In fact, approximately an equal amount of IR and visible light hits the earth's surface from the sun. So, what does it have to do with communications and healthcare? On a related front, IR detection is widely used in night vision, which is imperative in search and rescue. A popular example of IR in wireless communications is remote control for home appliances. When we pick up the remote control to adjust the volume of a stereo, the controller emits an IR signal that carries the instruction to the stereo's sensor.

IR is classified into three different categories by the International Commission on Illumination (CIE), of which near-infrared (IR-A) is used in night vision applications; whereas wireless communications usually use short-wavelength infrared (IR-B). It is worth noting that IR-B is widely used in long range optical communications, although we will not go into the details here, as we are looking at wireless networking here. IR wireless standards are governed by the Infrared Data Association (IrDA) for devices that use the successive "on" and "off" of an IR light emitting diode (LED) for communication. At the receiver, a silicon photodiode converts the received IR pulses to an electric current replicating the sequence of "on" and "off." It is a very mature technology that has been used for decades and is very easy to implement with virtually no interference issues, although it does not have the ability to penetrate walls. Another major issue is that it requires direct LOS and the transmitter must be aligned fairly close to the center of the sensor with only $\pm 15°$ offset possible. Although current IrDA compatible devices support only up to 16 Mbps, the introduction of Giga-IR offers a theoretical speed of up to 1 Gbps. It is often use in small ECG fragment transmission.

2.2.3 Wireless Local Area Network (WLAN) and Wi-Fi

The IEEE 802.11 standards are very widely used in wireless home networks, providing a low-cost and convenient way for Internet access. Unlike Bluetooth and IR, WLAN requires some efforts in setting up initial configurations before a communication link can be established. Popular IEEE 802.11 standards include a/b/g/n; these standards define the specifications for the physical layer ("PHY" which defines how raw data bits are transmitted over the air) and the WLAN's medium access control (MAC) layer, which provides addressing and channel access control procedures that allow several devices to communicate with a single access point). The standards provide details of these layers so that devices can be designed with full compliance to ensure operability. Apart from 802.11a, which operates at 5 GHz, the remaining three standards are 2.4 GHz. In this frequency band, significant interference can be caused by appliances of similar frequencies, such as cordless phones, microwave ovens, and also Bluetooth devices. Its coverage varies greatly depending on whether it is used for indoor or outdoor operation, ranging from 50 to 300 m, respectively.

A basic WLAN consists of at least one access point (AP) and the mobile client (MC). The MCs are essentially any mobile device that seek to maintain a wireless connection to the network via the AP. APs are placed in various locations throughout the coverage area to form the wireless network infrastructure. In its most primitive configuration, there is one AP in the center and one or more MC operating around it. The network coverage area can be increased by installing more APs. A wireless relay can be installed to further extend the range. When there are multiple APs within the proximity, an MC selects the closest AP whose signal strength is strongest for communication.

Information security is always a great issue, owing to its popularity and sharing of the unlicensed ISM band. This is covered in Chapter 6.

Wi-Fi provides a unified standard derived from IEEE 802.11 WLAN by the Wireless Ethernet Compatibility Alliance (WECA) for different types of wireless devices. Wi-Fi is sometimes referred to as "wireless Internet." Wireless devices are served by an access port, also known as a "hotspot."

Both Wi-Fi and Bluetooth have many similarities, but there are also many differences due to tradeoffs between coverage, data speed, and power consumption hence device size and cost. Owing to its popularity in home networking, Wi-Fi technology is very commonly used in off-site patient monitoring for people recovering at home as existing home networks can be utilized with minimal alternation.

2.2.4 ZigBee

These are small digital devices for wireless personal area networks (WPANs) complying with the IEEE 802.15.4 standard. Easy to implement and very low power consumption, they are not intended for intensive data transfer, because of their slow speed, and are primarily used for wireless control and monitoring. Currently, there is no global standard operating frequency: 868 MHz in Europe, 915 MHz in the United States, 950 MHz in Japan, and 2.4 GHz in most other parts of the world. In a way it may be viewed as a simplified version of Bluetooth and is often used in system-on-chip (SoC) implementations. It is so cheap that a transceiver is available for less than $1 per unit and is regularly used in safety precautionary devices, such as smoke detectors, and air conditioning control. It is also used in body area sensor networks (see Section 3.1 for details). Communication network is served by a Zigbee Coordinator (ZC) and accessed through a Zigbee Router (ZR), which effectively relays data between devices.

2.2.5 Li-Fi

Using visible light for data communication, a common consumer LED lightbulb can serve as a transmitter. This is similar to optical communication in that a light source is used to send data out. The main difference between Li-Fi and traditional fiber optic communication is that there is no cable, hence it is wireless. Li-Fi and Wi-Fi are fairly similar in that both transmit data electromagnetically. The main difference is that Wi-Fi carries data through radio waves, whereas Li-Fi puts data in visible light waves. In the receiving end, a photo-detector picks up the light signals with appropriate signal processing system to extract the data from the received light.

Li-Fi's operation is quite straightforward in that the data stream is sent into an LED lightbulb for transmission which is simply embedded in the light beam as we see it illuminating the surrounding environment. The 0s and 1s that represent the information being sent are presented by the very rapid dimming of the LED lightbulb. The use of visible light implies that the Li-Fi signal does not penetrate through walls so it is only suitable for deployment within a room.

2.2.6 Cellular Networks

Mobile phone networks are commonly known as "cellular networks" because the coverage area is composed of radio cells each served by a base transceiver station (BTS). Functions of the BTS may diverge considerably as determined by the service operator and the cellular technology. The coverage area can be enhanced by the establishment of more cells. In addition to improving coverage, the use of cellular composition also expands capacity and lowers transmission power requirements. The ability of users moving across cells with continuous connection is one of a cellular network's key features, provided by "handover algorithms." There are different technologies currently used in different parts of the world. We will give a brief account on those still widely used today, while omitting obsolete systems such as the "first generation" analogue AMPS from the early 1980s followed by the digital time division multiple access (TDMA) cellular technologies.

CDMA1900 (1.9 GHz): code division multiple access 1.9 GHz. This old digital cellular communication system is still used in the USA, as some operators are licensed to operate at 800 MHz as legacy systems rolled out prior to the FCC approval for 1.9 GHz. CDMA supports multiple simultaneous base stations on the same frequency channel.

2.5G (900 MHz): GSM Phase 2+ as defined by the European Telecommunications Standards Institute (ETSI). A system widely used in most parts of the world offering ease of roaming across countries with one single cellular phone. GPRS (general packet radio service) is an extension of 2.5G that supports a wide range of multimedia services at fairly slow speeds of up to 114 Kbps. The type of services supported is governed by an access point name (APN) defining services such as wireless application protocol (WAP) access, short message service (SMS), MMS, point-to-point (PTP), as well as Internet access.

3G (1.8 GHz): third-generation (3G) technology improving upon the previous 2.5G with a maximum speed of 14.4 Mbps. The main features of 3G include video calling and mobile TV broadcast. There are different interface systems defined by the ITU (International Telecommunications Union) within the IMT-2000 standards as 3G networks. The most significant ones are Mobile WiMAX and UMTS (Universal Mobile Telecommunications System), which is also known as W-CDMA, where "W" denotes "wideband." The former is named under the Worldwide Interoperability for Microwave Access (WiMAX) and is developed from the IEEE 802.16 Broadband Wireless Access (BWA) standard (see below), whereas the latter is a much more mature and widely used technology and direct successor to 2.5G that basically evolves from previously available mobile technology. An improved version, widely referred to as 3.5G and launched in 2006, is High Speed Downlink Packet Access (HSDPA) that supports over 20 Mbps.

Between 2.5G and 3G, there are popular technologies often classified as "2.75G," a term not often seen, but the following expressions should be very familiar to readers: CDMA2000 and EDGE (Enhanced Data rates for GSM Evolution). These are developed from CDMA1900 and GSM Phase 2+. There are misconceptions of these being classified as 3G because of their increased data rates supported versus 2.5G systems.

PHS (1.9 GHz): Personal Handy-phone System used exclusively in Japan for its high portability resulting from low power consumption and lack of a SIM card. The system was mainly designed for voice calls with data support of up to 256 Kbps. PHS is being progressively replaced by 3G networks.

4G (ISM band ranging from 700 MHz to 2.6 GHz depending on national legislations): Long-Term Evolution (LTE) is defined by the ITU in the IMT-A (International Mobile Telecommunications Advanced) standard of 2008. In contrast to 3G's two parallel infrastructures of packet switching and circuit switching (for backward compatibility), 4G only supports packet switching that also provides IPv6 support.

5G (rollout around 2019–2020): developed on the basis of the ITU's IMT-2020 specification. As with all previous generations, 5G is set to increase bandwidth while reducing latency. 5G is divided into three distinct service categories, namely enhanced mobile broadband (eMBB), ultra-reliable low-latency communications (URLLC), and massive machine type communications (MMTC) (Lien et al. 2017). eMBB is simply for consumer electronics devices like smartphones and tablet computers; URLLC is for industrial applications such as smart ambulances and post-disaster recovery, whereas MMTC are primarily for sensors, for example smart pills, health monitors in smartwatches, and smart clothing. In order to support data rates that are far higher than existing

4G cellular networks, 5G is designed to utilize much higher carrier frequencies in the range of 24–86 GHz, and use microcells since these frequencies are prone to rain-induced attenuation (Fong et al. 2003c). This is effectively to say that, at these high frequencies, radio signals will be significantly weakened by rainfall, hence the transmission distance should be shortened.

2.2.7 Broadband Wireless Access (BWA)

BWA supports a diverse variety of services, owing to its ultrahigh speed. Usually used for medium to long range distribution, the carrier frequency can be anywhere from a few GHz to 40 GHz depending on local regulations. The development of BWA is governed by the IEEE 802.16 Working Group on BWA Standards. Note that IEEE 802.16 does not specify frequency bands or equipment certification requirements. WiMAX, fixed or mobile operating in 2.4–5 GHz ISM band, is compliant with both IEEE 802.16e and ETSI HiperMAN wireless metropolitan area network (MAN) standards covering tens of kilometers, and has become very popular in recent years with a high degree of interoperability primarily. Local Multipoint Distribution Service (LMDS), being a prevalent BWA deployment, is intended for fixed network deployment implying that mobility support is very limited. The main difference between Fixed WiMAX and LMDS is the operating frequency that leads to a significant increase in channel bandwidth. LMDS is capable of supporting over 512 Mbps for carrying vast amounts of data. Since the radios have a 90° field of vision, it is possible to set up four radios for 360° coverage.

The properties of LMDS make it particularly suited for telemedicine backbone support. The term "backbone" refers to the medium that provides a main trunk line for interconnecting different local area networks (LANs) as well as equipment together over a large area. For example, a hospital may have several buildings having a network backbone in a hospital interconnecting different entities together, as shown in Figure 2.6.

Mobile WiMAX is also considered as 4G cellular communication as with LTE. It is an extension of the IEEE 802.16 standard as a subset of BWA and therefore not covered in Section 2.2.6. Subsequent generations beyond 5G are expected to further integrate cellular and BWA networks, and a framework has been suggested on integrating 5G microcell with satellite links, thus allowing connections to be initiated from virtually anywhere around the world (Liang et al. 2018).

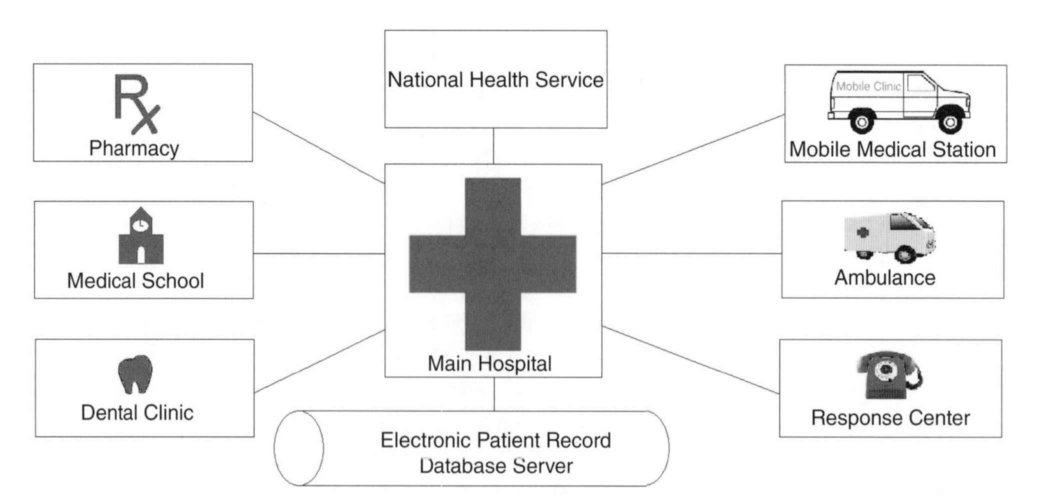

Figure 2.6 Network infrastructure linking the hospital to many entitles, both fixed and mobile services can be simultaneously supported.

2.2.8 Satellite Networks

These are sophisticated and expensive networks, as placing a satellite precisely above the earth is a very costly exercise. Its operating principle is, however, reasonably straightforward. A communication satellite (comsat) is laid in a pre-determined orbit above the earth. The choice of orbit depends on the desired coverage area. The comsat serves as a PTP microwave radio relay that provides a radio link between two earth stations. Satellites are frequently used in wide area networks (WANs). Despite being vulnerable to environmental interference such as solar storms, it is very reliable and provides very high speed links. Although such properties may appear suitable for remote robotic surgery given the vast amount of data that need to be transmitted, its inherent long propagation delay will likely affect real-time operations. For this reason, satellite networks are mainly used for remote recovery.

2.2.9 Licensed and Unlicensed Frequency Bands

We learned from the above discussion that some networks operate in licensed bands while others are unlicensed and shared with many other users. So, what are the differences that might affect telemedicine operations? Both licensed and unlicensed-band equipment can operate cooperatively for any telemedicine application. Finding out which is a better choice may depend on different situations (Dekleva et al. 2007). First, an unlicensed network does not incur any implementation delay and cost on acquiring a license, an unlicensed connection can be easily established, and anyone can do it with no restriction on the type of radio device used. Access by anyone means devices are at risk of security breaches and interference. Licensed networks operate within designated bands with exclusive use so equipment can be highly customized to exact requirements. Interference protection (there is no assurance of an interference-free environment) and guaranteed bandwidth availability are key features of using licensed frequency bands. So, generally, there is a compromise between cost and convenience versus security and operating environment.

By now we have looked at several different types of wireless networks that can be utilized for a wide range of telemedicine applications. Each has its own advantages and disadvantages. Careful selection based on their performance and properties will ensure that the type of services possible can be vast. Very often, the use of existing network is a desirable choice, owing to cost effectiveness and reduction in implementation time. As communication technologies advance and more choices become available in the near future, telemedicine is set to become even more reliable and accessible to more people with different needs.

2.3 M-health and Telemedicine Applications

Ever since the first mobile device was launched several decades ago, power consumption has been a perpetual issue that limits portability and operating time. Recent advances in battery miniaturization and SoC have made ultraportable and wearable devices readily available over recent years. SoC integrates all components of an electronic circuit into a single integrated circuit (IC) chip, as shown in Figure 2.7, where the working prototype of a wearable is all packed into a single chip for manufacturing. It is apparent that the prototype is not in a suitable form to be worn by a patient, whereas the production chip can be embedded in a small wristband that serves the same health monitoring function. In addition to having a substantial size advantage, SoCs consume much less power, as the electronics are integrated on a single electronic substrate.

Figure 2.7 System-on-chip (SoC) implementation for mobile health monitoring.

Mobile health (m-health) can be supported by sensors that monitor various health parameters, from general exercise monitoring of heart rates and steps taken to more sophisticated rehabilitation support, like knee recovery, or chronic disease management, such as blood glucose measurement. A wide range of possibilities exists through the use of appropriate sensors. As observed from the earlier illustration in Figure 1.3, where health information can be collected for monitoring and diagnosis through telemedicine, one common issue is the use of appropriate casing materials for the successful transmission of the health data from the sensor to the outside world (Li et al. 2018). The casing must provide adequate protection to the electronic components while minimizing absorption of radio waves and be able to withstand shock and vibration when the device is being worn.

Ensuring that the received data correctly represent a patient's state of health is vitally important in the use of telemedicine. M-health possesses significant challenges in that motion artifacts can have a substantial impact on sensor performance (Goverdovsky et al. 2015). As the device is moved because of the patient undertaking daily activities, such movement can significantly affect the data being taken so that the relative position of all sensors to the location where measurement is taken should be fixed or appropriate compensation must be applied to the received data to minimize the impact of movement on the measured data. Different types of sensors can be prone to varying uncontrollable factors – e.g. skin impedance can be affected by sweating, whereas optical sensors can be affected by ambient light. As shown in Figure 2.8, the oxygen saturation meter is designed so that the patient's finger is placed in the measuring clip where the foam surrounding the optical sensor serves both as a light seal against ambient light and to relieve the pressure exerted on the finger.

2.4 The Outdoor Operating Environment

As signal strength weakens over distance traveled (attenuation), the effects of electrical noise radiated by other nearby devices can be very significant. The noise can be so severe that the transmitted signal can be lost or corrupted, causing the received data to be useless. In addition to noise and

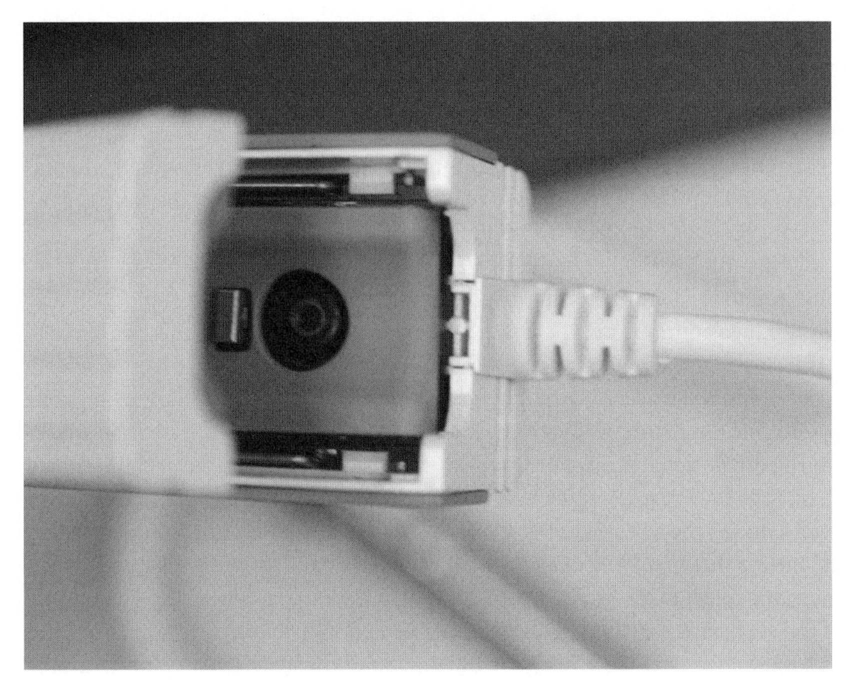

Figure 2.8 A small oxygen saturation meter that utilizes a pair of light source and sensor to measure the amount of oxygen within the bloodstream as the varying amount of light is absorbed as it penetrates through the patient's finger.

attenuation, signal distortion can be an issue as it travels through metal conductors. Distortion can take various forms depending on what lies along the signal path, but generally speaking the signal's shape is distorted, for example, when a square wave no longer retains its smooth pulse.

Although these signal propagation issues also exist indoors, there are more uncontrollable factors in the outdoor environment that make many signal degradation factors more severe.

The benchmark by which signal loss is measured in a transmission link is the loss that would be expected in free space, i.e. the loss that would occur along a path that is free of anything that might absorb or reflect signal energy. When a propagating radio wave hits a physical obstacle, it is subject to the following phenomena, as illustrated in Figure 2.9:

- *Diffraction*: the signal splits into secondary waves. It happens when the propagating signal hits a surface that has sharp edges. The waves produced by the surface are present throughout space, and a certain portion may penetrate behind the obstacle thereby causing power loss, giving rise to the bending of waves around the obstacle.
- *Reflection*: the signal reflects back to the transmitting antenna, just like light being reflected by a mirror. It happens when the propagating wave hits a physical object that is much larger than the wavelength of the carrier.
- *Scattering*: the signal reflects with different components spread in different directions as being diffused upon hitting an obstacle. Contrary to diffraction, scattering is an issue when the object that the propagating wave hits is small compared to the wavelength, such a rough surfaces, dust, and air pollutant particles, or other irregularities in the channel. When the signal scatters in all directions, it effectively provides additional energy as perceived by the receiver. This leads to the actual received signal being stronger than that affected by reflection and diffraction.

All these will result in loss of signal strength, collective known as "fading." Such effect can be compensated by using multiple antennas to pick up different components of the same signal arriving from different directions, such a technique is known as "space diversity." How it works is very simple, since different components are subject to different time delay, phase shift, and attenuation. One antenna may experience too severe fading that cannot effectively pick up the signal, whereas the use of more antennas will improve the chances of picking up a better version of the same signal.

In outdoor signal propagation, very often the signal must clear large physical obstacles, such as buildings and trees. One thing to remember is that a visual LOS observed, when looking from the position of an antenna toward another antenna, does not necessarily imply a radio LOS also exists especially in long range communication. Whether this is true depends on the clearance of the "Fresnel zone," this is because a radio wave needs some space to reach the receiving end as it is obvious that the wave cannot "squeeze" through a small hole drilled in a wall. A Fresnel zone is defined as a long ellipsoid that stretches between the two antennas. The Fresnel zone marks through a region that the direct signal propagates. A physical obstacle, such as the airplane shown in Figure 2.9, can fly outside the Fresnel zone while causing scattering to the signal (in addition to reflecting as shown) radiated by the antenna. This results in additional signal components reaching the receiver without altering the direct line of sight (LOS) signal. In contrast, if the obstacle moves within the Fresnel zone, the direct LOS signal is no longer that for free space. As a result, the Fresnel zone is a measure of the localization of the region along a given path that a radiated signal reaches the receiver. This is a spheroid space necessary for the wave to propagate toward the receiving antenna centered along the direct straight line path between the antennas. For example, suppose the signal frequency is 30 GHz. By applying the familiar (2.6) learned in high school physics:

$$v = f \cdot \lambda; \quad \lambda = v/f \tag{2.6}$$

Given that the speed of radio wave propagating through free space is approximately 3×10^8 m/s, the wavelength λ would be $3 \times 10^8/30 \times 10^9 = 0.01$ m or 1 cm. So, half wavelength will be 5 mm. So, the wave reaches the receiver by the direct straight line path, and it also reaches there within a spheroid area of 5 mm. It is said that at least 60% of the first Fresnel zone should clear any physical obstacle in order to achieve propagation characteristics comparable to that of free space. Also, the terrain profile around the spheroid area needs to be taken into account to estimate the path loss or attenuation. This can often be estimated using well-established models such as the Longley–Rice

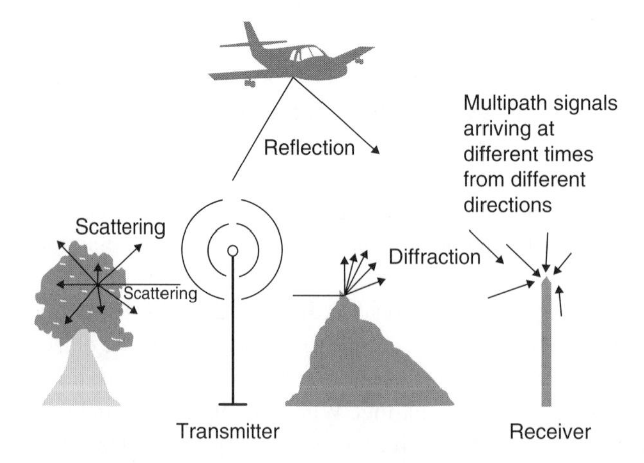

Figure 2.9 Propagating wireless signal degrades, owing to different phenomena.

model (Hufford 1999), where the median transmission loss is predicted using the path geometry of the terrain profile and the refractivity of the troposphere. An urban factor (UF) then accounts for additional attenuation due to urban clutter surrounding the receiving antenna. The model is effective as an irregular terrain model (ITS). However, it does not take into consideration the effects of buildings and foliage. When optimizing the propagating path for long range communication, usually when exceeding 5–8 km, the earth's curvature also needs to be taken into consideration.

The transmission loss depends on how much power reaches the receiving antenna. Attenuation is always an important consideration as the signal will eventually become too weak to be picked up by the receiver. Weather conditions, such as rain, fog, or snow, can severely affect the range and reliability of wireless systems. The effect of rain-induced attenuation can be very severe, especially in tropical regions where consistent heavy downpour in excess of 100 mm/h can last for hours. The dB/km measurement of attenuation indicates the power loss in dB for every kilometer of distance travelled. The actual impact is determined by several factors, primarily the rain rate and carrier frequency. In general, the heavier the rain and/or the higher the frequency, the more power is lost per kilometer. As a general guideline, rain-induced attenuation is not a significant problem for systems operating under 10 GHz, or when the rainfall rate is below 20 mm/h.

To see how severe this problem is, we take a look at Figure 2.10, which compares the attenuation for 10 and 50 GHz. Note, incidentally, that the plots evaluate vertically polarized signals; signals of horizontal polarization always undergo a higher degree of attenuation than vertical polarization under identical conditions. The difference between two polarizations also increases as the rainfall rate and/or frequency increases as shown in Figure 2.11. The effects of heavy rainfall on radio propagation path reduce the system availability because rain causes cross-polarization interference that subsequently reduces the polarization separation between signals of vertical and horizontal polarizations as the signals propagate through rain. The extent of radio link performance degradation is measured by cross-polarization diversity (XPD), which is determined by the degree of coupling between signals of orthogonal polarization (Fong et al. 2003a). Yamada et al. (2019) give a comprehensive definition of XPD as a measure of the strength of a co-polar transmitted signal that is

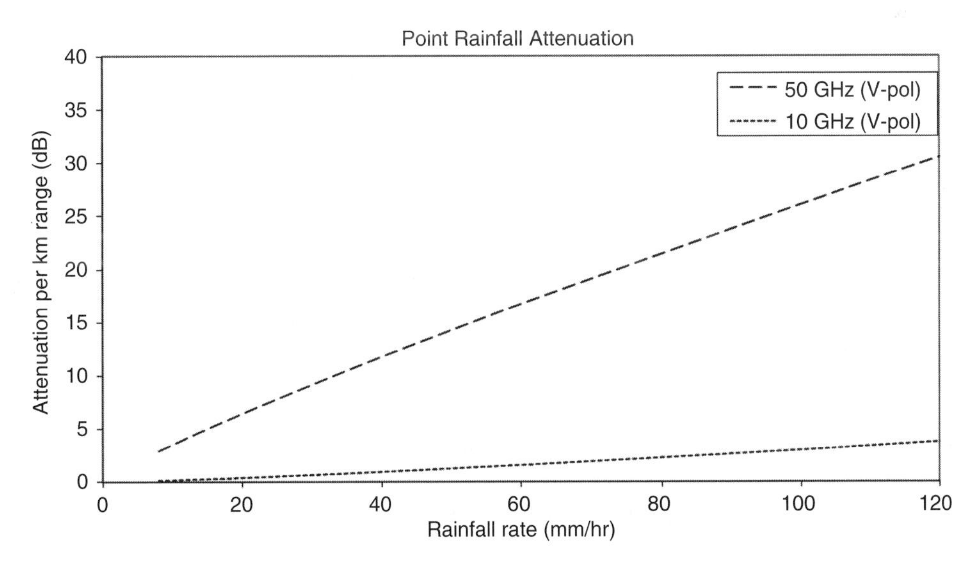

Figure 2.10 Effect of rain attenuation at different rainfall rates. The signal is weaker to a greater extent as the rain gets heavier. Also, higher carrier frequencies result in more severe attenuation at the same rainfall rate and distance traveled.

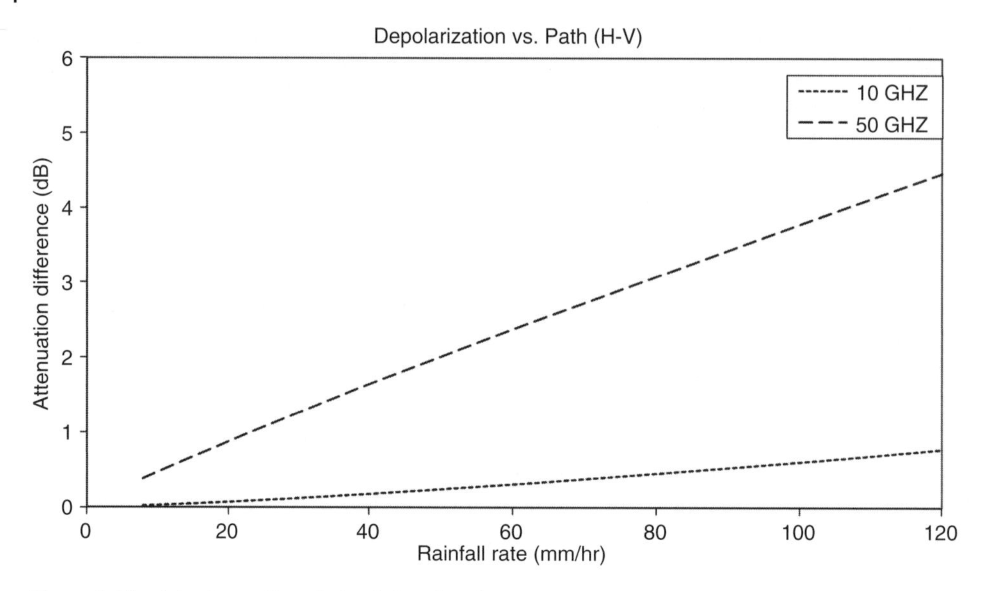

Figure 2.11 A horizontally polarized signal undergoes more severe attenuation than a signal of vertical polarization when the rainfall rate and distance traveled are the same.

received cross-polar by an antenna as a ratio to the strength of the co-polar signal that is received, which typically results in a 10% reduction in coverage, owing to cell-to-cell interference. While it may make logical sense to refrain from using horizontally polarized signals to avoid excessive power loss, we shall see in Section 3.2 why in many systems both are used simultaneously.

The issue of channel degradation due to rain must be thoroughly addressed in telemedicine because very often accidents occur as a direct result of heavy rain. So, telemedicine systems that assist emergency rescue operations must maintain an adequate level of quality. Optimizing the appropriate system margins would maximize the availability of radio link in such situations (Fong et al. 2003b).

"Multipath fading" is a phenomenon resulting from multiple components of a signal reaching the receiver from different directions of arrival (DOA) at different times due to reflection through different physical obstacles along the propagation path subject to different amounts of delay, and is illustrated in Figure 2.12. The shortest path between the transmitter and receiver is the straight line unobstructed path having LOS. When the propagating signal hits an obstruction, it will be spread out to take multiple paths, resulting in different traveling times to arrive at the receiver causing varying amounts of time delay. Multipath is generally an issue with signals below 10 GHz, whereas attenuation caused by rain is the most important consideration factor at frequencies above 10 GHz. Lower frequencies are therefore preferred in tropical regions where heavy and persistent rainfalls are expected. Otherwise systems at higher frequencies operate in a less congested part of the spectrum with more available bandwidth.

Another potential issue with wireless communications that may cause delays is Doppler spread, where fluctuations are caused by the movement of the transmitter, receiver, or the any physical objects between them. Doppler spread is particularly relevant in vehicular communications, where fast movement can severely affect signal reception.

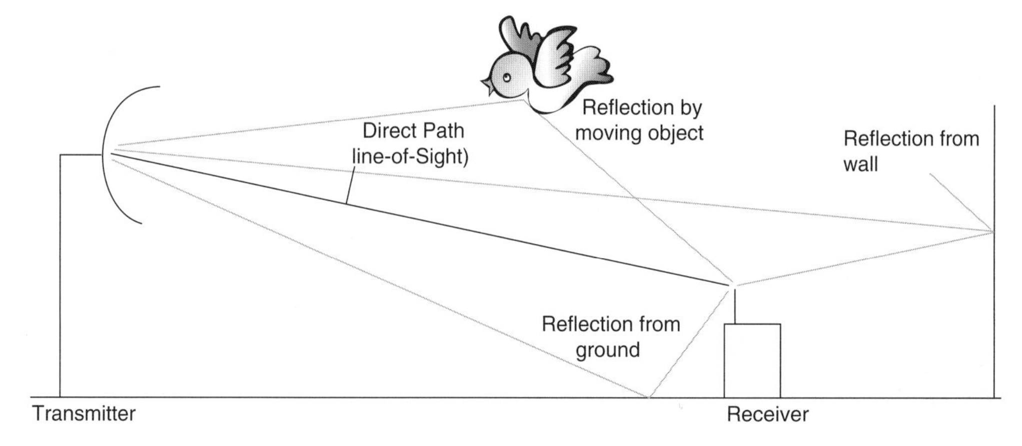

Figure 2.12 Multipath fading caused by different components of the same signal arriving at different times through different propagating paths. Some paths are longer than others and therefore take longer to reach the receiver.

2.5 RFID in Telemedicine

RFID (radio frequency identification) is an old technology that appeared as early as World War II but has only been widely used in many applications for everyday life over the past decade or so. As its name suggests, RFID is all about identifying an object using radio frequency signals. So, it is often perceived as an "electronic barcode" system. There are many different forms of RFID currently used throughout the world. Essentially, RFID involves "tags" that identify an object and "readers" that read and identify the tags. These come in various forms, portable or fixed readers, active or passive tags, meaning whether an internal battery is needed to power a given tag in order to respond to a reader. The battery serves the sole purpose of providing a longer read range that allows an active tag to be read from a longer distance. So, how does a passive tag respond without a battery since it does not have any power source? The answer is actually quite simple as it gets the necessary power from the reader while it receives the reading signal from the reader. Such a signal, carrying a certain amount of energy with it, hits the coiled antenna inside the tag thereby induces a magnetic field that energizes the electronic circuit containing information embedded within the tag including a unique identification number.

Advantages of passive tags are obvious: they are extremely small, cheap, and durable. These tags can be manufactured for less than 10 cents each and can be mass produced because a tag merely consists of a printed antenna and a small chip wrapped in paper. Figure 2.13 sketches the typical layout structure of an RFID tag. However, a short reading range may not be the only problem in telemedicine applications. It would not have the ability to supply power to biosensors if one were to be attached to the tag.

Major problems with RFID reliability are tag collision and reader collision. Tag collision occurs when multiple tags are energized by a single reader so that the tags respond at the same time causing read failure, whereas "reader collision" refers to situations when the coverage area of one RFID reader overlaps with that of a nearby reader. Another major issue is lack of security since tag signals can be picked up by any reader within range.

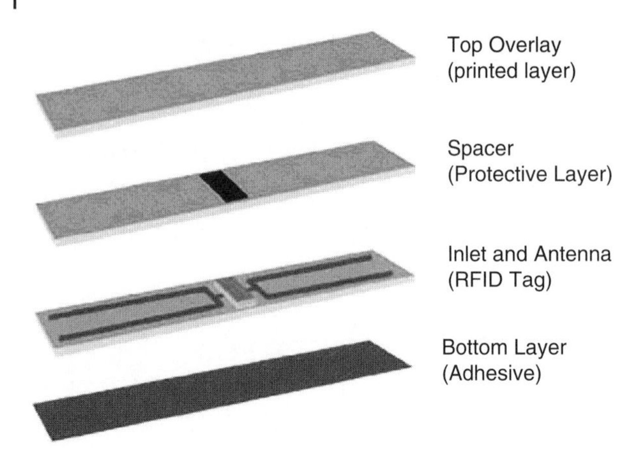

Figure 2.13 An RFID tag.

RFID systems operate in a number of different frequency ranges. Their propagation properties can severely affect the operation in different telemedicine applications. With LF (low frequency: 135 KHz) and HF (high frequency: 13 MHz) systems, signal reflection severely reduces the transmitted signal power, as shown in Figure 2.10. UHF (ultrahigh frequency: 900 MHz) systems can suffer from signal absorption by water, making it unsuitable for applications involving placements of tags on the human body.

RFID is used in many medical applications, for example it is very widely used in drug dispensary, linking patients with restricted or controlled drugs. Tracking babies and other patients as well as medical equipment is another key area of RFID usage. The list of applications is seemingly endless. An area that warrants more thorough discussion is implantation with medical devices within a human body. This is a challenging yet important application for devices such as the biventricular pacemaker and the glucose meter. UHF is not suitable, because of the composition of water in human tissue. Owing to the lack of security features and the high cost of readers, HF is a clear choice for surgical implantation. To implant a biventricular pacemaker, leads are implanted through a vein into the ventricle and the coronary sinus vein to regulate the ventricle. Since it is intended to serve patients suffering from serious heart failure symptoms, any irregularity must be reliably reported via a telemedicine network without delay in order to minimize the risk of sudden cardiac consequences. Further, patients with inadequate ejection fractions may require an implantable cardioverter defibrillator (ICD) in conjunction with the pacemaker to ensure sufficient heart pumps per beat are maintained. Since an ICD functions by rhythm detection and shocking the heart, such action can affect the operation of an associated RFID tag. Also, each tag associated with an implanted device may risk tag collision as they are so closely placed to each other. Obstacles along the signal propagating path include the lung's anterolateral surface of the inferior lingular segment, followed by rib bone, and finally all the way through skin that consists of epidermis, dermis, and subcutaneous fat leaving the body at the chest. There are many layers of barriers that would affect the signal path.

Capacitance between a tag and its housing can severely impact the antenna's tuning. To combat this problem, a tag must be tuned away from resonating at the reader's frequency to diminish mutual coupling with the other tag. The read range can therefore be improved by using an RFID tag with a tunable antenna.

The above case study may sound rather complicated. Let's look at a less demanding example of an implantable glucose meter for diabetes monitoring documented by Christiansen et al. (2018) that does not have any immediate life threatening consequences in the event of communication failure. The RFID tag would be responsible for transmitting the glucose meter data away from the body for subsequent analysis. Since the data storage capacity of the tag is no more than 2 KB as with any typical passive tag, the data needs to be sent away as soon as they are received from the glucose meter before it becomes full. Here, the part of the RFID tag is rather similar to a cellular phone connected to a laptop computer as a wireless modem via a USB (universal serial bus) cable. In this analogy, the mobile phone acts as a point of sending data to the outside world.

Having understood the tag's function, we need to look deeper into the mechanism involved. As a wireless transmitting device, it acquires the necessary energy from the incoming wave radiated from the external reader; the received energy must be sufficiently strong to power up the chip. When the signal is sent back from the tag's antenna, it must be efficient enough to ensure that the transmission power is adequate for the reader. So, here is the challenge: on one hand we need to ensure the tag is implanted as closely to the person's skin as possible to minimize the signal propagating distance. On the other hand, we also need to avoid immediate contact between the antenna and any internal tissue that may severely shield off the signal. Any housing for the tag that seals it off to avoid direct contact with tissue will certainly have an effect on signal penetration. The choice of material therefore becomes a critical issue in this scenario. Friedman (2001) describes various materials suitable for implantation. Polyvinylchloride (PVC) insulator of approximately 10 μm is suggested to be an optimal wrap for providing a reasonable separation between the tag and surrounding tissue without significant impact to signal propagation.

So, the communication aspect is more or less resolved, but what about integration with the glucose meter? The system does not seem to have many components, but the technical issues can be quite challenging as it entails biocompatible interface, glucose sensing, and a device to convert the captured reading into an electrical signal that can be written into the RFID tag for subsequent transmission away from the human body to the reader. We also mentioned that mechanisms must be in place to ensure that the previously stored data is sent away and the tag's memory content emptied before the next set of readings comes in. So, the device that connects the glucose meter and the RFID tag must be capable of programming the tag's memory in addition to generating the signal from the captured readings. Also, this device must be very small and power consumption must be so low that one energization activity by the reader can produce and store sufficient energy to last until the next energization, i.e. the next read operation. Ultimately, what needs to be achieved is to download the collected data for subsequent analysis and storage. This is mainly determined by the optimal design of an efficient antenna and related circuitry for the chip so that the data can get through to the outside world from inside the body.

RFID is capable of more than acting as an identifier, it is also a very small and economical piece of implantable object that supports short range wireless communications. It is so versatile that its application area is virtually unlimited and is certainly an important piece of tool for telemedicine.

Another important application of RFID is patient tracking (Cao et al. 2014). Hospital staff can obtain near real-time location information of individual patients from the tag locations. These inexpensive RFID tags can be worn by the patient either as a wristband or embedded in clothing. One of the major issues with hospital implementation is that the number of tags within a confined area increases, the likelihood of a "tag collision" increases (Xiao et al. 2018). This is particularly problematic when other RFID systems for medical resource management coexist. Mainstream anti-collision methodologies include random Aloha, binary search, and Hybrid with the common goal of coordinating tag transmission for collision avoidance (Zhu and Yum 2011). Each of these uses different mechanisms to read tags at different times. However, the binary search method is not suitable for clinical deployment when many patients are simultaneously tagged, owing to the time needed for execution (Ullah et al. 2012).

References

Cao, Q., Jones, D.R., and Sheng, H. (2014). Contained nomadic information environments: technology, organization, and environment influences on adoption of hospital RFID patient tracking. *Information & Management* 51 (2): 225–239.

Christiansen, M.P., Klaff, L.J., Brazg, R. et al. (2018). A prospective multicenter evaluation of the accuracy of a novel implanted continuous glucose sensor: PRECISE II. *Diabetes Technology & Therapeutics* 20 (3): 197–206.

Dekleva, S., Shim, J.P., Varshney, U., and Knoerzer, G. (2007). Evolution and emerging issues in mobile wireless networks. *Communications of the ACM* 50 (6): 38–43.

Fong, B., Rapajic, P.B., Hong, G.Y., and Fong, A.C.M. (2003a). Factors causing uncertainties in outdoor wireless wearable communications. *IEEE Pervasive Computing* 2 (2): 16–19.

Fong, B., Rapajic, P.B., Fong, A.C.M., and Hong, G.Y. (2003b). Polarization of received signals for wideband wireless communications in a heavy rainfall region. *IEEE Communications Letters* 7 (1): 13–14.

Fong, B., Rapajic, P.B., Hong, G.Y., and Fong, A.C.M. (2003c). The effect of rain attenuation on orthogonally polarized LMDS systems in tropical rain regions. *IEEE Antennas and Wireless Propagation Letters* 2 (1): 66–67.

Friedman, C.D. (2001). Future directions in biomaterial implants and tissue engineering. *Archives of Facial Plastic Surgery* 3 (2): 136–137.

Garratt, G.R.M. (1994). *The Early History of Radio: From Faraday to Marconi*. London: The Institution of Engineering and Technology.

Goverdovsky, V., Looney, D., Kidmose, P. et al. (2015). Co-located multimodal sensing: a next generation solution for wearable health. *IEEE Sensors Journal* 15 (1): 138–145.

Hecht, J. (2004). *City of Light: The Story of Fiber Optics*. Oxford University Press on Demand.

Hufford, G. (1999). *The ITS Irregular Terrain Model*. Boulder, CO: National Telecommunications and Information Administration.

Li, C., Kang, Y., Wu, T. et al. (2018). Numerical analysis for human perception of temperature rise on the fingertips during usage of a mobile device. *Bioelectromagnetics* 39 (2): 164–169.

Liang, X., Jiao, J., Wu, S., and Zhang, Q. (2018). Outage analysis of multirelay multiuser hybrid satellite-terrestrial millimeter-wave networks. *IEEE Wireless Communications Letters* 7 (6): 1046–1049.

Lien, S.Y., Shieh, S.L., Huang, Y. et al. (2017). 5G new radio: waveform, frame structure, multiple access, and initial access. *IEEE Communications Magazine* 55 (6): 64–71.

Shannon, C.E. (1948). A mathematical theory of communication. *Bell System Technical Journal* 27 (3): 379–423.

Sogo, O. (1994). *History of Electron Tubes*. IOS Press.

Tikkanen, J. (2005). *Wireless Electromagnetic Interference (EMI) in Healthcare Facilities*. BlackBerry Research White Paper.

Ullah, S., Alsalih, W., Alsehaim, A., and Alsadhan, N. (2012). A review of tags anti-collision and localization protocols in RFID networks. *Journal of Medical Systems* 36 (6): 4037–4050.

Xiao, F., Miao, Q., Xie, X. et al. (2018). Indoor anti-collision alarm system based on wearable internet of things for smart healthcare. *IEEE Communications Magazine* 56 (4): 53–59.

Yamada, S., Choudhury, D., Thakkar, C. et al. (2019). Cross-polarization discrimination and port-to-port isolation enhancement of dual-polarized antenna structures enabling polarization MIMO. *IEEE Antennas and Wireless Propagation Letters* 18 (11): 2409–2413.

Zhu, L. and Yum, T.S.P. (2011). A critical survey and analysis of RFID anti-collision mechanisms. *IEEE Communications Magazine* 49 (5): 2–9.

3

Information and Communications Technology in Health Monitoring

In Chapter 2, we learned that many alternative types of wireless networks are currently available for telemedicine services. These networks have very different properties and are designed for different situations. There is no simple answer as to what type of network is best for telemedicine, as different applications may have very different needs. Having looked at a variety of technologies, we have seen propagation as being one major issue that all wireless networks face. We have discussed why wireless telemedicine is far more popular than wired systems. Wireless networking is the underlying technology that enables the connection of healthcare in terms of both people and resources. Technological advancements over decades have enabled secure and reliable networks to provide services for life-critical services. In this chapter, we look at various situations where wireless telemedicine helps patient recovery and rehabilitation. We shall see how these can be accomplished and what technical challenges exist. In the past decade, radio frequency identification (RFID) has been extensively used in the management of both patients and supplies that range of medication to consumables (LeMaster and Reed 2016). More recently, the popularity of the Internet of things (IoT) and connected devices in various health management applications have paved the way for people and medical resources to be easily tracked down and monitored in a cloud environment (Darshan and Anandakumar 2015).

Specific network design is motivated by the application it provides such that it needs to fulfill the requirements to reliably transmit the type of information involved. For example, to monitor a ventricular tachycardia (VT) patient requires the regular transmission of electrocardiograph (ECG) and heart rate information to ensure any risk of ventricular fibrillation will be promptly detected; this may require resolving QRS complex separations of at least 0.05 seconds. In any telemedicine system, we must ensure the communication network used is capable of supporting the required data rate.

In this chapter, we begin by looking at the technologies and challenges of setting up a body area network (BAN) suitable to be used both by monitored patients and by health professionals. We shall then look at some major applications of remote patient monitoring utilizing wireless communication technology. Readers should be reminded that the examples given may have alternative solutions; these are by no means the only deployment option. Rapid technical advances in the healthcare and medical industry make it virtually impossible to provide an in-depth analysis of every single methodology available. The primary objective of looking at these examples is to grasp a good understanding of how telemedicine technologies support rescue missions in different situations and what challenges may be faced.

Telemedicine Technologies: Information Technologies in Medicine and Digital Health, Second Edition.
Bernard Fong, A.C.M. Fong, and C.K. Li.
© 2020 John Wiley & Sons Ltd. Published 2020 by John Wiley & Sons Ltd.

3.1 Body Area Networks

BAN, also known as personal area network (PAN), has become possible in recent years because the technology has allowed tiny radio transmitting devices to be securely installed on a human body. In addition to its increasingly popular deployment in healthcare, BAN is also used in many computing and consumer electronics applications, owing to its flexible deployment options. These devices are so small that some can even be implanted inside the body. It provides the underlying technology for monitoring various signs of the body and to automatically issue an alert should an abnormal behavior be detected. It also provides a convenient means of logging daily activities and determining whether a user has met a pre-determined target for workout during a session. It therefore supports people who require medical attention and those looking to monitor their levels of fitness. Biosensors are attached to the user's body for remote health monitoring offering extremely high mobility; a BAN typically consists of two major components: "intra-BAN" for internal communication around the body, where sensors and actuators are connected to a mobile base unit (MBU) that serves as a data processing center. The MBU can be just about any consumer electronics device that we carry on a regular basis, like a cellular phone, in-car hands-free kit, or the wireless modem that we use for connecting our laptops to the Internet; and: "extra-BAN" for external communication between components surrounding the body and the outside world. This is normally a telemedicine system that conveys the collected data for processing and analysis.

In general, BAN devices have properties of very low power consumption usually below 10 mW and a low data throughput of around 10 Kbps. There are a number of issues that BANs have. First, data security is a particular issue since no data protection mechanism is employed in most situations. QoS (quality of service) assurance for individual devices must be provided to ensure that all devices remain contacted. Coverage does have not to be vast, typically anywhere within 2 m from the base matching unit (BMU) should suffice. Antenna design as an integral part of a wearable sensor can be a very challenging task since it needs to provide omnidirectional coverage to ensure a high degree of mobility, and effects of absorption by human body on signal propagation need to be thoroughly investigated (Hirata et al. 2010). This is a particularly demanding mission for implanted devices.

Although there is currently no standards for BAN implementation, the IEEE 802.15 Working Group for wireless personal area networks (WPANs) has been working on allowing a broad range of possible devices to interoperate on various transmission media. Li and Kohno (2008) describe a number of prospects that may eventually lead to the standardization of IEEE 802.15 for BAN deployment. Different groups are in place for different media. For example, the most popular ones are IEEE 802.15.1 for Bluetooth and 802.15.4 for Zigbee.

Owing to the high degree of flexibility for different sensors to be attached, BAN is capable of monitoring suffers of asthma, diabetes, heart problems, etc., and logging and tracking of related data can be easily accomplished to detect any potential issues. In areas where many patients may be monitored in close proximity, such as hospitals and clinics, one major design challenge to overcome is the ability to distinguish each BAN system associated with each patient so that data collected will not be mixed up. Although many BANs can be built upon off-the-shelf sensors, a number of concerns exist for the sensors:

- Standards: functional specifications, operating environments, communication protocols, operating range, security, and privacy.
- Electromagnetic compatibility (EMC): amount of electromagnetic radiation induced, susceptibility to interference.

- Calibration: procedure and frequency for calibration, precision.
- Integration: connections, database linkage, mounting, and placement.

Let us go further into these design apprehensions. Currently, there are no standards that govern the development of BAN biosensors, and guidelines on power source requirements or communication protocols specifying how data is transmitted do not exist. Performance and reliability of sensors differ when used in different situations. For example, implantable sensors may not be suitable for operating above a certain altitude or when the person is submerged in water while participating in activities such as swimming and boat repairing. How far can a person move away from the point where data is collected needs to be specified to ensure that data can be successfully collected if the device does not have any internal data storage buffers. As with almost all healthcare systems, data security and privacy is an important topic to address. This is covered in detail in Chapter 6.

EMC is applicable to all wireless transmitting devices in most countries, and different countries may have different regulations. In cases where health monitoring devices can be brought to different countries, they must be designed to comply with all relevant regulatory governance concerning EMC. Calibration is an important process for all precision instruments in ensuring that the data captured is within the specified accuracy limits. It is possible to incorporate self-calibration and diagnostic functions for ease of maintenance. When this cannot be accomplished, there will be a need to specify how frequent calibration is necessary to maintain accuracy, and whether calibration can be performed by the user. Finally, how each sensor is connected to the MBU and how it is securely installed on the user must be thoroughly addressed to ensure reliability. Sensors can be implanted inside a human body given the appropriate protective housing; many of them are attached to the body on a temporary basis, while some are embedded on clothing, as investigated by Park and Jayaraman (2003) and Winters et al. (2003). To ensure maximum mobility, the sensors must be lightweight with a small form factor (physical size and shape); the weight and form factor is primarily determined by the internal battery installed within. These sensors must therefore be designed to be exceptionally power efficient to minimize size and maximize durability. Also, frequent replacement or battery charging will make usage inconvenient and impractical.

To better understand how BAN operates, we look at an example in Figure 3.1, which shows the infrastructure of a basic BAN that consists of sensors for monitoring a cardiac patient under

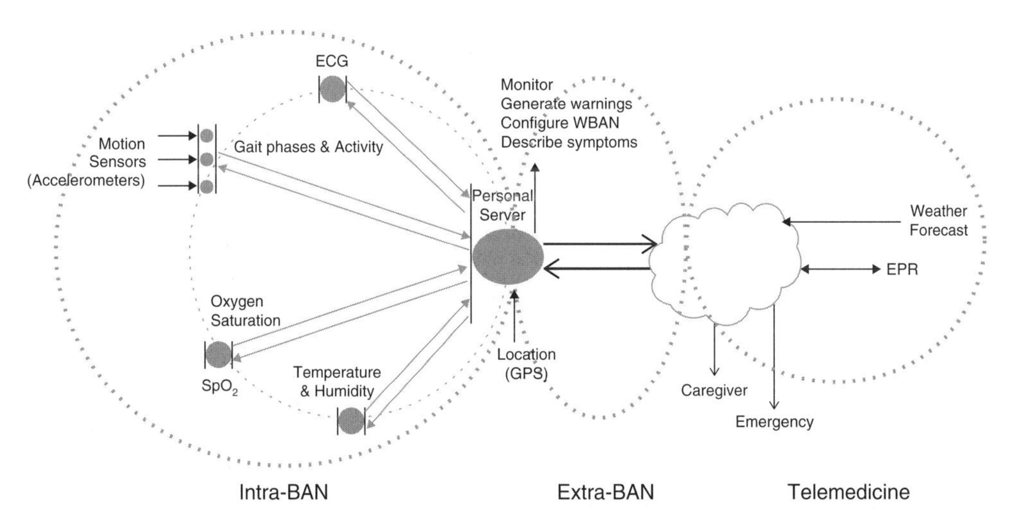

Figure 3.1 Body area network (BAN) connecting a range of devices/systems within and surrounding a patient's body to the outside world via a telemedicine link.

supervised recovery. Sensors are present for collecting ECG data, oxygen saturation, motion sensing for gait phase detection, and body and ambient temperature. Each sensor is connected to the MBU via a wireless link and data is sent at regular intervals. The patient's location can be tracked by a GPS (global positioning system) or through the position of an Internet access point (AP). The MBU conveys data captured from each sensor to an existing home wireless local area network (WLAN) that is linked to the telemedicine system. The electronic patient record can be automatically updated with the received data. In the event of an imminent medical condition being detected, an alert will be generated. As the patient's location is known, the necessary attention can be provided. The patient's environmental conditions, such as ambient temperature and humidity, can also be known and recorded. Quantitative analysis of various conditions and patterns for issuing appropriate recommendations can be easily achieved. Data can also be stored anonymously for the purpose of research so that the effects of each parameter to a given medical condition can be analyzed. Legal regulations in most countries may restrict access to patient-identifiable information.

The human body can affect BAN signal transmission. This is an important issue to be aware of and so close attention must be paid to where a sensor is placed and in what direction it faces when on the patient's body (Wang et al. 2009). When the person moves, some sensors may face nearer to the MBU, while others may be moved further away. Welch et al. (2002) discuss attenuation and delay induced by the human body as signal degradation factors caused by absorption, reflection, and diffraction. The electric properties of human tissue (i.e. the electric conductivity and permittivity) can be used to determine the behavior of radio signal propagating through the human body. Generally, relative permittivity decreases when the conductivity increases with the increasing signal frequency. A detailed description of these electric properties of the human body and their effect on wave propagation is given in Means and Chan (2001). Measurement, often done with appropriate simulating models instead of employing human subjects, is usually necessary to verify the network performance during the design stage to ensure reliability.

Connected biological and physiological sensors to intelligent consumer healthcare and medical devices through IoT open up vast opportunities for health service enhancement. Preventive care can be readily supported through anomaly detection when a measured parameter reaches a certain predetermined threshold. Comprehensive monitoring covers not only the patient but also the ambient environment that could affect the patient's health as an integral part of a smart home system (Fong and Hong 2012). Smart wearables can also be connected to an IoT network to obtain more comprehensive information about the patient's well-being (Yang et al. 2017). In addition to general health monitoring and assessment, Thapliyal et al. (2017) propose that IoT-enabled wearable smart health devices can also assist with psychological health improvement by relieving stress. IoT supports comprehensive health services through multiple communication links ranging from connectivity across the Internet via cellular networks (see Section 2.2.6 for details), as well as short range across Bluetooth/NFC (near-field communication) protocols. In low-power-sensing networks where a relatively small amount of data is transferred across short ranges, WPANs such as ZigBee (Section 2.2.4) are a good option optimized for efficiency and power consumption (Lee et al. 2016).

3.2 Emergency Rescue

Accidents do happen anytime anywhere regardless of how careful people are. Mishaps can be caused by nature, intentional or unintentional manmade commotion, machinery failure, or a combination of these. In the event of an accident that leads to injury, the utmost priority is always to provide appropriate treatment at the very earliest opportunity. Traditionally, the course of seeking help can be a very lengthy process. Minimizing the time to provide treatment is often the best way to save life, and telemedicine offers the solution for emergency rescue in this respect. How wireless technology can help is very obvious. An example can be as simple as the popularity of cellular phones in the past two decades, which has made a vast difference because people can immediately use their mobile phone to call for an ambulance from virtually anywhere, and so the amount of time saved compared to the era when it was necessary to look for a fixed line telephone can potentially be the difference between life or death to an injured person. This, of course, is only made possible so long as the cellular phone is within a service coverage area, thus extending coverage will improve the chance of saving someone. When used in conjunction with GPS, the caller's location can also be automatically reported. Wireless communication and multimedia technologies are combined in many ways for emergency medical services (EMS) as a diverse range of highly mobile devices become available.

Telemedicine can do far more than this. Ansari et al. (2006) outline how wireless telemedicine can be used in an emergency situation. Cellular phones equipped with cameras can do much more than just call an emergency center for assistance. Among various examples, Martinez et al. (2008) report the use of cellular phones in remote diagnosis, for example by their ability to transmit information pictorially about color change (which is often used as a disease marker). In this particular case, test strip images are sent indicating the presence of certain kidney diseases. To serve this purpose, the cellular phone's camera will suffice so long as the "color depth" is adequate to distinguish between different color changes reflecting the properties of the fluid under test. Here, the color depth is determined by how many binary bits are used to represent each primary color (namely red, green, and blue) of a given pixel in the image. A camera whose color depth is n bit is capable of capturing an image of 2^n different levels of shades of each primary color. Since the indication of the presence or absence of a substance does not require the distinction of subtle color change, an ordinary cellular phone is good enough for such an application. However, in other situations such as capturing images showing a wound, the required image quality may far exceed what a cellular phone's built-in camera can capture. It is therefore necessary to use more sophisticated devices in emergency rescue missions.

There is a huge range of information that telemedicine systems can capture remotely. Figure 3.2 shows the framework of an emergency rescue system capable of providing paramedics with a convenient medium for sending a large amount of information about an injured person to the hospital so that necessary preparations can be done prior to the patient's arrival. We look into the details of this system. In situations where fire engines are also required on site, direct communication can be provided for linking paramedics and firefighters to facilitate collaborative operations. Each paramedic carries a number of wearable devices, including camera, sensors, and communications equipment. Similarly, a firefighter can also wear tracking devices, gas detectors, and oxygen level indicators. The ambulance serves as an AP for all paramedics and provides two-way communication between the paramedics and the hospital, where paramedics can retrieve a patient's medical history from the electronic patient record stored in the hospital so that information such as allergy and health conditions can be known when first aid treatment is provided, as described in the next section.

3.2.1 At the Scene

A WLAN serves the proximity of the ambulance while attending the scene of an accident. It simultaneously connects devices carried by several paramedics so that data can be sent to the hospital.

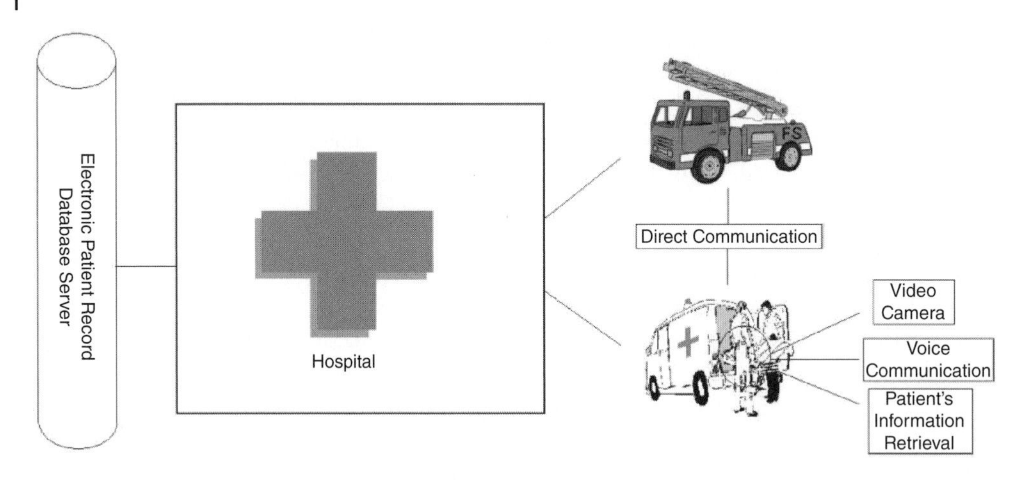

Figure 3.2 A simple emergency rescue system where multiple vehicles on the scene can coordinate the rescue operation with direct local area communications. Each vehicle is also simultaneously connected to supporting personnel.

Conversely, information about a patient can also be retrieved from the electronic patient record stored in the hospital's database.

Covering the area surrounding the accident, unless the recovery maneuver takes place deep within a high-rise building or a densely vegetated forest where the ambulance is unable to be parked nearby, a typical IEEE 802.11n WLAN will normally suffice. One major advantage of such a network is that line of sight (LOS) is not necessary to maintain a connection.

This network can perform the following: wearable camera captures high resolution images showing details of the injury sustained by the patient, image processing algorithms that estimate the amount of blood loss performed by object extraction that approximates the volume of spilled blood corresponding to the loss, various sensors acquire respective vital signs such as heart and respiratory rate, blood oxygen saturation (SpO_2) level, etc. In providing immediate healing, paramedics may need to rapidly retrieve the medical history of the injured patient, information such as drug allergy and any spreadable disease are vitally important so that necessary precautions can be taken. Video conferencing technology also makes consultation with specialists much easier especially when the patient's conditions are sent to the specialist in order to provide remote advice.

So, a vast amount of data is captured by different devices of each paramedic. There are various issues that need to be addressed. First, identification of each set of data must be clear in situations where more than one patient is treated on-scene, i.e. a given set of data belongs to which patient must be readily distinguishable. Whether all paramedics are well within the network's coverage area, it is necessary to ensure that paramedics can move freely near the ambulance with the assurance that they remain connected at all times. Data security must be addressed to ensure that patient's information will not be stolen by unauthorized personnel nearby, and at the same time not interfered or tampered with during transmission in an uncontrolled environment. To get some idea about how much data is collected by each paramedic, Table 3.1 lists an example of a paramedic carrying the devices described above. This may not seem to be a high demand compared to our daily usage of the Internet. One important difference to remember is that in the event of a failure we can easily reload a page while surfing the Internet, but in life-saving telemedicine applications the circumstances may not permit a second round of data capture and retransmission should a problem arises. So, adequate resources must be provided to ensure a certain margin for error is accommodated.

Table 3.1 Data throughout requirements for some telemedicine applications.

Source	Format	Approximate data rate	Compression
Video camera	FHD 1980 × 1080 25 fps	50 Mbps	Yes
Still image (each)	6000 × 4000, JPEG	6 MB	Yes
Voice	3 KHz bandwidth, 32 KHz sampling	128 Kbps	Yes
ECG monitor	12-lead ECG	12 Kbps	No
Ambient sensor	Binary data stream	<2 Kbps	No

Figure 3.3 Taking a closer look at the ambulance shown in Figure 3.2, there are a number of issues that may affect data communication around the ambulance.

In Section 2.4 we discuss the issues associated with outdoor wireless communications. In situations where paramedics attending to an accident scene will often find diffraction and reflection the utmost degradation factors to the connectivity of their devices. To see what the possible issues are, we zoom into this portion of Figure 3.2 to show the surrounding of the ambulance in Figure 3.3. As the paramedics are likely to move around during the rescue operation, there is no assurance that the transmitting devices will maintain a clear LOS to the AP antenna at the ambulance. The amount of diffraction primarily depends on the geometry of the obstructing object, and the properties of the signal, such as the amplitude, phase, and polarization of the carrier wave at the point of hitting the object. Sometimes the wave may also be partially diffracted during the process of reflection. The extent of reflection and diffraction are primarily dependent on the material of the object it hits, and generally affected by the polarization and the incident angle.

3.2.2 Smart Ambulance

One of the key challenges for smart ambulance design is to install the technology while meeting the demands of ergonomics, EMC, and ease of cleaning (Fong et al. 2018). In the most primitive form, a smart ambulance remains connected to the response center and information such as consultancy and electronic patient records (EPRs) can be readily retrievable by on-scene paramedics (Fong et al. 2017). Additionally, a local IoT network provides connection between ambulance instrument and

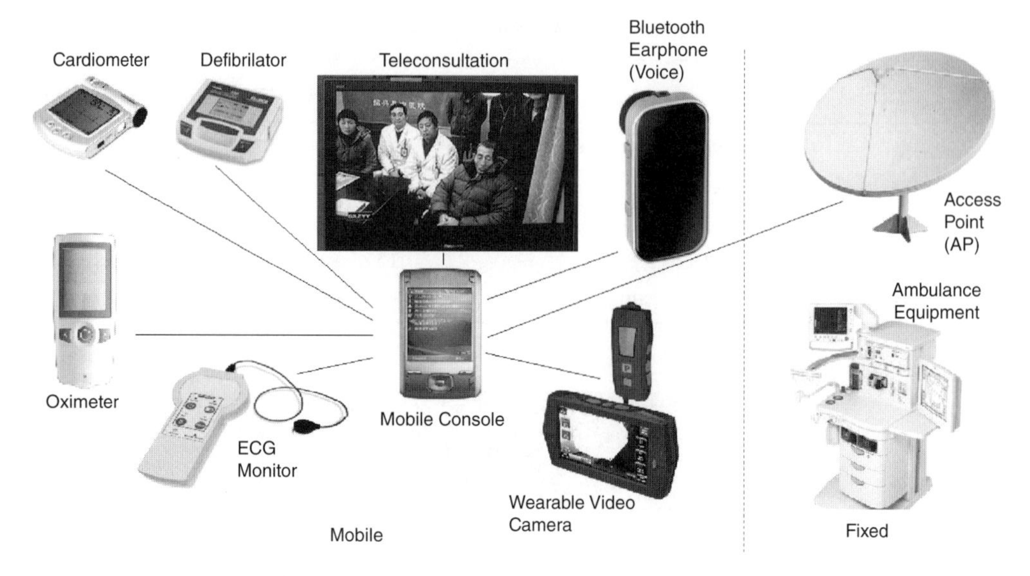

Figure 3.4 Wireless devices serving a paramedic on the scene.

paramedics, as shown in Figure 3.4, where a telemedicine link connects the ambulance to the hospital. At the rescue scene, the smart ambulance is equipped with the necessary equipment to support the paramedics.

A number of wearable devices may be carried by a paramedic depending on the nature of rescue and types of information sought. Figure 3.4 shows a collection of wireless equipment that a paramedic wears when attending to an injured patient. Owing to design considerations for small wearable devices, it is often advantageous to set up a BAN using Bluetooth instead of connecting individual devices directly to the ambulance as part of the local area network (LAN). In this particular example, a customized personal digital assistant (PDA) acts as an MBU that collects data from all sensors and cameras worn and it is connected to the ambulance network via a 2.4 GHz link. Chen et al. (2017) discusses situations where wires may sometimes be preferred in hooking up wearable devices. The paramedic's posture and the desired device interface may make wires preferable to options like Bluetooth or Zigbee. In general, small devices fitted in a pocket that requires minimal interaction during usage can be connected via wires. Wires are usually more reliable and data communication will not be affected by movement and orientation. It should therefore be a preferred option in situations where wires will not become tangled and the user's movement will not be affected in any way.

No matter what an individual device's function is, comfort and ease of use must be taken into consideration during design. Power consumption is an important factor to reduce size and weight. Also, reception properties in relation to movement and orientation need to be thoroughly studied for optimal operational reliability. Most of these wearable medical devices are highly customized, so very few off-the-shelf apparatus are available in the market. Ergonomic design is a vital attribute to ensure that, when worn, the device will not affect the paramedic's normal duties in any way while capturing data. Advances in programmable digital signal processing (DSP) chips enable one single type of processor to be tailored to drive virtually any sensor.

The specific supporting devices to be worn depend on the circumstances of each rescue mission. For example, illumination may be necessary for night operation and this requirement will draw more power, hence battery life may be significantly shortened. Most devices require waterproof housing for reliable operation in heavy rain. Secured mounting ensures that nothing will fall off while the paramedic runs. Many devices are available depending on the type of information required. It is even possible to implement physiological monitoring in disaster recovery where paramedics may be stretched to work nonstop for many hours.

3.2.3 Network Backbone

A backbone is a central part of any communication network that interconnects various pieces of devices, systems, and sub-networks. Its main purpose is to serve as a path for the exchange of information between different entities within the network. The backbone (see the example shown in Figure 3.2) is a 17 GHz wireless network described in Fong et al. (2005a). This is expanded in Figure 3.5 to show the relevant part of Figure 3.2 in more detail. The choice of carrier frequency primarily depends on the type of network being connected as well as government licensing legislations. This provides a two-way wireless link between the ambulance and a radio hub at the hospital. An IEEE 802.16 point-to-point network would work well when the ambulance remains stationary at the accident scene but its performance significantly deteriorates when the ambulance moves (Ansari et al. 2008). Mobile WiMAX would be a better option for moving vehicles if continuing assessment throughout the course of traveling back to the hospital is necessary. However, since most, if not all, vital information is acquired from the scene, mobility support is generally not essential for accident recovery.

Communication between the hospital and the devices featured in Figure 3.4 that the paramedic wears may not be available at all times. For example, tall structures may leave no LOS within the proximity of the accident scene or along the path between the scene and the hospital. From what we learned in Chapter 2, there are several issues to consider. The best way to ensure network coverage is by surveying the service areas covered by the hospital to establish a terrain elevation database that essentially consists of a computer elevation map. Its main function is to represent the terrain information to model the effects of buildings and trees in the township likely to affect communication with ambulances when serving different areas. Any given location z at a particular (x, y) position of the township is representing the relative altitude of the ground above a fixed reference, such as the rooftop of a high-rise building. A comprehensive database of these terrain points (x, y, z) can be viewed as a grid for the entire coverage area. The terrain database should cover anywhere that the hospital serves so that wherever an ambulance attends will be covered.

This communication link requires high reliability and availability while delay is not normally an important factor. Transmission must be error free, but since information about a patient does not have to reach the hospital in real-time, data retransmission is not an issue. In the event when

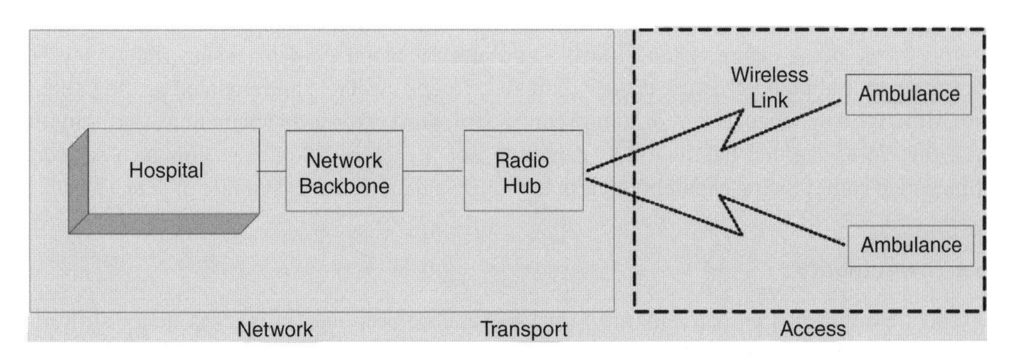

Figure 3.5 Emergency rescue network block diagram. A radio hub (which can be the base station of a cellular network) provides a link between individual ambulances and the hospital.

data is either lost or corrupted, they can be sent again. Retransmission guarantees successful data reception in the expense of time delay.

Heavy rain is well known to be a significant contributory factor to serious accidents. Indeed, rain not only increases the risk of an accident, it also affects the performance and reliability of radio links. Most notable problems caused by rain include attenuation and depolarization. The former weakens signal strength that leads to a reduction in coverage, whereas the latter has a very substantial impact on wireless links utilizing both vertical and horizontal polarization signals as depolarization may cause the two signals to overlap with each other, which eventually means there will be no signal at all.

The network backbone is a vital part of the telemedicine system providing a reliable link between personnel working on site and the hospital. Fong et al. (2005b) propose that generally, where licensing of radio frequencies permit, lower frequency of 10 GHz or below should be considered for tropical areas where heavy persistent rainfall of over 20 mm/h is frequently expected. Otherwise, anywhere between 25 and 40 GHz should be used for optimizing the backbone's performance and to avoid spectrum congestion.

3.2.4 At the Hospital

As information showing the extent of injury a patient suffers in an accident is sent to the hospital along with vital body signs, surgeons in the accident and emergency (A&E) unit can get a good idea of what to expect when the ambulance brings the patient in. An electronic patient record can be automatically retrieved so that the patient's medical history can be known. While the benefits to A&E personnel brought by telemedicine technology is clear, there are still a number of challenges that need to be dealt with. Earlier work by Benger (2000) identifies a number of potential problems that arise from the expansion of service capability, including human factors like convenience, reliability, and integrating telemedicine into current practice.

Surgeons and support staff need to get themselves familiarized with the system, what it delivers, and how it can be fully utilized. This may inevitably entail training to ensure that information delivered by the telemedicine system is correctly interpreted. Integration with existing medical systems may also demand special consideration as linkage to a proprietary system may involve compatibility and interoperability issues. Potential cause of interference is a topic that warrants investigation as transmitting devices are used in telemedicine systems, although Tachakra et al. (2006) report that no noticeable interference between the telemedicine transmitting devices and delicate medical instruments in A&E has been detected.

Telemedicine is capable of providing vital information about a patient prior to arrival. Information such as heart and breath, images showing extent of injury, vital signs such as heart and respiratory rates, pulse oximetry (SaO_2/SpO_2) and arterial blood oxygen tension (PaO_2) levels, and diastolic arterial blood pressure (DABP) can all be made available and updated as the patient arrives. Although a wide range of information can be sent, most of these do not incur a large amount of data; therefore, channel bandwidth is generally not a problem. Some systems also support real-time video conferencing capability that may require a data transmission rate of over 1 MB/s.

3.2.5 The Authority

Since e-health entails patient surveillance, privacy therefore become a primary issue for authorities concerned with any possible lawsuit that seeks damages for breach of security. Liability issue is therefore one impediment that may impact the popularity of telemedicine. Deployment crossing state boundaries can cause regulatory issues if a service spans different states with different legal and licensing directives. Initial deployment expenditure and lack of funding may also be an important issue limiting the exploitation of telemedicine for A&E as the cost benefits may not be obvious to officials even though the precious time saved in treating an injured patient that ultimately saves lives can be exceedingly significant. Authorities often make decisions based on a business point of view; sometimes the monetary investment is anticipated to yield financial returns within a specific timeframe. So, authorities need to be convinced of the perceptible advantages brought by telemedicine.

Technological challenges may not be as difficult to conquer as obtaining government endorsement in many cases. Setting up a comprehensive network for supporting emergency rescue may entail cooperation from various parties discussed in this section. Also, time needed to provide adequate training for healthcare professionals of varying capacities may be perceived as a time-consuming process for authorities. The long-term benefits brought to life saving are very obvious, yet gaining support on the financial and time investments needed is another issue that needs to be worked on.

The idea of supporting a wide range of healthcare services through telemedicine involves many government agencies and authorities, and many of these are country-specific. In the case of USA, medical devices are regulated by the Food and Drug Administration (FDA), the Federal Communications Commission (FCC) oversees radio spectrum allocation, whereas telecommunication policies are governed by the National Telecommunications and Information Administration (NTIA). The key point to note here is that, even for a very simple device, design and implementation could involve regulatory compliance with several authorities.

3.3 Remote Recovery

Wireless telemedicine facilitates healing just about anywhere, on land, at sea, as well as in the air. Over a decade ago, commercial airliners began linking their planes to MedLink, described in Mchugh (1997) as a service to complete healthcare coverage beyond the earth's surface, offering basic life support information to trained airline personnel to carry out basic medical emergency procedures and decide whether urgent medical attention is necessary. The underlying technology allows the airline industry to save the time and money involved in unscheduled stopovers to drop off passengers for immediate medical attention who may not need it.

Through video conferencing, experts in different countries can offer real-time medical advice to service personnel who may not even have any prior healthcare training. It is not only a matter of offering recommendations as what to do: telemedicine also enables the electronic medical records of a specific passenger to be retrieved so that any existing medical conditions can be known. In addition to assisting patients in the air, telemedicine makes recovery and healing available virtually anywhere. Remote recovery often involves swift discovery of the patient's exact whereabouts, and any potential hazards surrounding the patients must be known to avoid endangering rescuers. Technology allows telemedicine to assist remote recovery in these aspects. We shall look at three situations where telemedicine frequently helps save lives of both the general public and the professionals who risk their lives to rescue them.

3.3.1 At Sea

Maritime recovery is a challenging scenario since cellular communication networks cannot serve areas in the sea. In vast oceans where no land can be seen, communication is limited to satellite links. Although most modern vessels are equipped with precision GPS, this may not always work because the person who requires urgent rescue may be thrown overboard, or the vessel may have sunk or be without power. Technology makes recovery in many situations much easier than before. Finding a person is perhaps the very first thing rescuers need to do at sea. Video extraction technology used in conjunction with high-resolution video capturing makes locating a person or an object in the sea much easier. Coordination between rescue boats, helicopters, and a control center must be supported in real time as current drift can move the person to be rescued very quickly.

Satellite communication is often used in maritime rescue since this is the only means of providing comprehensive coverage across vast oceans. Since satellite communication utilizes millimeter-wave band in the GHz magnitude, retrieval of sunken vessels is virtually impossible, owing to absorption through water at these frequencies. Underwater wireless communication is far more challenging than communication through air. In contrast, acoustic waves propagate some five times faster in water than in air making acoustic pressure channels suitable for underwater communication. Long-range underwater acoustic propagation studies commenced as far back as the fifteenth century. As its name suggests, an acoustic channel involves an audible frequency range that spans between tens of Hertz up to around 20 KHz. So, wireless communication systems have very different requirements in terms of transceiver structure and antennas. As an acoustic wave propagates far slower than a millimeter-wave band through air, a long transmission delay is to be expected. In addition to a significant propagation delay, underwater communication also suffers from a high variation of multipath and narrow available bandwidth.

3.3.2 Forests and Mountains

Search and rescue in densely vegetated areas often cannot be performed visually. Although infrared cameras can help pinpoint the location of a person in some situations, it is by no means an all-round solution and its effectiveness is limited by many circumstantial factors. Rescue is made even more difficult since cellular phone coverage is highly unlikely to be available in remote forests and mountains.

Radio communication is extremely difficult in these areas. First, it makes no economic sense for operators to provide coverage to these areas given the enormously low subscriber density and utilization rate. An LOS link cannot be maintained, owing to dense vegetation; therefore, diffraction and reflection can be important degradation factors that affect successful communication. Remember, the basic mission of a radio link is to deliver sufficient signal power to the receiver so that some kind of meaningful information, such as the receiver's whereabouts or images showing the surrounding area, can be realized. The effects of physical obstacles like plants on wave propagation go back to the concept of clearing the Fresnel zone, as outlined in Section 2.4, which refers to the volume of space enclosed by an ellipsoid of the two antennas between the ends of a radio link. The radio link can be maintained if no objects within the area cause significant diffraction into the corresponding ellipsoid. Having said that, it does not necessarily imply that failure to clear the Fresnel zone will always result in a loss of communication. The actual network degradation experienced very much depends on the operating environment. The ground reflection path will sometimes be obstructed by the disturbance of trees and other plants, whereas ground reflections can be a major factor of path loss in plateaus of thick vegetation or lakes. Although the existence of an LOS path may not be likely, if it does have some gaps between the transmitter and the receiver, one direct path and a ground-reflected path may both exist. In such cases, the path loss would depend on the relative amplitude and phase relationship of the signals propagated through the two paths. The amplitude and phase of the reflected wave depend on a number of variables, including the conductivity and permittivity of the reflecting surface, frequency, angle of incidence, and polarization. They will overlap each other, and the effects can vary. The relative signal strength between the two paths would depend on the ratio between the ground-reflected paths having Fresnel clearance and the LOS signal path (if present). If the former undergoes little loss due to reflection, the two paths would have similar signal strengths. This situation may result in either a boost of up to 6 dB over the signal across the direct path alone or cancelation, resulting in additional path loss of 20 dB or more depending on the relative phase shift of the two paths. Two signals combined in phase (without a relative phase shift) would result in "constructive interference," while being out of phase (180° relative phase shift) causes "destructive interference," as taught in high school physics. Spread spectrum techniques and antenna diversity are often considered effective solutions to control this problem. In addition, attenuation from fog can be significant at frequencies above 20 GHz, and this can be an important consideration for communication systems in humid forests.

"Clutter," defined by *Wikipedia* as "excessive physical disorder," is a term often seen in wireless communications that refers to vegetation that affects signal propagation. Clutter usually causes attenuation and scattering when radio waves hit a surface resulting from variation of multipath, owing to movement of branches and leaves by wind. The extent of scattering usually depends on the density of leaves, their shape, and the amount of water held within each leaf. It is therefore extremely difficult to predict the propagation characteristics through forests.

3.3.3 Buildings on Fire

Among the types of awkward rescue operations discussed in this section, recovering people from a fire is generally the most challenging situation given the amount of time available to rescuers. Fire can spread very quickly and is almost always accompanied by thick smoke, impairing vision. The combined effect makes finding the exit path very difficult at times. Paramedics and firefighters alike require extremely reliable communication systems and tools that can bring them back to safety in the shortest time. The fact that people without special needs may not carry any transmitting identification devices makes finding people trapped more difficult in the event of a fire. It is therefore an unrealistic expectation to locate a missing person by using a radio under any assumption that a person sought wears some kind of radio transmitting device that is fully functional. This problem implies safe recovery can only rely on professionals risking their own lives to ensure any missing person is found at the earliest opportunity and to lead the person through a safe escape route to safety. Unfortunately, a floor plan may not always be available for rescue professionals upon entering the building. A building on fire can easily turn itself into a maze. Further, the path taken when entering the building may not necessarily be the shortest and safest to take for escaping. The entry path may also risk subsequent blockage by falling obstacles. Having said all this, how can technology assist in finding the optimal path?

Thick smoke can severely impair vision, making it virtually impossible to see one's surroundings. Likewise, radio links can be blocked by partitions composed of energy absorbing materials. Metal is a particularly "unfriendly" material for radio waves to pass through and it is used in buildings for a variety of reasons. Without experiencing it for oneself, it is difficult to understand what it is like for rescuers trying to find people in a smoke-filled building, battling with the heat and fear while dealing with their communication being intermittently cut off. Technology is here to assist their operation and to maximize the chance of a successful rescue. This is an area where telecommunications, in particular, can help save lives.

We learned in Chapter 2 that signal frequencies in excess of several GHz generally penetrate materials better than lower frequency waves. The reader may have experienced cellular phone

Figure 3.6 A well-equipped paramedic capable of carrying out a rescue mission in hostile scenes such as mines, buildings on fire, and at high attitudes as the paramedic's health and ambient environment is also continuously monitored and their exact location tracked.

service interruption when entering a lift (or elevator), as most lifts are made of steel enclosures that effectively act as a metal cage that "shields" the commonly used 900 MHz signal for cellular communications. For this reason, firefighters need something more reliable than a mobile phone to ensure that they can find the safest exit route. A comprehensively equipped paramedic is depicted in Figure 3.6. It shows a rescuer with different apparatus as those shown in Figure 3.4 and some with data transmission and reception capabilities. Each communicating device has its own function in ensuring the user's safety.

Another vital survival tool is the amount of remaining oxygen to ensure that breathing can be sustained until reaching a safe location. An alert must be generated to allow adequate time for escaping, and in the event of any mishap the rescue team must be able to bring in any necessary additional oxygen supply. In the process of issuing such an alert, it must be made in a subtle way to avoid putting unnecessary pressure on the firefighter to ease any additional anxiety. In addition to the oxygen supply status, the detection of any flammable or toxic gas and, if available, video footage showing the site's environment can be reported to an off-site control center or command post in order to build a better picture about what is going on inside a blaze. So, a reliable network that supports a range of communication needs is necessary to keep rescuers safe. A report by the TriData Corporation (2005) points out a number of deficiencies with the conventional VHF (very high frequency) radio of 30–300 MHz range used by US fire departments. More recently, the FCC assigned the 800 MHz band for public safety radio communication in an effort to reduce spectrum congestion that spanned across the range used by commercial broadcasters and so risked interference. It is reported that different radio channels are often used within a fire department. Interoperability is therefore said to be a major issue. So, is it really feasible to standardize firefighters' communication systems?

In hindsight satellite may sound good, owing to its vast coverage and excellent penetration properties. A satellite phone may be a good choice for the sole purpose of providing a medium for talking to off-site supporting personnel. However, the precision for location tracking is far from adequate for fire rescue within a building. GPS accuracy is affected by uncontrollable factors, including satellite placement and nearby buildings, both of which can increase the DOP (dilution of precision). Normally, GPS can only identify a location within a radius of several meters at best. This may mean a person sought after in an emergency may be mistaken as being trapped in an adjacent room that consequently increases the search time. Such precision deficiency may even lead to a search being conducted on the wrong floor of the building when 3D positioning is not used. Satellite is therefore not the suitable solution for fire rescue. Other solutions such as RFID for short range path recognition and marking can be explored since markers can be placed automatically along the entry route during an operation and should therefore be considered.

There are several fundamental requirements that need to be satisfied: equipment needs to be lightweight and easy to operate, with minimal intervention, possess precise position tracking, have good penetration through various materials commonly used in building construction, and be resilient to excessive heat. Until now there is no single technology that satisfies all these requirements. The most likely solution is therefore to integrate different solutions with a high degree of interoperability.

In this section, we have looked at three different demanding situations for rescue operations. Each of these has different fundamental requirements and problems. The only thing they all share is an exceptional level of dependability and ease of operation. Through technological advancements of different types of communication networks, more sophisticated wireless systems can be developed to meet growing needs in an effort to improve the chance of survival in difficult situations.

3.4 Smart Hospital

Information technology (IT) has increased automation and safety into hospitals over several decades. The list of improvements in the way a hospital runs introduced by IT is endless. Felt-Lisk (2006) gives an example of how six different areas of IT make a significant difference to a hospital. It is simply impossible to cover all of the different IT applications in one single book, and so we concentrate here on how communication technologies can help to modernize a hospital by first briefly reviewing a case documented by Williams et al. (2019) before delving deeper into the important role played by IT in the daily operations of the various departments of a hospital. The article starts by reporting a case where a physician received an automated warning when he requested a drug to be dispensed. An information system in the hospital had detected a possible risk of mixing this particular drug with one that the patient had previously taken. Such alerts prompt the physician to prescribe an alternative medicine as a remedy to eliminate the detected risk. This is just one of the many examples where the timely delivery of information easily saves lives. The article then describes physicians examining X-ray images and controlling a robot inside the hospital that can be accomplished remotely. All these plus many other tasks are made possible by telemedicine. Technical advances in telemedicine and digital health have come a long way over the past decade, and health monitoring devices have become much smaller and more affordable.

A hospital that provides comprehensive services may be composed of many departments with a central administration. Figure 3.7 shows a simplified version of a typical hospital having several departments all housed in a single complex, and all linked together by a network for information sharing and coordination. Obviously, each department would have its own requirements on the

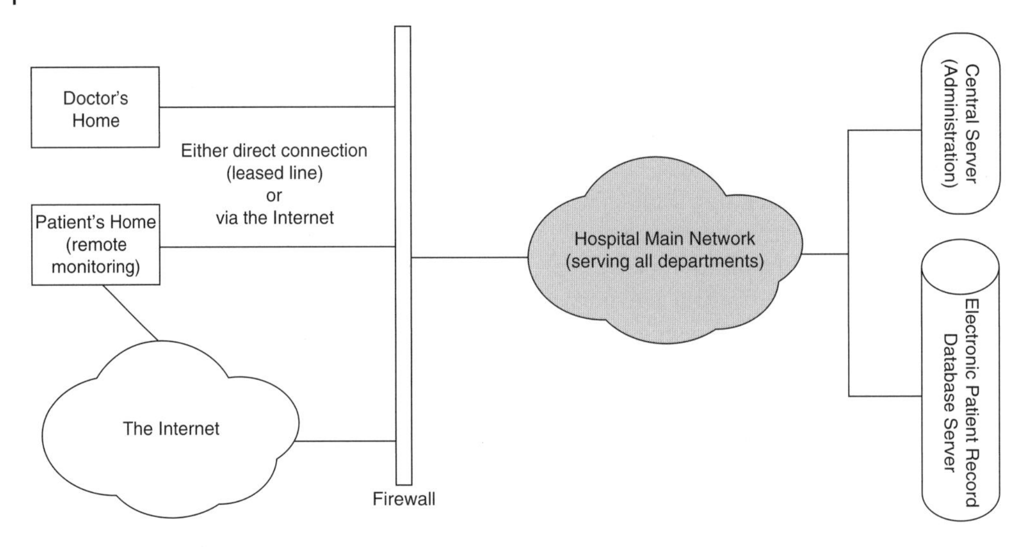

Figure 3.7 Block diagram of a typical hospital network.

type of information processed and the urgency of information retrieval prioritized. For example, an A&E nurse treating an injured patient would require information about the patient much more urgently than a pediatrician carrying out a general health assessment. Both of these involve the retrieval of medical history and updating new information, but the tolerance to delay and the ease of reading retrieved information are very different for these two examples. In Section 3.2, we discuss the importance of telemedicine on maximizing the efficiency of the A&E department by providing surgeons with the necessary information about an injured patient even prior to arrival. Here we look at three other examples where communication technology can help save costs and lives in a hospital. Since there are so many different situations where telemedicine finds its importance at a hospital, the examples below are:

- Cost saving measures in radiology with accurate and timely information.
- Precision control of robots for surgery.
- Reliable tracking of newborn babies to ensure misidentification never happens.

These examples are selected to illustrate different categories of wireless communication applications, namely quality assurance, remote sensing, and surveillance. Many other situations can utilize telemedicine with very much the same underlying technology.

3.4.1 Radiology Detects Cancer and Abnormality

Radiology is an important area of medicine for early diagnosis and treatment to ensure maximum chance of survival. This is an application where communication is critical both among hospital staff and patients. So, telemedicine extends beyond the hospital network. Delay in the delivery of correct information to the appropriate parties may lead to an unnecessary delay of treatment that may result in legal consequences. Radiology involves the accurate interpretation of medical images. The images by themselves frequently do not make any sense to the patients, and so contextual explanation is an important part of communication between the hospital and the patient. Images are therefore accompanied by reports. Figure 3.8 shows a block diagram of what the radiology information system entails. Read and write permission must be granted independently to each

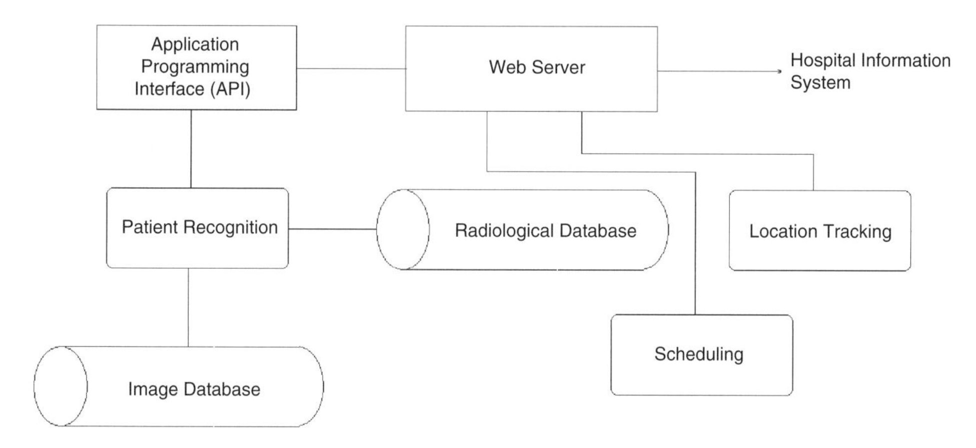

Figure 3.8 Case study: radiology information system.

hospital personnel involved to avoid any risk of unauthorized access and the undesired alteration of information. In this system, an SMS (short message service) text message is automatically sent to the patient as a reminder for a checkup through the scheduling module. When the patient arrives at the hospital, an RFID card carried informs the radiology department staff of the patient's arrival and their electronic patient record is automatically retrieved (it will also be used for tracking). Results are sent to the hospital information system for analysis so that any necessary action can be taken. Archives will be stored in separate databases for images and radiological data.

One major objective of effective communication is cost saving: ensuring the proper delivery of information can potentially save a large amount of money. As Brenner and Bartholomew (2005) report, the cost of an erroneous communication is, on average, over $200 000 per case. So, the entire process shown in Figure 3.8 from the radiologist capturing the image to delivering an extract of its associated report to the patient must be made error-free to ensure the proper communication of information is attained so that cases of indemnity payout are kept to an absolute minimum. This is best achieved by the proper design and maintenance of the related telemedicine system. Referring back to Figure 3.8 again, a radiograph is taken with a patient in the X-ray facility location with one radiologist serving a number of patients per session. These images are captured, digitized, and sent to specialists handling respective patients.

Next, we investigate what can possibly go wrong during the process. The worst possible calamity is mixing up images of different patients leading to an incorrect diagnosis of a healthy person with cancer while the cancer sufferer is wrongfully discharged as free of cancer. Obviously, such misdiagnoses will lead to negative psychological consequences to patients and their families, and healthy patients being unnecessarily operated on while leaving the other patient with their cancer undetected. The first line of safeguarding each image and ensuring that they are correctly referred to the respective patient is by the proper filing of each image throughout the entire process. With the correct procedures in place an strictly adhering to them, information management can help ensure images are well taken care of. Next, when each image is successfully passed on to the specialist, it is examined and any abnormal signs detected, either manually by the specialist or with automated feature extraction using image processing techniques. At an early stage of tumor formation, especially in the CIS (carcinoma in situ) stage, subtle signs may not be easily seen. This is a critical time to prevent the invasive phase when other treatment options are available. Noise free high resolution images without loss of fine details are therefore vitally important in the image transmission process. Radiographic images are usually transmitted digitally and image clarity very much depends on the

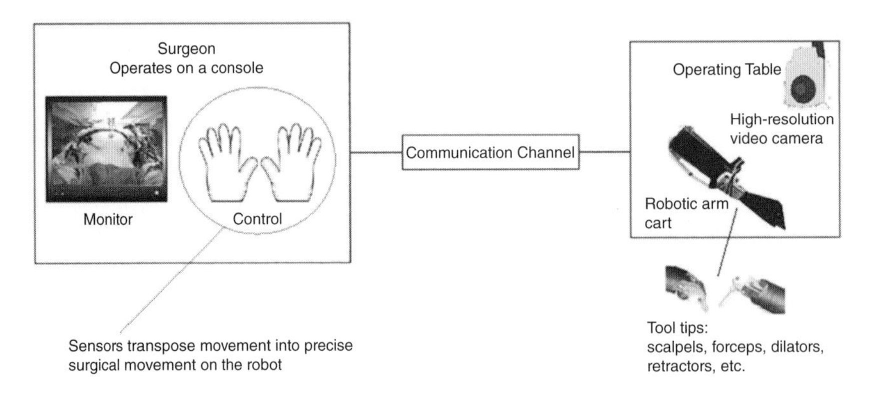

Figure 3.9 Telerobotic surgery.

bit error rate (BER), which effectively measures how many bits are sent when one bit is corrupted in the data stream. Ultimately, we never want to miss any subtle sign identifying the presence of a tumor because of a few bits missing from the digital image.

3.4.2 Robot Assisted Telesurgery

The term "telesurgery" refers to surgical operations being carried out by surgeons remotely without being physically situated in the operating theater. Robots are only capable of carrying out precise surgery because of the introduction of tiny sensors and actuators that were previously unavailable. These small actuators make very small movements that usually involve moving in all three dimensions. The primary function of an actuator is to initiate a robot's movement based on instruction given by the surgeon. Figure 3.9 shows how a surgeon can carry out an operation remotely by using a telemedicine system to control a robot in the operating theater. In addition to hand controls, Peters et al. (2018) reports that eye controlled robots make 3D mappings of tissue possible and automatically calculate the depth of tissue by tracking the controlling surgeon's eye movement to precisely track where the surgeon is operating.

Robotic telesurgery effectively brings a surgeon's professional techniques into an operating theater that does not have a surgeon physically present. However, in order to make this happen a large amount of data exchange is involved between the surgeon and the robot "acting on behalf." For a start, the surgeon needs a good view of what is going on inside the operating theater. Cameras are installed in the operating theater that must incorporate remotely controllable rotation and high-power zooming functions. Also, the video image captured must be displayed at the surgeon's side in real time without any noticeable delay so that any movement of the robot to take an action will not be delayed. Even a very small amount of time delay in the robot's action can lead to irreparable damage to the patient's body. Time delay (latency) is a big issue with long distance telesurgery. With telecommunications that span continents, transmission delay is an unavoidable issue. This is likely to be one of the most challenging issues for long distance operations.

User interface for control must be carefully designed to ensure that the entire system cooperates well with the surgeon. Voice activated control would ensure minimal disruption is caused during the operation. This involves speech recognition algorithms that not only correctly interpret each individual command issued by the surgeon but also to identify the voice of each individual person in the theater so that only the respective surgeon's command is acted upon. This is vital in ensuring that voice commands from other surgeons or supporting staff will not be mixed

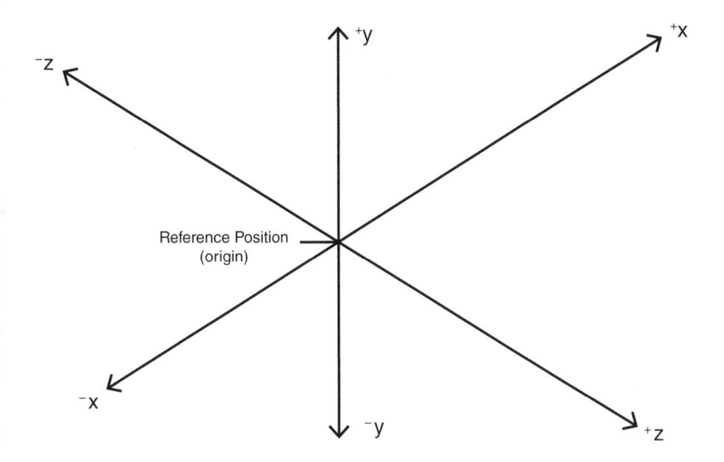

Figure 3.10 The 3D space is represented by "six dimensions," effectively the +/– of x, y, and z axis.

up and inappropriately acted upon. Robot control requires exceptionally high precision 3D hand movement manipulation commonly through a pair of virtual gloves. Sometimes the term "six-dimensional" is used to describe the sensors in these gloves. The "six dimensions" are just the positive and negative directions along any of the x, y, and z axes representing the 3D space away from any fixed reference point, as depicted in Figure 3.10. The sensor's movements drive the respective actuators in order to control the robotic hand that operates a surgical tool, including the change of different tools on the robotic hand. In addition to control signals, a voice channel should also be given for video conferencing between the surgeon and personnel inside the operating theater. So, telesurgery involves transmission of high resolution real-time video images and control signals of high precision. Minimizing time delay is a vitally important issue for the successful implementation of robot assisted surgery.

Portable robotic surgeons would be extremely useful in demanding remote rescue missions, such as the examples discussed in Section 3.3. In the worst case, only an expensive robot will be written off without putting any precious life into jeopardy. They can even go underwater if necessary (Kawaguchi et al. 2016). There are, in fact, many situations where robots go into dangerous situations during high risk rescues.

3.4.3 Ward Management Using RFID

We commence our discussion by looking at keeping track of babies in a hospital as an example of how technology can help prevent mistaking individual babies among a group (the word "tracking" in our example has nothing to do with any possible breach of privacy that involves surveillance). Cases of newborn mix up that often leads to avoidable yet substantial emotional damage payouts have been reported throughout the world. Most mix-up cases are indeed avoidable since they are direct consequences of irresponsible personnel failing to follow the necessary procedures for handling babies. The good news is foolproof technology is here to help eliminate such risk simply with a couple of tags that cost a few pennies (a mere 10 cents). By referring back to Section 2.5, where we talk about RFID, it is not difficult to understand how RFID tags can help identify each individual baby.

Figure 3.11 shows the layout of a typical hospital maternity ward where RFID readers are installed on both sides of the main entrance and each nurse carries a handheld reader. When

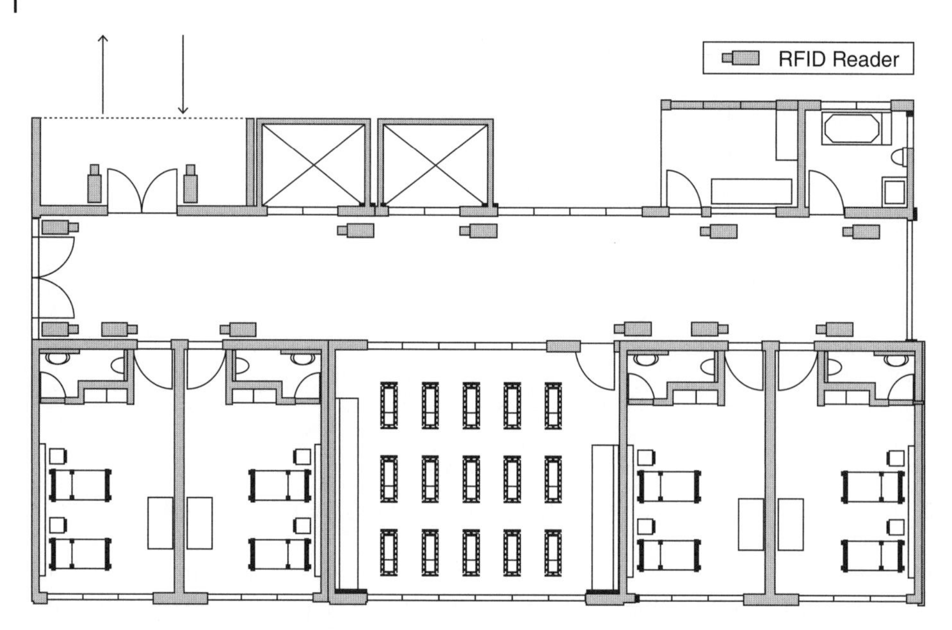

Figure 3.11 RFID readers installed in a hospital maternity ward to ensure that every babies' whereabouts is known at any given time. Anyone who carries out a baby from the ward will be automatically recorded.

a baby is born two RFID tags are attached with separate submersible bands. The band must be comfortable while not too loose to pose any risk of falling off. The reason for using two tags instead of one is solely for redundancy so that in the event of any unforeseen problem there will still be another one for identification or confirmation. Since RFID tags are small and light, they are highly unlikely to cause any discomfort to the baby or inconvenience those hospital staff or parents handling them. Also, RFID tags are fairly robust and can be submerged in water so that they need not be removed for a bath. A quick scan of the tag can affirmatively identify each baby even though they may look very similar to each other.

It is even possible to track the movement of each baby if active (battery operated) tags are attached so that every time the baby (and the tag) passes a reader that is installed in a given location it can record the whereabouts of the baby. This also prevents unauthorized carriage of the baby away from a certain area simply by triggering an alarm when the tag comes close to the exit. However, this alarm system does not have any proficiency to preclude tag removal unless the band is tough to make cutting difficult. So, to make it more secure additional circuitry can be included to activate an alarm when the tag is tampered with or when the tag remains stationary for a specified time, say one minute, that may reasonably assume the tag has been removed or else the baby's movement would confirm the tag remains intact. This works particularly well with babies because they tend to move frequently even when asleep. The use of active tags may cost more but the benefits they bring are obvious. Typically, a baby only stays in the wards for a few days before going home. For this reason, battery life is not a concern since even a very small embedded battery can easily power an active tag for over a couple of weeks continuously. Further, RFID usage poses no risk of excessive radiation even to a delicate newborn, since the intensity of electromagnetic radiation emitted is even lower than surrounding cellular phones that people carry.

A communication system is necessary for linking the alarm system and pagers carried by hospital staff together so that they can be automatically alerted. This can be easily set up with a console that stores information about all tags issued with a map of reader locations. Information about the type of alert together with the location of the nearest reader that picks up the RFID signal can be broadcasted to all staff pagers for necessary follow-up actions.

3.4.4 Electromagnetic Interference on Medical Instrument

The examples demonstrate that wireless telemedicine is a vital part of an efficient and reliable hospital, but what about possible EMI that may affect the operation of delicate medical instruments? Radio transmitting devices such as cellular phones may cause medical equipment to malfunction, and in some cases even critical life-supporting apparatus can be badly affected. As ambient environmental electromagnetic noise from various sources both within and outside the hospital site cannot be controlled, proper shielding of medical instrumentation therefore becomes the most effective way of ensuring reliability, irrespective of interference noise levels surrounding the area of operation. However, many instruments by themselves are sources of EMI. For example, cardiopulmonary resuscitation (CPR) draws a vast amount of electrical current that would generate an excessive amount of noise. The proper design of housing with appropriate metal shielding will be able to protect an instrument from external interference.

Operating theaters and ICUs (intensive care units) are most vulnerable to EMI, owing to the inherent nature of the instruments used in them. In these areas, it would be necessary to restrict the use of transmitting devices including cellular phones by anyone nearby. As for the venue itself, it is possible to set up the partition with an absorption chamber, with pyramidal polyurethane foam inside its walls. This is a layer of foam that effectively stops electromagnetic radiation from entering the venue.

3.4.5 Smart Wearable Integration

Smart wearables are capable of tracking many health parameters of a user. Some health problems could well be triggered from a source that is not detectable by the wearable device by itself because of its limitation in the types of sensors being installed. The tremendous growth of the consumer healthcare sector has improved the quality of life of people with chronic diseases such as diabetes and dementia (Wei 2014). Devices capable of providing health management are followed by consumers' desire for longevity and health, and a patient's state of health can be made readily available to hospital staff. A range of important diagnostic information such as heart rate, medication record, activities, and dietary logs can be downloaded and mapped to one's EPR.

Health monitoring through smart wearables also reduces the problems of seeking unnecessary treatment that in turn puts extra pressure on already stretched hospital resources. Smart wearables are set to tackle this challenge through taking care of a wide range of health management needs and are highly customizable to cater for individual circumstances. The endless possibilities of health management using smart wearables lead us to the next important topic: general health assessments.

3.5 General Health Assessments

Telemedicine for healthcare extends beyond medical protection for patients in need of special attention. It can also facilitate the general public for maintaining good health in many situations, indoor or outdoor, at rest or on the move. No matter where we are, technology is always helping us with optimizing our well-being. Information technology is found in many areas of health assessments in daily life, for example dietary monitoring for those who are concerned about weight gain, color matching for skin care, calculating the amount of calories burnt during a workout, the nutritional intake of a child, baby monitoring alarm, automated reminder for a dental checkup, etc. There is something for all ages. Strictly speaking, even ergonomic design factors of appliances that can affect our well-being due to usage can have a close relationship with IT and healthcare very simply because proper product design eliminates the risk of causing users to require medical attention.

Telemedicine finds its use in many situations, for example assisting with the reduction of obesity for subscribers to weight-loss programs. They can have their body weight automatically sent to the control center for record keeping and progress tracking. Before ending this chapter, we shall look at a number of situations where telemedicine help us in our daily life. We take the availability of such technology for granted when using it on a regular basis. Let us look at how they work by looking at some examples.

3.5.1 Case Study I: Fitness Monitoring for a Morning Jog

Since off-the-shelf foot-contact pedometers can only count the number of steps taken, accelerometers and gyroscopes are often used for motion monitoring. Bouten et al. (1997) report that a sampling rate of around 18 Hz is adequate for sampling human activities. Pappas et al. (2004) and (Bamberg et al. (2008) have conducted comprehensive studies installing shoe integrated gait sensors into running shoe insoles, as illustrated in Figure 3.12, with combinations of accelerometers, gyroscopes, electric field sensors, piezoelectric sensors, and resistive band sensors (Morris 2004). This set of sensors is installed to capture foot movement. A tiny transmitter can send the data out for analysis on the level of activity and track the state of the user. This mechanism can also track

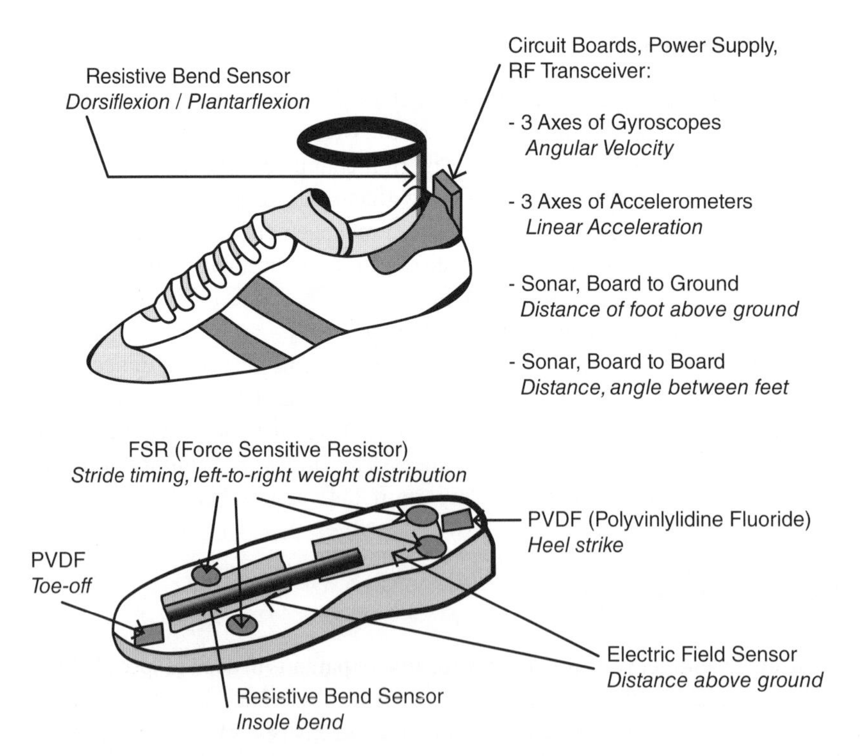

Figure 3.12 Schematic of the shoe-integrated gait sensors. Source: Reproduced with permission from Morris and Paradiso © 2002 IEEE.

uneven wear of the heel part of the sole and detect abnormal wear patterns and in so doing aid running comfort. Technology can help us keep track of how far and how fast we run, as well as reducing the uneven wear of our shoes.

Some people may jog with an intention of weight loss. This is also an area where telemedicine helps. Exercise accelerates digestion so one may feel hungry after running. Communication technology can help us activate a microwave oven so that it prepares our breakfast, at a certain predetermined stage of the jog, say the last 1 km before returning home. A signal can be automatically sent to the smart home control console to start heating the breakfast so that it is ready when the jog is over. This is just one simple task smart home automation can do. Anything else like a coffee machine can also be activated in very much the same way. After programming on the console, either programmed in advance or remotely via a cellular phone, no further action is necessary unless manual override is desired.

A morning jog not only strengthens our muscles, it is also a gentle cardiovascular workout that optimizes both respiration and blood circulation. It also helps ease any digestive problems that may be accumulated because of a busy work schedule. While we can feel the benefits ourselves, technology lets us quantitatively realize the difference and keep a record of our progress by logging our daily activities, such as length of jogging route, number of steps taken, duration, and heart and respiratory rates. A wearable pulse meter can be bought cheaply and some can even be part of a wristwatch. This little device helps keep track of our heartbeat while we jog. Technology can also help us monitor what we eat and automatically generate a nutritional report of each meal throughout the day. So, the after-jog breakfast can be prepared according to the amount of calories burnt. By linking the health monitoring devices to a home computer, the user can check the improvement of health on a daily basis and retrieve a meal recommendation list that is generated from the captured data for optimal nutritional balance.

Although running in an outdoor environment in fine weather may be more pleasurable than in a gym, sometimes a gym workout is more desirable as different types of equipment offer full body exercising – and it is weatherproof. So, this brings us to the next case study: gym health monitoring.

3.5.2 Case Study II: Gym Workout

Many gyms offer free Wi-Fi Internet access even though a physical workout does not normally need it. We may not surf the Internet in the gym but a wireless network does offer a range of possibilities for keeping track of our activities there. Just like what we can wear for a morning jog, small sensors can be worn throughout the body for reading different signs depending on the nature of the exercise. Walking or running on treadmills or steppers applies the same technology as that used for morning jogging – almost identical devices except that the downloading of captured data is much easier in this case as the gym wireless network can readily support continual downloading of data so that no memory storage is needed within the BAN of sensors and related electronics. Figure 3.13 shows a block diagram of a gym equipped with commonly found apparatus, for example a treadmill, stepper, weights rack, leverage bench press, elliptical trainers, exercise bikes, and seated row machine. Although there are many types of apparatus, how technologies facilitate health assessment can be very similar if we group them according to the way they are used. For example, in the health assessment perspective, the weights rack and leverage bench press are similar in nature as they both involve using the upper part of the body to lift a certain amount of weight. Weight-lifting is intended for muscle building. The result is best judged by the growth of muscle that can be detected by examining the change of body shape as progress is made. A quick scan of the appropriate part of the user's body can be sent for storage that enables subsequent comparison of body shape change

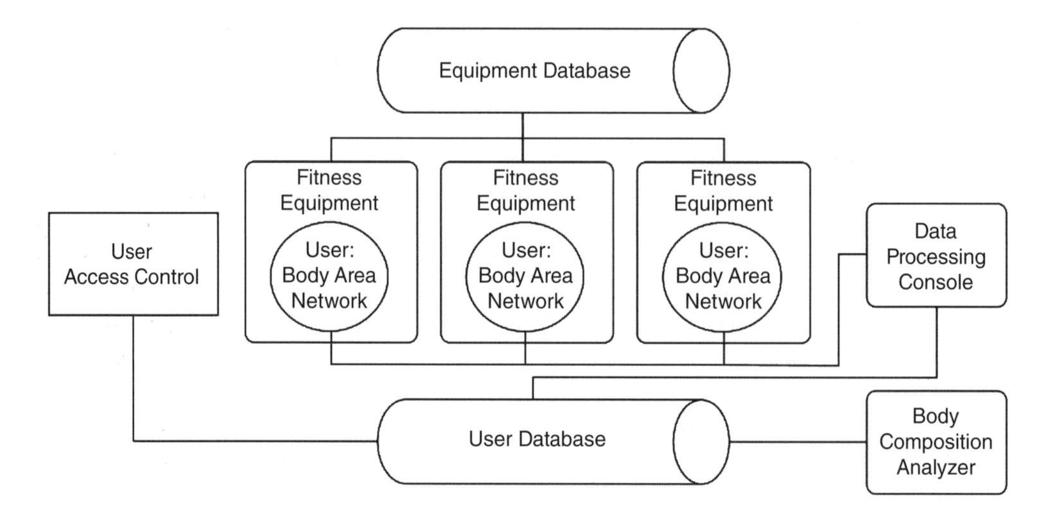

Figure 3.13 A gy network that simultaneously providing health and safety tracking for multiple users.

when the user returns in the next session. Technology can help beginners by offering guidance on the proper way of handling dumbbells so as to avoid injury. This can be done by projecting an image to illustrate the correct procedures so that the user can follow step by step.

Multiple users can be identified in many ways. The most convenient is perhaps an RFID sticker on the shoe with readers on the mat associated with each piece of equipment. How it works is very simple: once a user steps on the mat the unique identification number will be recorded, and when the equipment is started readings will be captured and marked as belonging to the identified person. In addition to serving the purpose of health tracking, the system can also be used for billing purposes if usage is charged on a per-use basis. Upon completion of a session, a user can choose either to download the data onto a removable storage device to bring home or to have it sent home via the gym network. Proper user identification procedures should ensure data associated with each individual user will not be mixed up and their privacy assured.

3.5.3 Case Study III: Swimming

Difficult challenges need to be tackled in underwater wireless communication, as we explain in Section 3.3.1 where we discuss the difficulties with submerging a transmitting device into water. Despite the challenges, wireless communications can help save life above and beyond health assessment capabilities similar to that in a gym environment. This is particularly advantageous at beaches where lifeguards may not be able to keep an eye on all of the swimmers. Any small waterproof transmitter can be used to call for help in the event of an accident. Warning can also be issued from land in a sudden emergency situation such as sighting of shark. All it needs is a waterproof receiver that can pick up broadcast signals from the shore.

So, a system that allows small transceivers to be brought with a swimmer can potentially save lives. Since the body does not go far below the water surface and the distance away from the shore does not normally exceed a couple of hundred meters, water absorption does not necessary block off radio waves completely. One important issue to bear in mind is the material used in waterproof housing, since this will also have an effect on signal absorption. Another fact worth noting is that wave penetration properties differ between saltwater and that with traces of bleach in a swimming pool. Since the data rate generally does not exceed 1 KB/s, an underwater wireless acoustic network

(UWAN) should do the trick. The major drawback of such networks is severe propagation delay that can be as much as 1 s/km. This must be taken into consideration during the system design stage.

Setting up a UWAN can be quite complex. Swimmers move around inside water hence it is highly unlikely that the transceiver will remain stationary. To illustrate the effect of changing speed we shall look at some basic mathematics here. Given the combined speed of swimming and water flow v at an angle relative to the acoustic signal propagation direction θ, the effective acoustic propagation speed v' is:

$$v' = v \cdot \cos\theta \tag{3.1}$$

Logically, the effective propagation speed v' increases if the combined speed v is moving toward the same direction as signal propagation, whereas v' decreases when v moves in the opposite direction to propagation. The water flow will result in a slight bending of a narrow acoustic beam in the same direction, but its effect is reasonably insignificant. The propagation speed changes significantly when entering a different medium, namely from water into air or vice versa. This effect is due to refraction as the dielectric constant changes just as light bending from air through water or glass. So, refraction will change the direction of the propagating signal. Note, incidentally, that the term "refraction" is also used in optometry when it refers to the examination of an eye in the process of evaluating whether a spectacle prescription enhances vision. Such application is sometimes known as "refractometry."

In addition to refraction, reflection will also occur when the signal hits the boundary between two media, resulting in a portion of signal being reflected back into the water from the surface without going into the air, as shown in Figure 3.14. In shallow water, as in the case with most beaches and swimming pools, reflection from the bottom will also induce a multipath effect. The received signal *r(t)* can therefore be expressed mathematically as:

$$r(t) = \sum_{n=1}^{N} \alpha_n \cdot s(t + \tau_n) \tag{3.2}$$

where the attenuation coefficient α_n denotes the reduction in signal strength due to absorption attenuation that effectively turns the signal energy into heat and loss due to reflection, which is both frequency and distance dependent; and the original transmitted signal *s(t)* is subject to a delay of τ_n resulting in $s(t + \tau_n)$. *N* is the number of incident acoustic signal paths caused by the multipath effect. In shallow water at short range, it will likely be $n = 3$ since there are three signal paths: direct LOS between the transmitter and receiver, a reflection from the surface, and a reflection from the

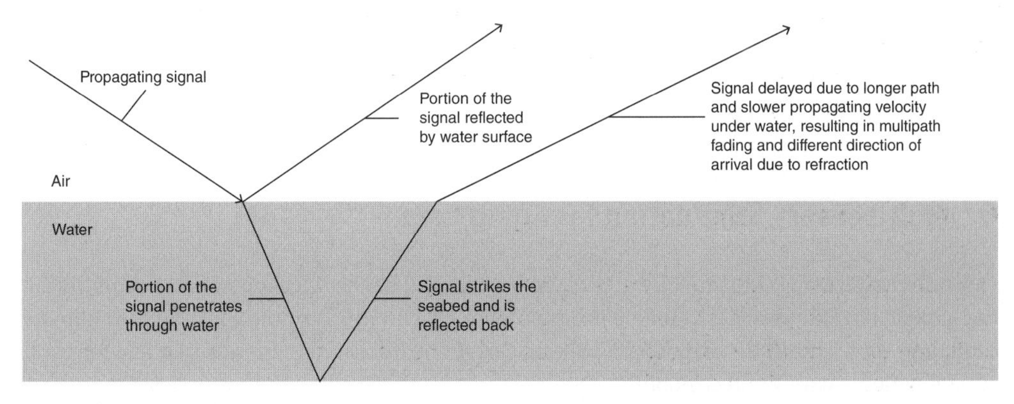

Figure 3.14 The water's surface causes reflection and refraction, which makes health tracking in water particularly challenging.

bottom. n and τ_n generally increases when the depth and range increase as more reflections will occur and the time for the signal to reach the receiver increases. The reflection loss due to water surface and the bottom can be very different since the bottom may have deposits that make it far from even. The molecular movement of the water surface caused by the propagating signal is very small (the carrier wave is highly unlikely to carry sufficient energy to cause significant movement to water); therefore, only a very tiny fraction of the signal will be transmitted from water into air. Virtually the entire signal will be reflected back into the water. Also, acoustic pressure does not couple well with air, just like an "impendence mismatch" with an electrical current hitting a load. A similar situation applies from air into water. This is exactly the reason why when we hold our heads underwater we can hardly hear anything from above. This coupling problem does not generally exist with the bottom since deposited particles are "more friendly" with the movement of water molecules. With better coupling, a certain portion will be reflected back into the water, while some will be absorbed. This is good news for communication, since absorption will have a negative effect on multipath. The bottom effectively acts blocks some of the reflected signals, thereby reducing n. The actual effectiveness will depend on the composition of the deposit.

Till now we have looked at signal propagation relative to time. Before we end our discussion, let's turn our attention briefly to the effects with respect to distance. Consider the signal $S(d)$, where d is the distance traveled. Obviously, the signal S weakens as d increases. Their relationships can be expressed in basic mathematics as:

$$S(d) = S(d = 0) \cdot e^{-\alpha d} \tag{3.3}$$

Since attenuation is usually expressed in dB, we can represent the signal loss L (not to be confused with the notation in Eq. [2.2], which denotes the number of levels there) as:

$$\mathrm{L} = 20 \cdot \log_{10}\left(\frac{S(0)}{S(d)}\right) = 20 \cdot \log_{10}\left(\frac{S(0)}{S(0) \cdot e^{-\alpha d}}\right) = 20 \cdot \log_{10}(e^{\alpha d}) \tag{3.4}$$

This can be simplified as:

$$L = 20 \cdot \log_{10}(e)[\alpha d] = 20 \cdot [0.434][\alpha d] = (8.86\alpha) \cdot d \tag{3.5}$$

The above discussion gives an insight into the complicated situation of applying telemedicine to healthcare in an underwater environment. Readers who are interested in finding out more about data communication through water are encouraged to refer to Etter (2003) for details on underwater wireless communications.

All the above case studies support general health assessments while undertaking regular exercise. What they have in common is that the health and activity data gathered is transferred and processed via a smartphone, which effectively serves as a centralized console for multiple purposes (Piwek et al. 2016). The integration of different sensors in a smartphone can support continuous monitoring of many health parameters.

3.6 Multisensory Stimulation for Aging Care

We have discussed using consumer healthcare devices for general health assessment. Keeping track of the patient's health provides valuable information for facilitating remote consultation. Remote consultation serves multiple purposes in making certain health services more accessible for rural areas and patients with limited mobility.

In addition to reducing the needs for hospital visits, Rosenbloom et al. (2017) propose that teleconsultation can facilitate cognitive care for dementia patients. Senior citizens with cognitive

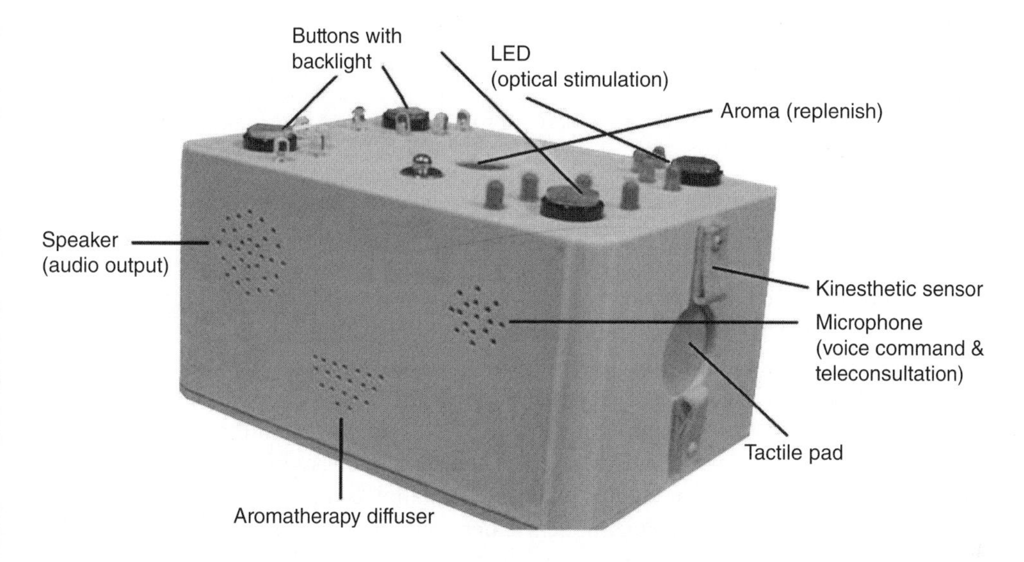

Figure 3.15 A multipurpose sensory stimulator for senior citizens.

impairment such as dementia experience difficulty in expressing their needs and uncomfortable health conditions. Independent living can be supported through multisensory therapy (Staal et al. 2007). The integration of multisensory therapy in telehealth enhances both safety and health enrichment. Technical advances have made the miniaturization of multisensory stimulation devices traditionally designed for Snoezelen rooms and clinics to portable units so that now senior citizens as well as developmental coordination disorder (DCD) patients can bring these devices with them. (Figueroa 2017). The three functional features are: (i) health monitoring, (ii) alerts and reminders, and (iii) assistive communication.

A portable multisensory stimulator like the device shown in Figure 3.15 provides various stimuli including cognitive, visual, olfactory, tactile as well as auditory stimuli to an elderly user, with components such as optical fiber, aroma diffuser, tactile boards and audio input/output for hearing stimulation and teleconsultation thereby reduces the impact of dementia on an older patient (Lorusso et al. 2017). Coupled with a smart home control system, stimulation can be supported as follows:

Visual

- Display moving images through variation of light colors, patterns, cognitive stimulation through vintage picture synchronized with auditory stimuli by means of audio commentary.
- Optic fibers displaying different light patterns.
- Cognitive stimulation through counting the number of light sources.

Olfactory

- Aroma.

Tactile

- Touch-sensitive temperature varying pad.
- Tactile pictures projected on the wall.
- Vibration, sequence/touch buttons with different textures.

Auditory

- Music with varying moods.
- Sounds of nature and animals.

Kinesthetic

- Interactive sound/ lighting for arm movement, bilateral-like clapping).
- Grasp/squeeze/stepping/kicking with targets.

All these can be highly customized according to the patient's condition. Most importantly, the multisensory therapy can be designed and customized by caregivers remotely and any anomaly can be detected and an alert sent to the caregiver for necessary attention. The health monitoring feature collects information about the patient's health conditions, such as blood pressure, body temperature, and SpO_2 readings through appropriate sensors, medication, and nutritional intake, and fall history. Such clinical information will be analyzed on a regular basis for monitoring purposes (Abawajy and Hassan 2017). In addition, the clinical information can be connected to and shared with healthcare facilities (e.g. general practitioners or hospitals) via any telemedicine network. This feature is particularly suitable for older patients with cognitive impairment users who are recovering at home after hospitalization (e.g. after hip fracture surgery) while still under close surveillance by hospital staff. In addition, this feature can help reduce demand on hospital resources as well as travel time for the patient.

A range of features can be supported, including alerts and reminder (e.g. drug intake, flush toilet after use, safe use of gas stove, bring key and wallet when leaving home); assistive communication.

In this chapter, we have looked at a number of situations where telemedicine can help save lives. It can also be used in applications for general health monitoring, so its benefits extend to healthy people, too. Wireless communication systems can face difficult challenges in some harsh environments. Barriers such as water and vegetation can significantly affect system reliability; therefore they are not 100% problem free, even though technological advances have made them far more effective than ever before.

References

Abawajy, J.H. and Hassan, M.M. (2017). Federated internet of things and cloud computing pervasive patient health monitoring system. *IEEE Communications Magazine* 55 (1): 48–53.

Ansari, N., Fong, B., and Zhang, Y.T. (2006). Wireless technology advances and challenges for telemedicine. *IEEE Communications Magazine* 44 (4): 39–40.

Ansari, N., Zhang, C., Rojas-Cessa, R. et al. (2008). Networking for critical conditions. *IEEE Wireless Communications* 15 (2): 73–81.

Bamberg, S.J.M., Benbasat, A.Y., Scarborough, D.M. et al. (2008). Gait analysis using a shoe-integrated wireless sensor system. *IEEE Transactions on Information Technology in Biomedicine* 12 (4): 413–423.

Benger, J. (2000). A review of telemedicine in accident and emergency: the story so far. *Emergency Medicine Journal* 17 (3): 157–164.

Bouten, C.V., Koekkoek, K.T., Verduin, M. et al. (1997). A triaxial accelerometer and portable data processing unit for the assessment of daily physical activity. *IEEE Transactions on Biomedical Engineering* 44 (3): 136–147.

Brenner, R.J. and Bartholomew, L. (2005). Communication errors in radiology: a liability cost analysis. *Journal of the American College of Radiology* 2 (5): 428–431.

Chen, M., Ma, Y., Li, Y. et al. (2017). Wearable 2.0: Enabling human-cloud integration in next generation healthcare systems. *IEEE Communications Magazine* 55 (1): 54–61.

Darshan, K.R. and Anandakumar, K.R. (2015). A comprehensive review on usage of Internet of Things (IoT) in healthcare system. In: *2015 International Conference on Emerging Research in Electronics, Computer Science and Technology (ICERECT)*, 132–136. IEEE.

Etter, P.C. (2003). *Underwater Acoustic Modelling and Simulation: Principles, Techniques and Applications*, 3e. Taylor & Francis.

Figueroa, R.C. (2017). Personal spaces for multisensory stimulation as support to rehabilitate patients with cognitive disabilities. In: *Proceedings of the XVIII International Conference on Human Computer Interaction*, 56. ACM.

Fong, B. and Hong, G.Y. (2012). A prognostics framework for health degradation and air pollution concentrations. *Journal of Advances in Information Technology* 3 (1): 64–68.

Fong, B., Fong, A.C.M., Hong, G.Y., and Ryu, H. (2005a). Measurement of attenuation and phase on 26-GHz wide-band point-to-multipoint signals under the influence of rain. *IEEE Antennas and Wireless Propagation Letters* 4: 20–21.

Fong, B., Fong, A.C.M., and Hong, G.Y. (2005b). On the performance of telemedicine system using 17-GHz orthogonally polarized microwave links under the influence of heavy rainfall. *IEEE Transactions on Information Technology in Biomedicine* 9 (3): 424–429.

Fong, B., Situ, L., and Fong, A.C.M. (2017). *Smart Technologies and Vehicle-to-X (V2X) Infrastructures for Smart Mobility Cities. Smart Cities: Foundations, Principles, and Applications*, 181–208. Wiley.

Fong, B., Fong, A.C.M., and Li, C.K. (2018). *Internet of Things in Smart Ambulance and Emergency Medicine. Internet of Things A to Z: Technologies and Applications*, 475–506. Wiley-IEEE Press.

Felt-Lisk, S. (2006). *New Hospital Information Technology: Is It Helping to Improve Quality?* Washington, DC: Mathematica Policy Research.

Hirata, A., Fujiwara, O., Nagaoka, T., and Watanabe, S. (2010). Estimation of whole-body average SAR in human models due to plane-wave exposure at resonance frequency. *IEEE Transactions on Electromagnetic Compatibility* 52 (1): 41–48.

Kawaguchi, M., Shimada, M., Ishikawa, N., and Watanabe, G. (2016). Underwater robotic suturing. *Minimally Invasive Therapy & Allied Technologies* 25 (3): 129–133.

Lee, W.C., Faan, H.H., Tsang, K.F. et al. (2016). RSS-based localization algorithm for indoor patient tracking. In: *IEEE 14th International Conference on Industrial Informatics (INDIN)*, 1060–1064. IEEE.

LeMaster, N. and Reed, D. (2016). Interdependence in the healthcare industry: how a provider and a supplier collaborated to achieve mutual success. *Management in Healthcare* 1 (3): 217–223.

Li, H.B. and Kohno, R. (2008). Body area network and its standardization at IEEE 802.15. BAN. In: *Advances in Mobile and Wireless Communications* (eds. I. Frigyes, J. Bito and P. Bakki), 223–238. Berlin: Springer.

Lorusso, R., Gelsomino, S., Parise, O. et al. (2017). Neurologic injury in adults supported with veno-venous extracorporeal membrane oxygenation for respiratory failure: findings from the extracorporeal life support organization database. *Critical Care Medicine* 45 (8): 1389–1397.

Martinez, A.W., Phillips, S.T., Carrilho, E. et al. (2008). Simple telemedicine for developing regions: camera phones and paper-based microfluidic devices for real-time, off-site diagnosis. *Analytical Chemistry* 80 (10): 3699–3707.

Mchugh, T. (1997). MedLink bails out in-flight emergencies, *Phoenix Business Journal* (21 November). http://phoenix.bizjournals.com/phoenix/stories/1997/11/24/focus5.html (accessed 20 January 2020).

Means, D.L. and Chan, K.W. (2001). Evaluating compliance with FCC guidelines for human exposure to radio frequency electromagnetic fields, additional information for evaluating compliance of mobile and portable devices with FCC limits for human exposure to radiofrequency emissions. FCC Supplement C (Edition 01–01) *OET Bulletin* 65 (Edition 97–01), 1–53.

Morris, S.J. (2004). A shoe-integrated sensor system for wireless gait analysis and real-time therapeutic feedback. Doctoral dissertation. Massachusetts Institute of Technology.

Pappas, I.P., Keller, T., Mangold, S. et al. (2004). A reliable gyroscope-based gait-phase detection sensor embedded in a shoe insole. *IEEE Sensors Journal* 4 (2): 268–274.

Park, S. and Jayaraman, S. (2003). Enhancing the quality of life through wearable technology. *IEEE Engineering in Medicine and Biology Magazine* 22 (3): 41–48.

Piwek, L., Ellis, D.A., Andrews, S., and Joinson, A. (2016). The rise of consumer health wearables: promises and barriers. *PLoS Medicine* 13 (2): e1001953.

Peters, B.S., Armijo, P.R., Krause, C. et al. (2018). Review of emerging surgical robotic technology. *Surgical Endoscopy* 32 (4): 1636–1655.

Rosenbloom, M.H., Barclay, T.R., Dorwart, A. et al. (2017). Cognitive screening results from the medicare annual wellness visit in a primary care practice. *Alzheimer's & Dementia: The Journal of the Alzheimer's Association* 13 (7): P825.

Staal, J.A., Matheis, R., Collier, L. et al. (2007). The effects of Snoezelen (multi-sensory behavior therapy) and psychiatric care on agitation, apathy, and activities of daily living in dementia patients on a short term geriatric psychiatric inpatient unit. *The International Journal of Psychiatry in Medicine* 37 (4): 357–370.

Tachakra, S., Banitsas, K.A., and Tachakra, F. (2006). Performance of a wireless telemedicine system in a hospital accident and emergency department. *Journal of Telemedicine and Telecare* 12 (6): 298–302.

Thapliyal, H., Khalus, V., and Labrado, C. (2017). Stress detection and management: a survey of wearable smart health devices. *IEEE Consumer Electronics Magazine* 6 (4): 64–69.

TriData Corporation (2005). Current status, knowledge gaps, and research needs pertaining to firefighter radio communication systems. A report prepared for the NIOSH. http://www.cdc.gov/niosh/fire/pdfs/FFRCS.pdf (accessed 20 January 2020).

Wang, Q., Tayamachi, T., Kimura, I., and Wang, J. (2009). An on-body channel model for UWB body area communications for various postures. *IEEE Transactions on Antennas and Propagation* 57 (4): 991–998.

Welch, T.B., Musselman, R.L., Emessiene, B.A. et al. (2002). The effects of the human body on UWB signal propagation in an indoor environment. *IEEE Journal on Selected Areas in Communications* 20 (9): 1778–1782.

Wei, J. (2014). How wearables intersect with the Cloud and the Internet of Things: considerations for the developers of wearables. *IEEE Consumer Electronics Magazine* 3 (3): 53–56.

Williams, P.A., Lovelock, B., Cabarrus, T., and Harvey, M. (2019). Improving digital hospital transformation: development of an outcomes-based infrastructure maturity assessment framework. *JMIR Medical Informatics* 7 (1): e12465.

Winters, J.M., Wang, Y., and Winters, J.M. (2003). Wearable sensors and telerehabilitation. *IEEE Engineering in Medicine and Biology Magazine* 22 (3): 56–65.

Yang, C., Puthal, D., Mohanty, S.P., and Kougianos, E. (2017). Big-sensing-data curation for the cloud is coming: a promise of scalable cloud-data-center mitigation for next-generation IoT and wireless sensor networks. *IEEE Consumer Electronics Magazine* 6 (4): 48–56.

4

Data Analytics and Medical Information Processing

In Chapter 3 we look at a number of situations where telemedicine and related technologies can save lives, which was impossible only a few decades ago. Telemedicine covers just about all corners of the globe. Its comprehensive range of service facilitates everything from search and rescue operations to general health monitoring. This involves medical information being captured and converted into the digital domain. Numerous advantages exist for handling digital data instead of leaving everything in the original analog form. Mosco (2017) describes the ease of transmission, processing, and subsequent storage with digital data compared to analog data manipulation. So, what do these long strings of 0s and 1s representing medical data have that differ from anything else digital in daily life, like TVs and cameras? One thing they have in common is that, in all these applications, information is sent and processed in "binary bits," i.e. we only deal with 1s and 0s. However, the requirements for capturing and handling medical data are quite different from those for general purpose consumer electronics devices. For a start, medical information is often specifically related to a single individual. A person's medical history must be kept strictly confidential at all times. Compare the consequence of losing a few songs on an MP3 player to losing the analysis results following a medical test. The maximum liability of the former is probably downloading from the server again by paying a nominal fee, whereas the latter can lead to lengthy legal proceedings and damage claims, and the patient may lose precious time receiving prompt treatment, in addition to the impact on the medical institution's reputation. The fundamental difference in requirements extends to the way information is processed and the tolerance to faults, errors, and omissions. Looking at the above comparison again, data misinterpretation may lead to momentary disruption to music playback or degradation in sound quality and no consequence will result thereafter once normal playback is resumed in a few seconds' time. The consequences of losing or corrupting medical data can be dire, including the possibility of failure to diagnose life-threatening conditions.

The course of making use of medical information, just like most information systems, as illustrated in Figure 4.1, begins with data acquisition from data sources. In the case of telemedicine, the majority of data comes from patients that involve a diverse range of data types from biosignals to surveys about daily activities. Once captured, the data needs to be transmitted to an appropriate location for processing in order to make sense of what the data implies. Next, processing entails technologies in different areas such as signal processing, multimedia, and data mining; how the data is processed depends on the nature of the data and its related application. Having analyzed the data such that any necessary actions can be taken in response to the given situation, the data needs to be archived as this can be very useful in a number of ways; for example, a patient who is allergic to certain substances needs to be known prior to providing treatment. Data can also be used anonymously for statistical analysis of virus mutation and spread pattern in the study of disease control, and government agencies can use the anonymous data for regulatory planning, etc.

Telemedicine Technologies: Information Technologies in Medicine and Digital Health, Second Edition.
Bernard Fong, A.C.M. Fong, and C.K. Li.
© 2020 John Wiley & Sons Ltd. Published 2020 by John Wiley & Sons Ltd.

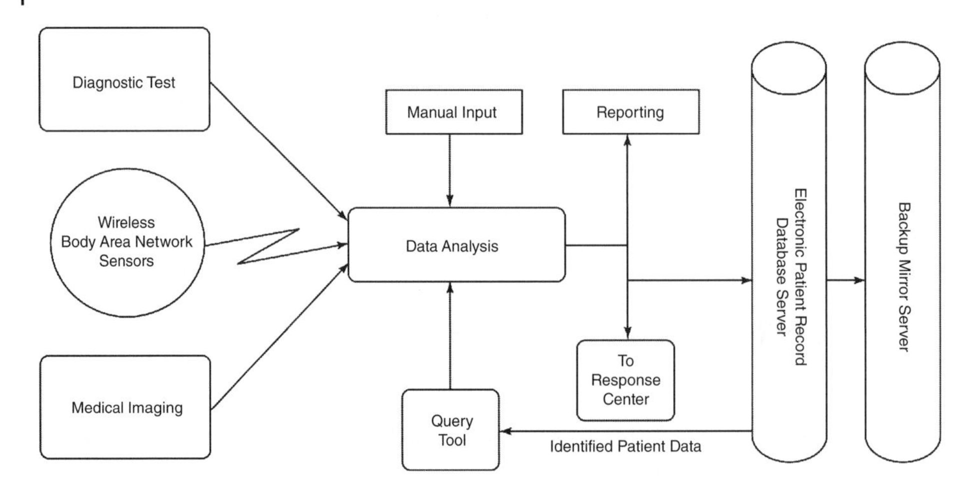

Figure 4.1 Block diagram of a medical information system.

So, an effective way of storing a massive amount of data and speedy retrieval of relevant data is also an important topic to study. The main purpose of this chapter is to walk through the entire process of medical information processing. We conclude the chapter by taking a look at an electronic drug store that utilizes medical information for the efficient and safe dispensing of medication as an example to demonstrate the importance of technological advances in medical information technology for assisting patients with special needs so that medication becomes risk free and easy to access.

4.1 Noninvasive Health Data Collection

There are all kinds of data to be collected from a patient, from head to toe, within and around the body. We concentrate our discussion on biomedical data related to the human body and leave the qualitative survey (verbal and written) collection topic behind in order to deliberate on the technical aspects. So, we look at what kind of information about a patient can be collected and see how it can be collected. We also look at an overview of any necessary precautions in the process of collecting such data. The human body is so complex that it would be impossible to cover every single measurable parameter in a single book. Our main objective here is to look at some commonly used attributes and to get a good understanding of what is involved while processing medical information.

The obvious candidates are the vital signs of a human body. These are signs that determine the health status of an individual. Indeed, a person without any one of these may not even be alive. So, these are important signs to collect and monitor; we look at some properties of these signs and how they can be collected below. Some of these signs are inherently known to present circadian rhythms in a 24-hour behavioral cycle with fluctuation due to temporal regulation of the ambient environment and activities. Given the comprehensive types of health signs that can be measured, we only concentrate our discussion on those that can be easily monitored using noninvasive methods.

4.1.1 Body Temperature

(Normal Range: 36.1–37.5 °C)

The "normal" body temperature of a person varies based not only on the surrounding environment but also to a greater extent on where within the body the temperature measurement is taken. Mackowiak et al. (1992) reveal that even gender plays a role in the mean body temperature that is considered normal. Body temperature measurement is the principal factor that tells whether a person suffers from hyperthermia or hypothermia upon exposure to extreme conditions. The former is above 40 °C that may result in severe dehydration caused by excessive sweating, whereas the latter is below 35 °C after exposure to "freezing" conditions in the cold. Both can be fatal if medical attention is not given promptly. Abnormal body temperature can also indicate fever, which may lead to permanent organ impairment or even mortality. The precise measurement of body temperature and monitoring its changing pattern is therefore an important issue to consider.

There are many methods of measuring body temperature with varying precision and the time required for measurement. Measurement can be taken from many points of the body, most commonly the armpit, mouth underneath the tongue, ear, or rectum. These positions are listed in ascending order of nominal temperature that spans across the 37.6–38.0 °C range. A number of factors that affect the reading taken are outlined in the study by Sandsunda et al. (2004). The age of the subject also makes temperature measurement less predictable, children playing hard may generate a considerable amount of heat inside the body as an absolutely normal response, while older people may not have the adequate energy to generate as much heat under normal situations. To illustrate the extent of normal body temperature variation during the day, we take a look at sample reading from three perfectly healthy persons in Figure 4.2, of a child, an adult, and an older person aged 5, 35, and 70, respectively. Although the activities taken by each subject varies during the day, the circadian rhythms of everyone involved appear fairly consistent. The significance of this behavior tells us that body temperature measurement for what considered as "normal" can be quite capricious.

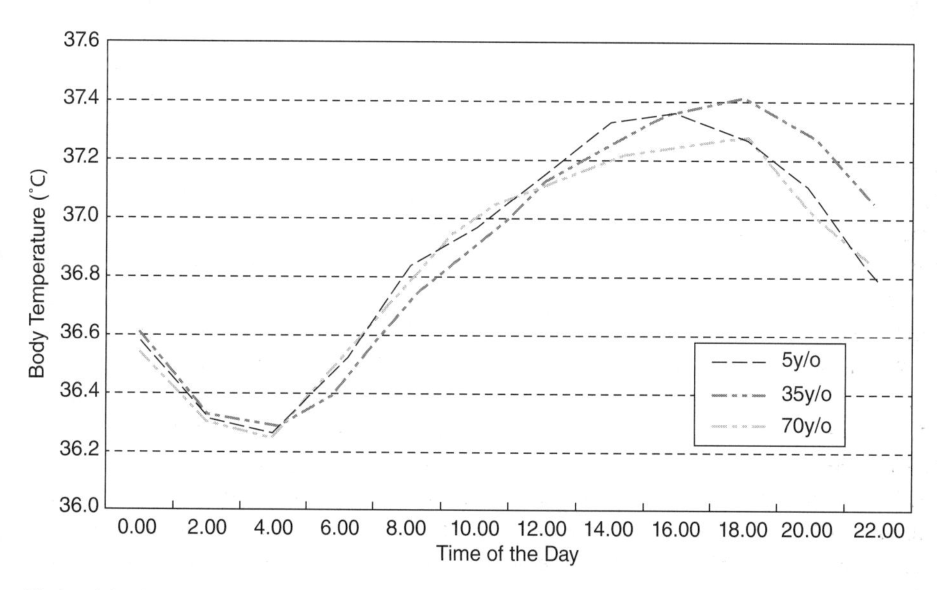

Figure 4.2 Normal variation of body temperature throughout the day.

Body temperature measurement can be accomplished in several ways and each method can be affected by different environmental variables. For example, the very traditional way of oral measurement by putting a thermometer into the mouth can result in very significant deviation following the consumption of either a hot or cold drink. Likewise, a reading taken under the arm can be greatly affected by sweat and ambient temperature change. Therefore, more reliable methods have been developed using advances in technology. For example, tympanic temperature can be measured economically and reliably with an infrared ear thermometer that operates by measuring the amount of infrared energy radiated from the subject's eardrum. Ear measurement is intrinsically reliable since the eardrum is situated very near the hypothalamus, the core temperature regulator of the human body. This method is fairly fast with a reading obtained in about 0.1 seconds, and the small portable thermometer is very affordable for consumer use. With appropriate wireless technology, the reading can be automatically transmitted to a nearby workstation for patient record updating.

The infrared ear thermometer is good for measuring individual subjects' body temperatures as a probe needs to be placed inside the ear for each measurement. To monitor and prevent the spread of certain diseases, body temperature monitoring is sometimes imposed at checkpoints where people come and go. For example, during the severe acute respiratory syndrome (SARS) and avian influenza pandemics, we saw border control authorities imposing body temperature checks in many countries. To ensure the smooth flow of people, a color image of each subject is captured by a contactless heat sensing camera that instantaneously shows the core body temperature as soon as the subject walks past the camera. Infrared thermal imaging is commonly used for this purpose where abnormalities of body temperature can be revealed by a change of color in the image. Although representation of color does not offer high measurement precision, it is a fast and convenient method that can be programmed to trigger an alarm if a certain color that represents a certain preset threshold temperature is detected among a group of people that enter the camera's operating area. However, its reliable use demands precise calibration, both in terms of the process of performing instrumental calibration and the calibration stability that determines how frequently the device needs to be calibrated again. Also, its reliability can be significantly affected by surrounding error sources such as radiation and heat generating machinery.

Forehead and spot infrared thermometers are also commercially available, although not widely used, for good reasons. Forehead measurement can be greatly affected by ambient temperature as well as the use of fever-lowering medication such as acetaminophen or ibuprofen, whereas spot infrared involves the use of a laser beam that can be potentially hazardous if accidentally pointed at a subject's eye.

Accurate detection of high body temperature in infants is a particularly important issue as permanent disabilities can result if treatment is not provided at once. Cranston et al. (1975) describe the human body response to infection that results in a fever. The cause of temperature elevation can be simply wearing too much as many parents tend to overprotect little babies. Sometimes this can be a normal response to a vaccination, or more serious case caused by viral infection that requires immediate medical attention. Technology can help parents monitor their newborn child with a small heat-sensing camera in the event of suspecting a fever. The system generates an audible alarm should the baby's body temperature exceed 38.0 °C and automatically alerts the clinic should the temperature reach 38.9 °C, which requires immediate medical attention. This is an example where a simple thermometer can be linked to a telemedicine system for improved healthcare monitoring.

Technology advances provide more precise means of measuring body temperature than traditional mercury thermometers with added feature enhancements, such as automatic updating of patient records and alerts for temperature exceeding a certain preset threshold. An analysis of variations in temperature can also suggest a possible cause that prompts medical attention; different measurement methods should be optimized between speed, precision, and ease of operation. Different methods are optimized for specific applications and operating environments.

4.1.2 Heart Rate

(Normal range at rest: 60–100 bpm)
The measurement and subsequent analysis of the heart rate is useful in many applications, from life-threatening conditions such as abnormal behavior due to heart failure to general fitness assessment in gymnasiums, as discussed in Section 3.5. Not as homogeneous as body temperature, the heart beat daily pattern of the human body also exhibits a certain degree of circadian rhythm, as shown in Figure 4.3. It shows that generally, neglecting any irregular, heavy activities, the heart beats almost 30% higher during the day than sleeping at night and a daily average of around 70 beats per minute (bpm) over a range of around 58–82 bpm. Note, incidentally, that under normal circumstances a female subject generally beats some 5% faster than a male subject in identical situations. Obviously, more blood is pumped across the body while a person moves around than when sleeping. A reading taken once every two hours would eliminate any sudden impulse caused by exercise or a nightmare during the night. The purpose of obtaining a set of heart rate readings is usually associated with the study of certain activities. It would have a much better margin of error than body temperature measurement, which is more prone to uncontrollable environmental conditions.

As the measurement unit "bpm" suggests, the heart rate is measured by the number of beats within any one-minute period. The easiest way is by counting the number of pulses over one minute. In theory, this can be read from anywhere of the body with an artery running near the skin. Most commonly, measurements are taken at the radial and carotid artery (wrist and neck, respectively). Some fitness equipment may use the brachial artery (elbow) for ease of access during a workout. Gymnasium equipment is widely fitted with heart rate sensors, such as the one illustrated in Figure 4.4. As we use the same kind of measuring method on a regular basis, it appears to be

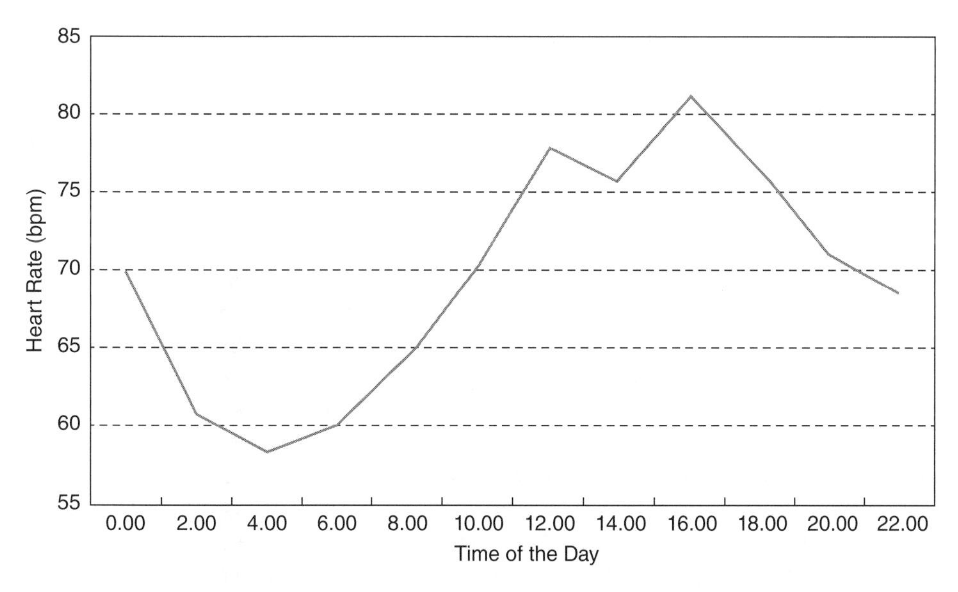

Figure 4.3 Circadian rhythm of heartbeat.

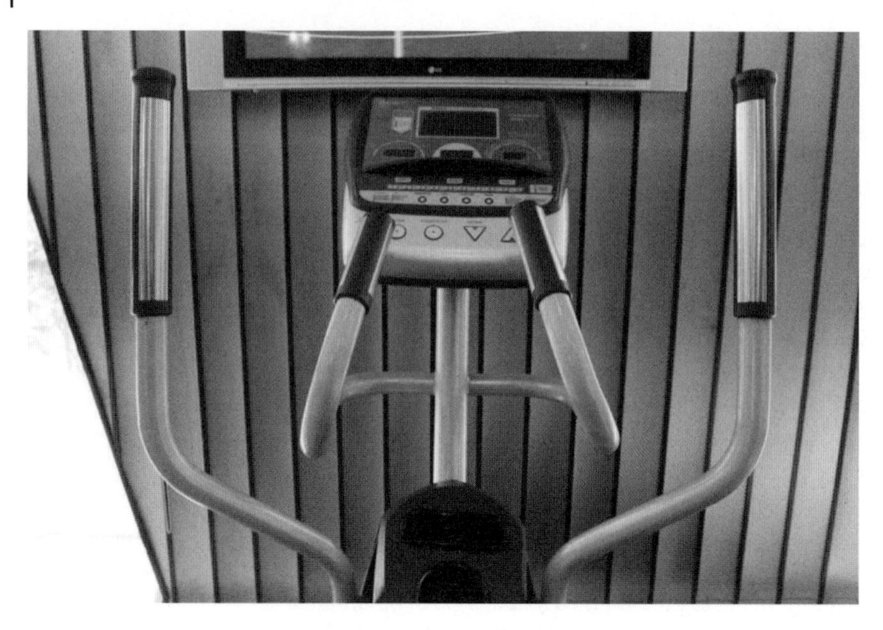

Figure 4.4 Heart rate sensors mounted on the hand grips.

extremely effortless for us to use. However, its design means the user needs to grab the sensor fairly tightly in order to get an instantaneous reading of the heart rate while exercising. We may overlook two important limitations: restriction of movement while gripping the handle and running (as in the case of a treadmill), and lack of means for keeping a log on our health condition (an instantaneous reading may not convey a lot of useful information about our health status). To improve this, a simple wearable counter can be deployed as follows: an electrical signal is induced through the heart muscle during contraction as the heart beats. A wearable transmitter that picks up the signal can be placed near any of the three locations mentioned above. The transmitter then sends an electromagnetic (EM) signal corresponding to each pulse to a receiver that counts over a certain period of time, say five seconds, and displays the heart rate by normalizing it to the number of beats per minutes; in the case of counting in five seconds this would mean multiplying the counts by 12 for an estimate of the number of beats per minute. One of telemedicine's many features is keeping track of health. Londeree and Moeschberger (1982) suggest the maximum heart rate for a given age decreases by 1 bpm for each increase of age by one year. A user can program the fitness monitoring device that alerts them to slow down when their heart rate reaches a certain predetermined level to ensure a safe workout. Devices designed for use by older people should avoid measuring at the neck as inappropriate exertion of force during the process may risk light-headedness, which may have serious consequences. For infants of less than one year old, their normal heart rates are notably higher than toddlers, hence the range of measurement will be very different. So, different requirements are present for application in different areas.

Ultimately, telemedicine technology should assist with attracting the necessary attention when a sudden change of heart rate may indicate a serious medical situation. There are numerous possible causes to heartbeat aberration: hypothyroidism and influence by medication are common causes to lowering heart rate. Conversely, factors such as heavy exercise, stress, disease, and the effect of stimulants such as coffee and alcohol can rapidly increase the heart rate. A simple automated system such as that illustrated in Figure 4.5 can help ensure older people who live alone are monitored at all times. This system is easy to set up and does not require any user interaction. Quite simply,

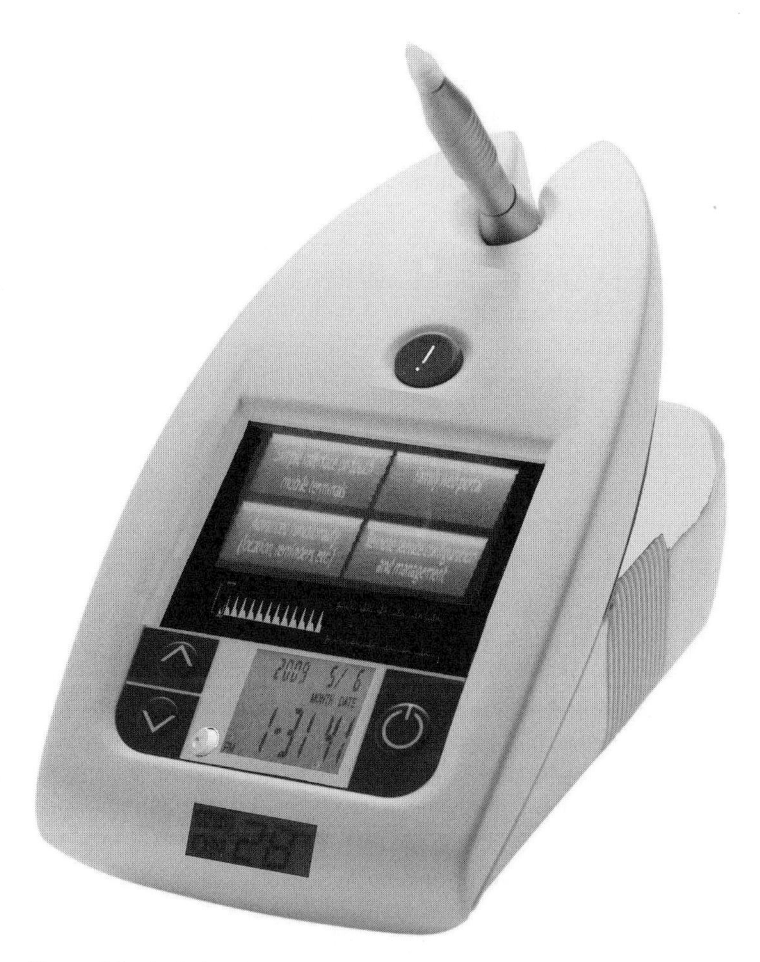

Figure 4.5 Assistive device with environmental sensing and communication capabilities for older people. A call button is also available in case of emergency.

a small pulse counter is placed at the back of a wristwatch that continuously monitors the user's heart rate when worn. If the reading falls outside the predetermined nominal range, it will send a signal to the responding unit that in turn alerts the service center via a telemedicine network. Supporting personnel will attempt to place a phone call to find out if the user is undertaking normal activities, and immediate attention will be brought to the user in case the call is not answered. One important consideration with such a system is that the threshold must be set with sufficient margin to minimize the chance of a false alarm while any serious problems will not go undetected. However, a weak pulse cannot be detected simply by counting the number of beats. An abnormally weak pulse may be due to potentially fatal causes such as blood clot or heart and peripheral arterial disease. It is therefore necessary to deploy more refined methods for measuring heart beats for detecting palpitations.

Electrocardiography (ECG)/electroencephalography (EEG) (see Section 4.2) can also be used for measuring heart rate. They provide accurate measurements and also provide an indication about the rhythm corresponding to heartbeat pattern as well as strength. Such additional information can be useful in detecting any signs of heart disease and abnormalities of blood vessels. More sophisticated methods of analyzing heart beats are therefore necessary as counting the number of beats per minute may not be able to provide sufficient information to determine whether a blockage of blood vessels has occurred. Palpitations can be chronic or acute with varying consequences and each has different requirements for detection. Malik (1996) describes a number of alternatives, such as statistical, time, and frequency domain methods. These rely on a series of ECG recordings that may span a prolonged period of time, such as over 24 hours. The main purpose is to distinguish problems from beats irregularities owing to normal variants.

Sometimes, a patient's medical history can reveal impending problems. Also, there are situations where chemical analysis may be involved when identifying ingestion-related causes of palpitation. For example, detecting the presence of substances that may influence heartbeat will normally require laboratory diagnosis before any appropriate adjustments to the measured data can be made. Both situations require linking the testing site to respective departments of the hospital in order to facilitate outpatient palpitation monitoring. It may even be necessary to implant a device beneath the patient's skin when permanent monitoring for certain conditions is required.

4.1.3 Blood Pressure

(Normal systolic pressure range: 100–140 mmHg)

Blood pressure is a measure of the pressure (force divided by surface area) exerted on the walls of arteries. It makes blood circulate through the body so that oxygen and nutrients can be delivered to all organs. Compared to the two vital signs discussed above, blood pressure exhibits the least regular pattern over the day. An attempt to plot the blood pressure variation of a certain person during a typical day would most likely produce a somewhat messy chart from which little of significance could be learned. In general, it would be true to say that the blood pressure is normally higher when awake than asleep. This is, however, not necessarily always the case. For a normal healthy adult, the mean systolic pressure (peak pressure in the arteries) is around 120 mmHg, but this can vary about 20 mmHg above or below the mean during normal activities. Note, incidentally, that the diastolic blood pressure (minimum pressure in the arteries) of the same person is typically slightly over half that of the systolic value with a healthy range of around 80–90 mmHg. Due to the irregular pattern exhibited throughout the day, a spot measurement of an instantaneous reading may not be useful at all. The reader may remember their doctor using a "sphygmomanometer" to measure their blood pressure; it is a very simple, noninvasive measurement method using a cuff around the arm that is inflated by the doctor using a mechanical hand bulb pump to obtain a spot reading of our blood pressure. Before we look at how technology sets in, let's take a closer look at what blood pressure measurement is about. Quite simply, it is a measurement of the pressure exerted inside a blood vessel when the heart is beating and pumping blood through the arteries of the human body. Such measurement is known as the systolic pressure. This can essentially be done wherever there is an artery near the skin. The diastolic pressure, which also needs to be measured in most circumstances, is the pressure when the heart is at rest in between two consecutive beats. A hypertension condition is defined if any one of these parameters is too high. When measuring the two blood pressure parameters using the traditional method with a sphygmomanometer, a stethoscope is also used, for listening to the heartbeat so that readings can be taken at the appropriate times: either a pulse is heard for the systolic pressure or the absence of a pulse corresponds to the diastolic pressure. In this manual process, the reading is taken at the moment when synchronized with hearing hence there will certainly be a delay that can introduce error. This is just a spot reading taken at a certain time during a visit to the clinic and is therefore inappropriate for ambulatory blood pressure monitoring (ABPM) that involves continuing measurement

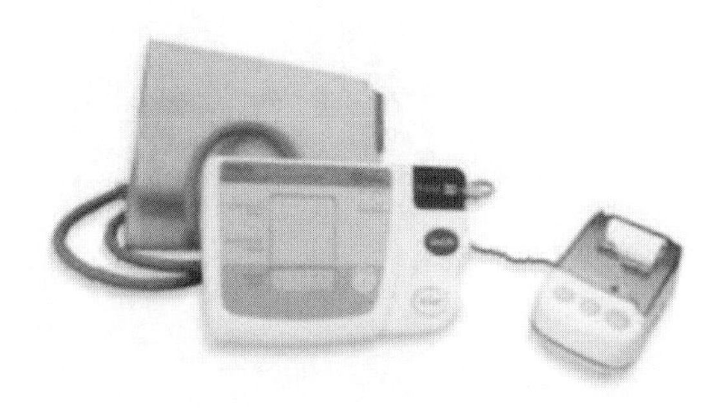

Figure 4.6 Blood pressure meter that can automatically send the measurement data to a smartphone or service center for storage and analysis.

throughout the day. This would require a wearable monitor that collects blood pressure readings throughout the day and is able to transfer the data to an external device for analysis by medical personnel. ABPM is usually deployed on a temporary basis for circumstances such as abnormally high blood pressure under the influence of certain prescribed drugs, or for patients subject to prolonged anxiety undergoing psychological treatment.

So, there are circumstances that continuous monitoring becomes necessary. This is where technology makes it possible and the subject feels comfortable with the wearable device, which is particularly useful for hypertension patients who are resistant to pharmacotherapy. The entire process involves reading, scanning, and the analysis of captured data. Marchiando and Elston (2003) describe a number of methods for carrying out ABPM where appropriate measuring apparatus can be remotely linked to the hospital for off-site measurement. This is particularly useful in monitoring cardiovascular patients where accurate measurement that reflects readings from normal daily activities is necessary. As Pickering (1999) explains, many patients tend to become nervous and this drives up the blood pressure reading unintentionally during a visit to the doctor. Remote measurement would ease tension and hence obtain a more accurate measurement.

Small wearable automatic blood pressure meters are readily available in the consumer electronics market in different forms for measurement taken at different locations on the body. In addition to the arm, measurements can also be taken at the wrist, leg, or even finger. An example of a small monitor is shown in Figure 4.6. Many similar devices are on the market priced below \$100. Its design is very simple: air is pumped into an inflatable wrap with a pressure sensing switch that acts as the manually inflated cuff of a sphygmomanometer, a quick succession of readings on the pressure exerted on the switch are taken with one high and one low reading that corresponds to the systolic and diastolic pressure, respectively. With an internal clock, the time of the reading can be recorded and the data stored and analyzed.

Telemedicine technology can do more than facilitating remote and periodic blood pressure monitoring. It can also help alert medical personnel when certain methods are not suitable to be carried out on patients with special conditions. For example, applying a noninvasive measurement with a sphygmomanometer to sufferers of sickle cell anemia is not recommended since excessive pressure applied to the patient's arm can lead to intravascular sickling resulting in intravascular thrombi, tissue necrosis, and hemolysis. To avoid this, the retrieval of their medical history from the electronic patient record can alert medical personnel to the existing condition prior to putting the patient at unnecessary risk by using a sphygmomanometer.

We have looked at a brief description on how technology can assist with blood pressure measurement and monitoring and will now move on to the next vital sign: the respiratory rate.

Figure 4.7 Telemedicine under water.

4.1.4 Respiration Rate

(Normal range: 12–24 breathes/min)

Among all body vital signs, respiratory rate is probably the most difficult to measure due to its significant variation over a very short period of time. Its pattern is somewhat related to changes in the heart rate, as the intensity of activity affects both parameters. Taking a deep breath may lengthen the duration of a breathing cycle, thereby reducing the respiratory rate while the heartbeat is much less affected. The respiratory rate is much lower than the heart rate – typically a healthy adult breathes around 12–24 times/min. The rate varies quite considerably with age: newborns may have over 40 breaths/min as normal behavior and the average rate for a toddler may be reduced to around 30. Although this parameter may provide less important information than those three listed above when determining the health state of a person, the accurate measurement of the respiratory rate would be most useful for activities such as diving where the respiratory rate would govern how long a diver can be submerged for. A range of equipment can be fitted to a diver, as shown in Figure 4.7, where the most important device is a button for seeking help. Used in conjunction with a beacon, the diver's position can be easily located. As this subsection is all about technology for measuring the respiratory rate, we concentrate our discussion on the part which measures the respiratory rate to provide a constantly updated estimate of how much oxygen is left before the diver must decompress and return to the surface. It also triggers a remote alarm to alert support staff on the shore and nearby divers in case a sudden abnormal respiration is detected.

Under normal circumstances, the respiratory rate is normally measured for patients with lung disease or taking medication that suppresses respiration. Also, asthma symptoms are closely linked to bouts of breathlessness, which can be readily detected by respiratory rate monitoring. "Tachypnea," the anomalous increase of respiratory rate, is an important behavior to be detected since it can be caused by serious problems such as pneumonia, fever, and congestive heart failure. Breathing is easy to be counted as it is usually slow and rhythmic: counting the number of expansions and contractions of the thorax can measure the respiratory rate. So, thoracic motion during breathing can be measured by placing a pressure-sensitive switch with a counter inside a vest. The chest expands, the thoracic diaphragm contracts, and the chest cavity shrinks as the diaphragm settles. The frequency of this repeated motion can be counted via the switch.

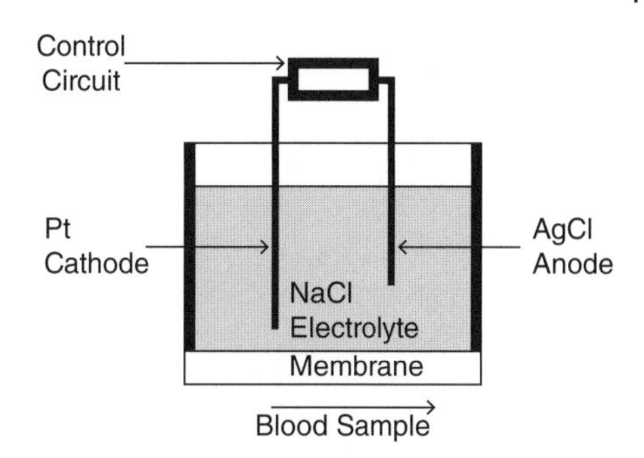

Figure 4.8 Partial pressure of oxygen in arterial blood (PaO_2) measurement.

4.1.5 Blood Oxygen Saturation

(Normal range: SaO_2: 95–100%, PaO_2: 90–95 mmHg)
Blood oxygen saturation measures the ability of the lungs supplying oxygen to the blood. In the blood, oxygen is carried chemically in hemoglobin and dissolved physically in plasma. Measurement is done to evaluate the oxygenation and saturation of hemoglobin in the blood. There are several parameters involved: partial pressure (in mmHg) of oxygen in arterial blood (PaO_2), which is an invasive method used to measure the arterial percentage of blood, whereas "SaO_2" and "SpO_2" refer to the direct and indirect measurement of the percentage of the blood oxygen saturation level, respectively. The former is measured by pulse oximetry and the latter is measured by arterial blood gas sampling. Although SaO_2 and SpO_2 may sound similar, these two parameters differ fundamentally. Conditions such as thrombolysis and influence by anticoagulant medications can significantly affect the readings obtained in an arterial blood gas sampling. These parameters are related to respiration as "inhalation" brings oxygen into the lungs, while "exhalation" brings carbon dioxide out.

PaO_2 is about gas measurement that can be measured by a polarographic oxygen electrode, as illustrated in Figure 4.8. It consists of a platinum cathode and a silver chloride anode where an electrical current is generated which is proportional to the oxygen tension. The blood sample is isolated from the electrode by a membrane to avoid protein deposition. The apparatus has to be kept in a temperature-controlled oven in order to maintain a temperature similar to that of the human body (of around 37 °C). Another precaution is to ensure that the membrane does not have any protein deposit that may accumulate on its surface over time.

Pulse oximetry is a noninvasive method of continual arterial oxygen saturation monitoring. Pulse oximeters are usually small portable devices that paramedics can carry when attending an accident scene. These can measure the arterial oxygen saturation (SaO_2) of a patient. In theory, the maximum amount of oxygen that the blood can carry can be calculated as:

$$SaO_2 = \frac{O_2\text{content}}{O_2\text{capacity}} \times 100\% \tag{4.1}$$

This would give some insight into what to expect on an oxyhemoglobin dissociation curve. A more accurate measurement of the actual value needs an oximeter that relies on a light source with red and infrared LEDs (light-emitting diodes; 600 and 800 nm wavelength, respectively) that penetrates certain parts of the body where a relatively translucent area of blood flow can be exposed to the light. Oxygenated hemoglobin absorbs infrared light, whereas deoxygenated hemoglobin absorbs red light, as illustrated in Figure 4.9. Measurement is very often taken from the finger or ear lobe. Light passes through the blood vessel, which absorbs a certain portion of red and infrared light

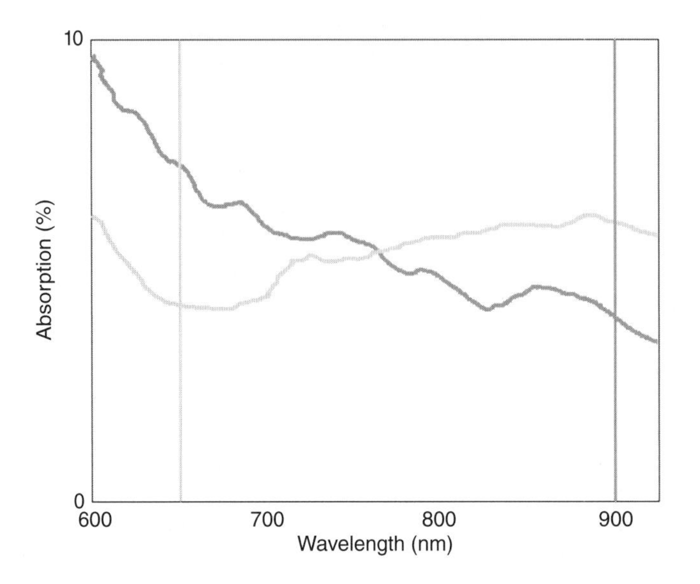

Figure 4.9 Infrared energy absorption by hemoglobin versus wavelength.

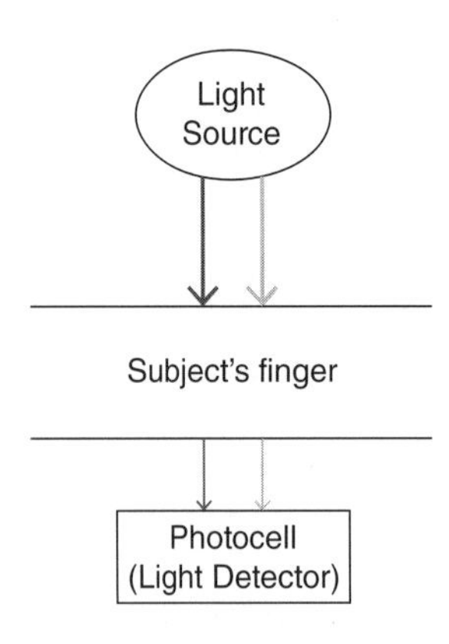

Figure 4.10 Pulse oximeter oxygen saturation (SpO_2) measurement.

beams. Whatever is left over is received by a photocell that can then deduce the red-to-infrared ratio of absorbed light through blood. This simple arrangement is shown in Figure 4.10, where a 100% SpO_2 yields a received light ratio of about 0.5. It should be noted that calibration is necessary due to the varying extent of light absorption by skin and tissue. Also, the amount of arterial blood flow varies due to heartbeat sequence, which may also affect the reading. It is therefore necessary to measure for a sufficient time covering two successive heart beats to obtain an average reading. The measurement of oxygen saturation at an accident scene is important for detecting hypoxia so that necessary emergency treatment is ready to be provided by the time the patient reaches the hospital. It is also worth noting that conditions such as tricuspid regurgitation, hypovolemia, or vasoconstriction affecting blood flow may impair the reading from an oximeter. As a final note, an oximeter cannot distinguish carboxyhemoglobin from normal oxygen-carrying hemoglobin in the event of carbon monoxide poisoning, and the reading obtained may be higher than what it should actually be.

4.1.6 Blood Glucose Concentration

(Normal blood glucose level: nondiabetics, fasting: 3.9 and 7.1 mmol/l)
Similar to the oxygen saturation described in Section 0, blood glucose level can also be measured optically (Ozana et al. 2015). In fact, both parameters can be measured with a near-infrared light of very similar wavelength (940/1050 nm). The most primitive setup consists of a pair of LEDs fixed toward a photodiode. The light goes through a translucent part of the patient's body such as a fingertip or earlobe. One LED transmits red light with a wavelength of 660 nm, while the other is 940 nm near-infrared. There is a substantial difference in the absorption properties of light at these wavelengths; it differs significantly as the light penetrates through blood. Measurement takes place at 2 Hz so that within one second the two LEDs alternately emit 1 pulse/s so that the photodiode reads the light of different wavelengths separately.

The theory is reasonably straightforward in that a mathematical equation can be mapped from the light absorption to a corresponding glucose concentration reading (Maruo and Yamada 2015). There are, however, practical issues in the design and implementation of such a measurement device. First, the received light signals fluctuate in time, because the amount of arterial blood under measurement increases as pulses with each heartbeat. This can be overcome simply by subtracting the minimum transmitted light from the peak transmitted light for each respective wavelength. Other problems, such as motion artifacts, are not as easily dealt with given the fact that the subject is highly unlikely to remain still throughout the course of measurement (Delbeck et al. 2018).

4.2 Biosignal Transmission and Processing

The main function of telemedicine is to provide medical services remotely. To serve this purpose, data must be transmitted from one location to another, such as from an accident scene or a patient's home to the hospital. Further, any data received needs to be processed before any useful information can be extracted for analysis and storage. There are so many types of relevant information. Some are fairly self-explanatory, like instructions for taking medication, whereas parameters like oxygen saturation may require expert analysis before the cause of any abnormalities can be established.

For any kind of data about a patient to be collected and processed, we need some kind of mechanism similar to that of Figure 4.11, which is expanded from the basic communication system shown in Figure 2.1, with the inherent additive noise in Figure 2.2 understood and omitted for simplicity. Here, we have a simple block diagram showing biosensors that capture data, such as those described in Section 4.1; the sensor network being connected to a transmitter, via an

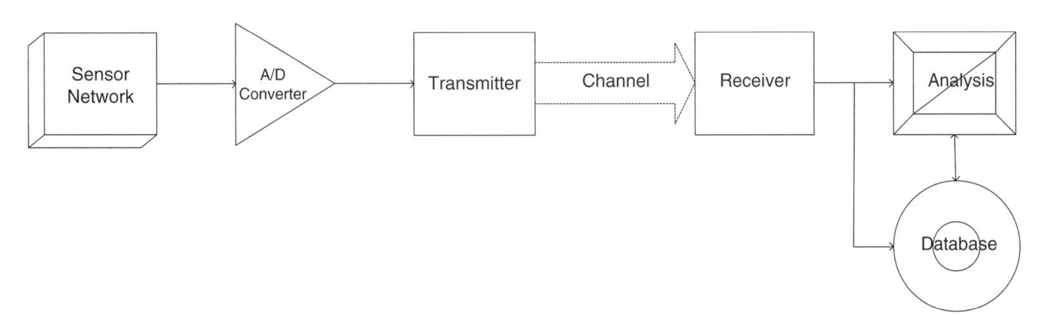

Figure 4.11 Block diagram for collecting patients' information.

analog-to-digital (A/D) converter, will send the collected data to a remote receiver. The purpose of converting the captured analog data into the digital domain is for transmission efficiency and security. While transmission efficiency is dealt with in this section, the topic of information security is addressed in Chapter 6. At the receiving end, the data will be analyzed and/or stored. Stored data can also be retrieved for analysis at any time.

This is a typical information system that deals with basic information theory. It would be virtually impossible to discuss the topic any further without revisiting the landmark work by Shannon (1948), which quantifies information in "entropy," a term that refers to a certain anticipated value in association with a dataset. In essence, Shannon entropy measures the maximum amount of information that can be sent across a given communication channel (Bousso 2017). The theory essentially describes the capacity of a given channel based on statistical model of the channel under the influence of additive noise during the transmission process. We shall not go into the mathematics behind this; readers interested in the underlying mathematical theories are advised to consult the comprehensive study by Cover and Thomas (2006) for details. Leaving the mathematics behind, the concept is fairly simple. We begin a brief discussion by referring to the basic communication system shown in Figure 2.1, whose transmitter consists of a discrete source S (with finite number of possible values per output sample) that produces raw data at a rate of R bits per symbol. The source has entropy:

$$H(S) \leq R \tag{4.2}$$

Shannon's theorem indicates that S can be coded into an alternative, but equivalent, representation at $H(S)$ bits per symbol. The original representation can be recovered in its original form by the receiver. This is theoretically possible as long as the transmission rate is above $H(S)$. Therefore, $H(S)$ is a measure of the actual information content in the output of S. Next, we also look briefly at "channel coding" by considering the transmission of a stream of information bits $b \in \{0, 1\}$ over a digital communication channel with bit-error probability (the probability of having an error bit per one million bits sent) q and capacity $C = C(q)$. A channel code consists of a block of k information bits and maps. These data bits are placed into a new block of n such that $n > k$ coded bits, c, hence introducing "redundancy." The "information content" per coded bit r is:

$$r = \frac{k}{n} \tag{4.3}$$

The coded bit sequence c is transmitted and a decoder at the receiver produces estimates $\widehat{b}$ of the original information bits, such that the probability of error is:

$$p_b = \Pr(b \neq \widehat{b}) \tag{4.4}$$

So, p_b can be minimized given that $r < C$. We can see from the above discussion that C is a measure of the channel quality, that is how noisy the channel is. With the basic concept of channel quality understood, we proceed to the topic of transmitting and processing medical information.

4.2.1 Medical Imaging

Medical imaging technology is very widely used in areas such as X-ray, body scan (whole or specific part), anatomy, remote surgery, and accident recovery. Here, we begin by looking at the simple flow chart depicted in Figure 4.12 which shows the process of medical imaging. In almost any situation, medical images are captured, sent, analyzed, and stored. In nonemergency cases, images are scanned and stored for later referral or kept for archival purpose. Whereas immediate attention

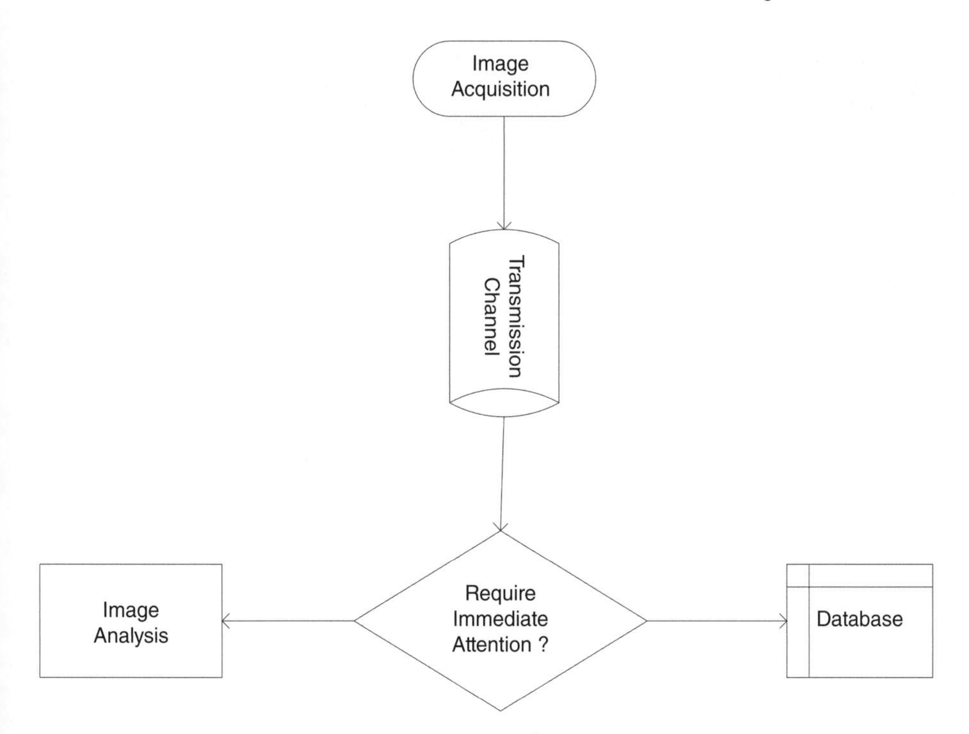

Figure 4.12 Image processing. The fundamental idea of medical imaging is to determine whether diagnosis or prognosis is required. Images that do not exhibit an anomaly can be stored in electronic patient records.

will be given once the image is obtained in the case of an emergency, these are situations such as MRI (magnetic resonance imaging) scans for surgery or photographs taken at an accident scene showing the wounds of an injured patient. Before going deeper into the transmission aspects we first look briefly at how various types of images are taken.

4.2.1.1 Magnetic Resonance Imaging

As shown in Figure 4.13, an MRI scanner looks similar to a tunnel and is about the length of an adult's body when lying flat, surrounded by a large circular magnet with a radio frequency (RF) coil and a gradient coil. The magnet generates a strong magnetic field that aligns protons within the hydrogen atoms. All the protons line up in parallel to the magnetic field like tiny magnets. The radio waves knock the protons from their position when the scanner operates by emitting short bursts of radio waves toward the subject. The subject slides into the scanner during the image acquisition process. When emission stops, the protons realign back into their original random orientations. During this realignment process, they too emit radio signals. The protons that locate in different tissues of the body realign at different speeds so that the signal emitted from different body tissues diverges, hence tissues of different properties can be identified by such variation of signal emission. From the radio signals, a spectrometer inside the scanner can produce an image of the body. An example of an MRI scan of a healthy human brain is shown in Figure 4.14. The key features are the different shades of gray that represent different parts of the brain.

4.2.1.2 X-ray

Similar to MRI scanner, an X-ray camera is also operated by a radiographer who controls how and where the image is taken. An X-ray image, commonly known as a radiograph, is usually taken for

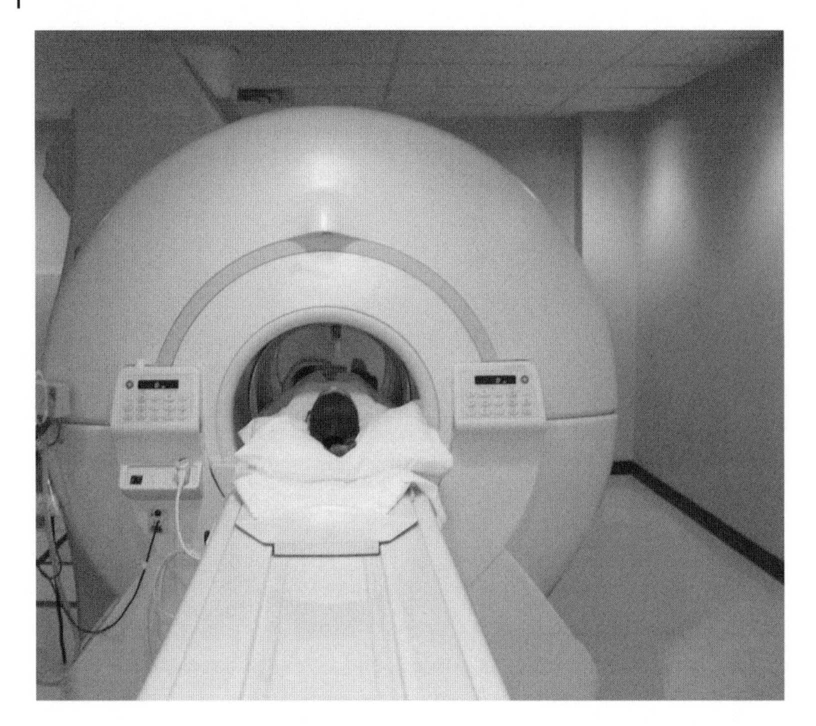

Figure 4.13 MRI scanner where the patient enters the scanner that allows a detailed scan of any part of the body to be carried out.

diagnosis purposes. X-ray radiography is perhaps the earliest medical imaging technology. It was introduced by Wilhelm Röntgen in 1895 (Koeningsberger and Prins 1988). A portable X-ray camera was made commercially available in the following year. This was about a century after Alois Senefelder invented lithography. While X-ray radiography has been very widely used in medical science for over 100 years, lithography never found an application in medicine. Indeed, the invention of X-ray was such an important event that it won Röntgen the first Nobel Prize for Physics in 1901.

X-ray incurs energy that is sufficient to ionize atoms resulting in positively charged ions that may damage human tissue. X-ray radiography relies on the capturing of radiated EM radiation whose frequency range, hence energy level from elementary physics as in Eq. (4.5), is way above that of visible light.

$$E = hf \tag{4.5}$$

$$h \sim 6.63 \text{ x } 10^{-34} \text{ (J s)}$$

The incident energy E, measured in electron volts (eV), is directly proportional to frequency f since h is Planck's constant that relates the energy in one quantum. This is the potentially harmful energy that can lead to health problems such as energy in excess of 1 KeV can change the chemical bonds of vital substances within the human body. Note, incidentally, that radio frequencies do not carry sufficient energy to alter an atom. For this reason, MRI is much safer than X-ray.

The physics behind X-ray radiography is actually quite simple. Consider the situation where an X-ray beam carries sufficient energy to "knock off" an electron within an atom causing it to ionize, as illustrated in Figure 4.15. An X-ray photon strikes an electron causing the electron to move from a higher energy shell into a lower energy shell closer to the nucleus. This process releases dissipation

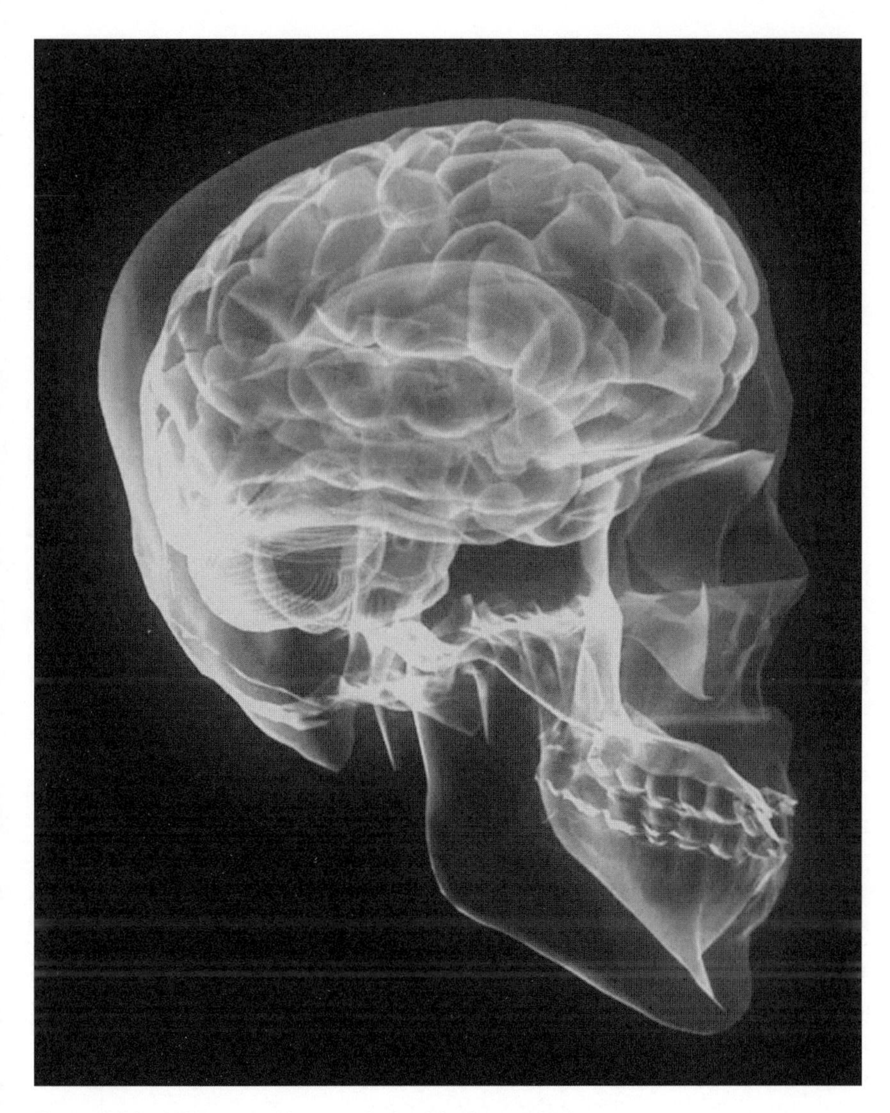

Figure 4.14 MRI scanned image of a healthy human brain.

energy that produces a photon. The photons produced during this process are known as fluorescent or characteristic energy.

To study X-ray image processing, we need to understand how a clear image can be produced. The above physical properties lead to Compton scattering when the incident X-ray photon is deflected from its original path due to an electron. Another Nobel Prize in Physics was awarded in 1927 to Arthur Holly Compton for the discovery of this phenomenon. Unlike the above situation, only a part of the photon energy is transferred to the electron during the X-ray strike. So, a photon is emitted with less energy through an altered path. The energy shift caused by a reduction of energy, hence wavelength change $\Delta\lambda$ (as in the simple relationship $v = f\lambda$), depends on the angle of scattering:

$$\Delta\lambda = \frac{h}{m_e v}(1 - \cos\theta) \tag{4.6}$$

$$\Delta\lambda = \lambda' - \lambda \tag{4.7}$$

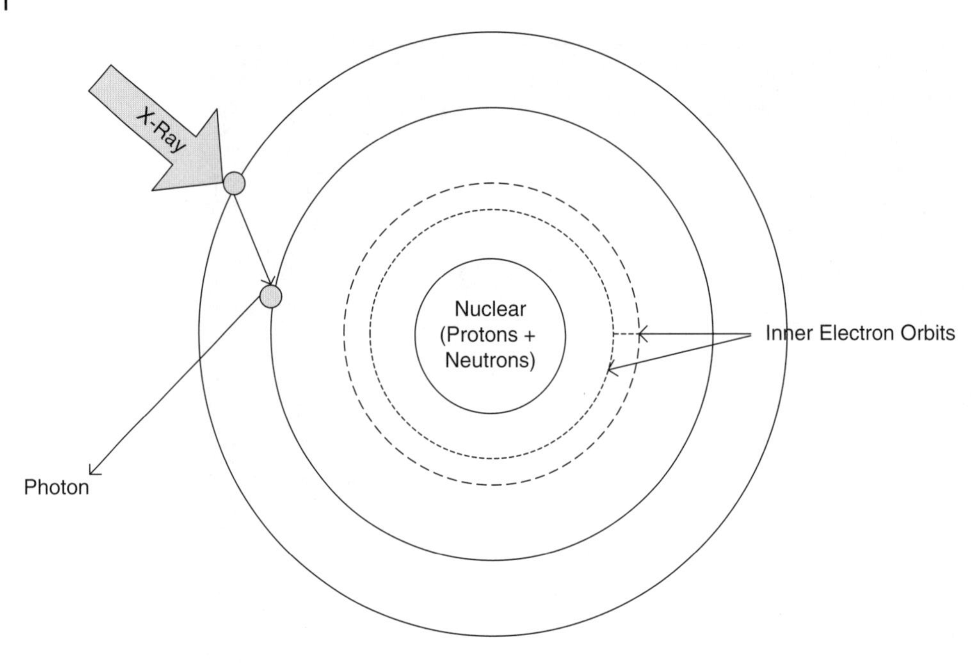

Figure 4.15 X-ray radiography.

The scattered photon has an energy E' relative to E is:

$$E' = \frac{E}{1 + \frac{E}{m_e v^2}(1 - \cos\theta)} \tag{4.8}$$

where m_e is the mass of electron, which is a constant, and θ is the photon's scattering angle, as shown in Figure 4.16. λ' and λ are wavelengths of scattered and incident X-ray photon, respectively. Energy is lost to an electron that is driven out from the atom. Compton scattering is an important topic to study since it is the major source of background noise on an X-ray radiograph. It is also the major cause of tissue damage. It is obvious from Eq. (4.7) that the scattered energy E' is independent of scattered angle θ if the incident energy E is low. So, scattered photons with higher energy will continue in about the same direction as that of the X-ray source.

There is a tradeoff between patient safety and the effectiveness of X-ray penetration that produces a clear image in the adjustment of X-ray dosage. The absorbed dose exposed to a patient is measured in terms of the energy absorbed per unit of tissue. Details on X-ray dose can be found from the RSNA (Radiological Society of North America) report of 2009. Other sources of interference include cosmic radiation, nuclear plants, and natural radioactive materials that exist almost everywhere. More details about the possible risk of excessive radiation dosage are discussed in Section 8.5.3.

Since X-ray images reveal abnormalities inside the body, small tumors are divulged somewhere inside the image with different shades of gray. Converting images into digital format can make transmission and storage far more efficient than with silver-based films. Therefore, preservation of tiny but important details requires digital imaging techniques that provide sufficient resolution and bit-depth that can distinguish any tumor from the background. Additive noise imposed on the image or transmission loss may completely ruin the usefulness of a radiograph. We look at the details in Section 4.2.2.

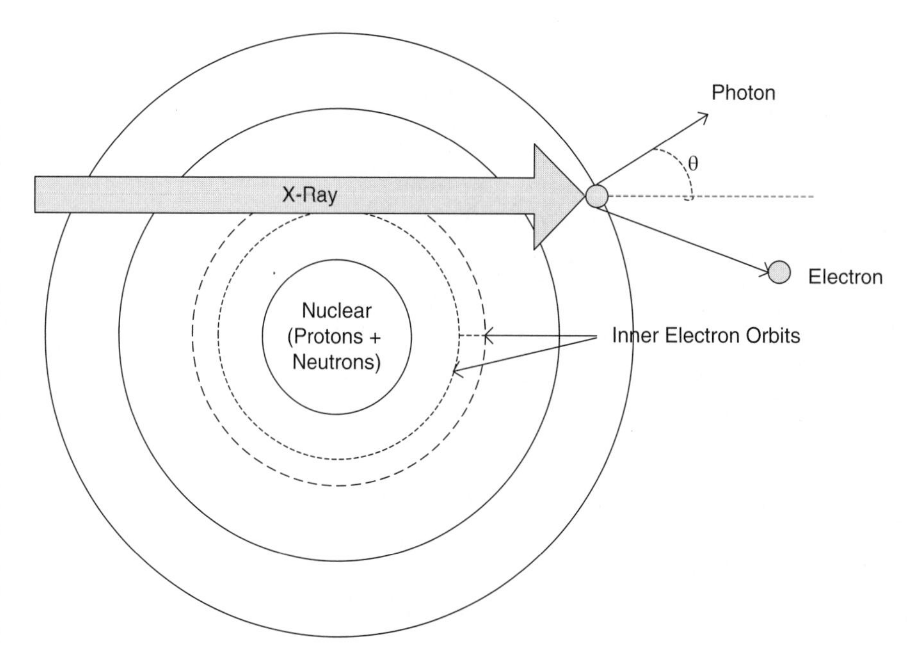

Figure 4.16 Photon scattering.

4.2.1.3 Ultrasound

Ultrasound measurement relies on several different properties of sound propagation; these include propagating velocity, attenuation, phase shift, and acoustic impedance mismatch. With variation of these properties while propagating through different substances, tissue structure characteristics can be analyzed (Tempkin 2009). It is a high frequency sonic signal above the audible frequency range that propagates through fluid and soft tissues. The ultrasonic signal is then reflected back as an "echo" to form an image. The denser the tissue it strikes, the more is reflected back producing a lighter image. Images of organs and structures with different shades of gray can therefore be created.

An image is formed by scanning a probe across the area of interest. This probe does not have to enter the body, and the entire process is carried out on the skin. The probe emits pulses of ultrasound and picks up the echo as the ultrasound signal is reflected back. We first take a look at how an image is generated by using an example of a heart scan that generates an "echocardiogram." The ultrasound signal penetrates through blood in the heart chamber and is reflected back when it strikes the solid valve. The presence and absence of tissue reflecting the signal produces a black and white image with varying contrast, as in Figure 4.17. A monochrome image that shows a healthy heart is formed. This is particularly useful in detecting any abnormalities that may lead to heart problems. Very similar techniques can be used in different areas, such as the detection of breast tumor and renal calculi (kidney stones) for cancer and hydronephrosis diagnosis at an early stage so that treatment can be provided before the condition deteriorates.

In addition to providing early treatment, an ultrasound scan is also very widely used on pregnant women to constantly monitor the development of their unborn babies in the womb. An example of a 21-week-old healthy growing fetus is shown in Figure 4.18. These seemingly blurry pictures convey important information, such as the gender of the child and whether all parts of the fetus are developing normally.

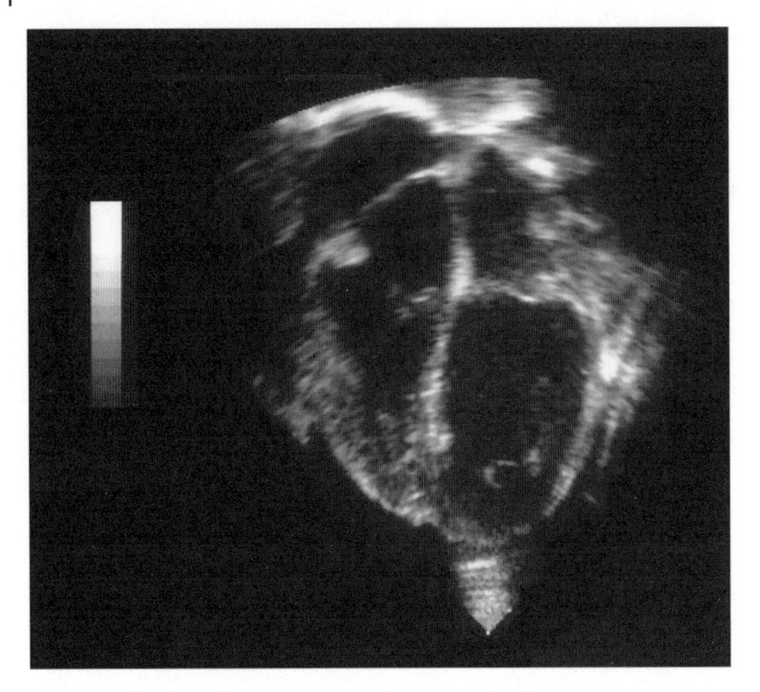

Figure 4.17 Ultrasound image of a healthy beating heart.

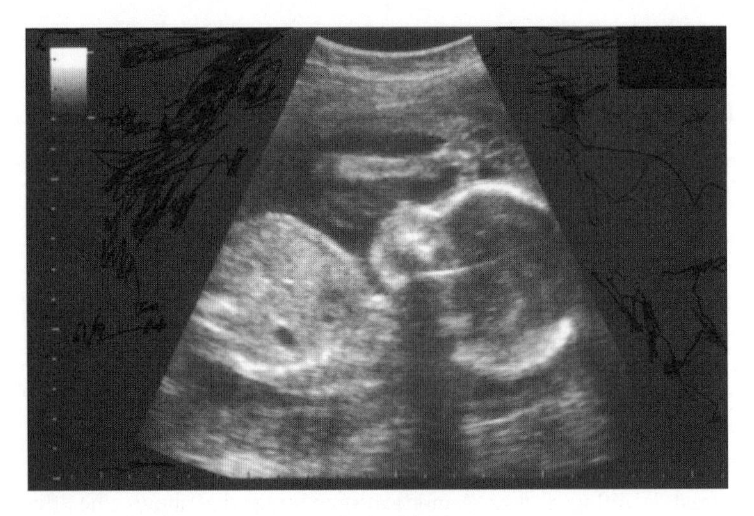

Figure 4.18 Ultrasound image of a healthy fetus.

4.2.2 Medical Image Transmission and Analysis

We study three major types of medical image acquisition technology above. We now move on to the topic of processing these images without going further into alternatives such as OCT (optical coherence tomography) and PET (positron emission tomography) as these modalities exhibit many similarities when compared with the image types that we have covered as far as image processing algorithms are concerned.

The technologies related to transmitting a medical image from one location to another may be very similar to that of general purpose photo transmission, just like snapping a photo with a camera-equipped smartphone and uploading a digital photo onto the Web. The procedures may be similar but the requirements are certainly very different in the sense that faithful reproduction is the key to an image's usefulness since the main objective of taking a medical image in the first place is likely to be for the identification of any subtle details embedded in the image. It may be a tumor hidden somewhere in a confined area of the image that needs to be identified. Also, many such images are monochrome so the different shades of gray can hold the key to diagnosis from the image.

To learn more about the successful transmission of medical images, we look at a case study where an X-ray radiography is sent from the radiographer's site to an expert for analysis by first referring to Maintz and Viergever (1998). Remember, an X-ray radiograph is a 2D depiction of 3D mapping that represents the attenuation of X-ray absorption properties of tissues that are exposed to a dose of X-ray. In this case, suppose an X-ray beam of intensity I strikes the tissue of a subject; the cross-sectional area of the X-ray beam and that of an atom within the tissue is A and S, respectively. The atom density of the tissue, namely the number of atoms per cubic centimeter of the tissue, is N. The total cross-sectional area of the atoms in this mass of tissue is therefore $N \times S$, and the total area of atoms hit by the beam is $A \times N \times S$. These parameters are shown in Figure 4.19. The rate of change of the beam intensity while penetrating through the tissue across thickness x is:

$$\frac{dI}{dx} = -NSI \tag{4.9}$$

This is very important in deducing the "attenuation coefficient" μ, which is a function of the photon intensity at any arbitrary position through the tissue x, namely $I(x)$, as:

$$I(x) = I \times e^{-\mu x} \tag{4.10}$$

So, the image formed on the radiograph is essentially a map of the photon energy across the area photographed with adequate contrast between bone and different types of tissues. For example, tumor will be highlighted on the radiograph with different shades of gray compared to bones and healthy tissues. In the sample radiograph of Figure 4.20, the left side of the patient shows abnormally dark cavity, indicating a decay of tissues inside the left lung. This particular radiograph

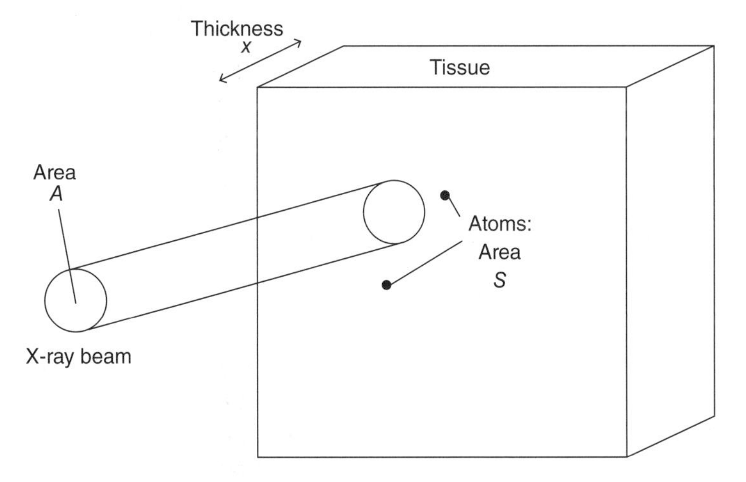

Figure 4.19 When an X-ray beam strikes the tissue.

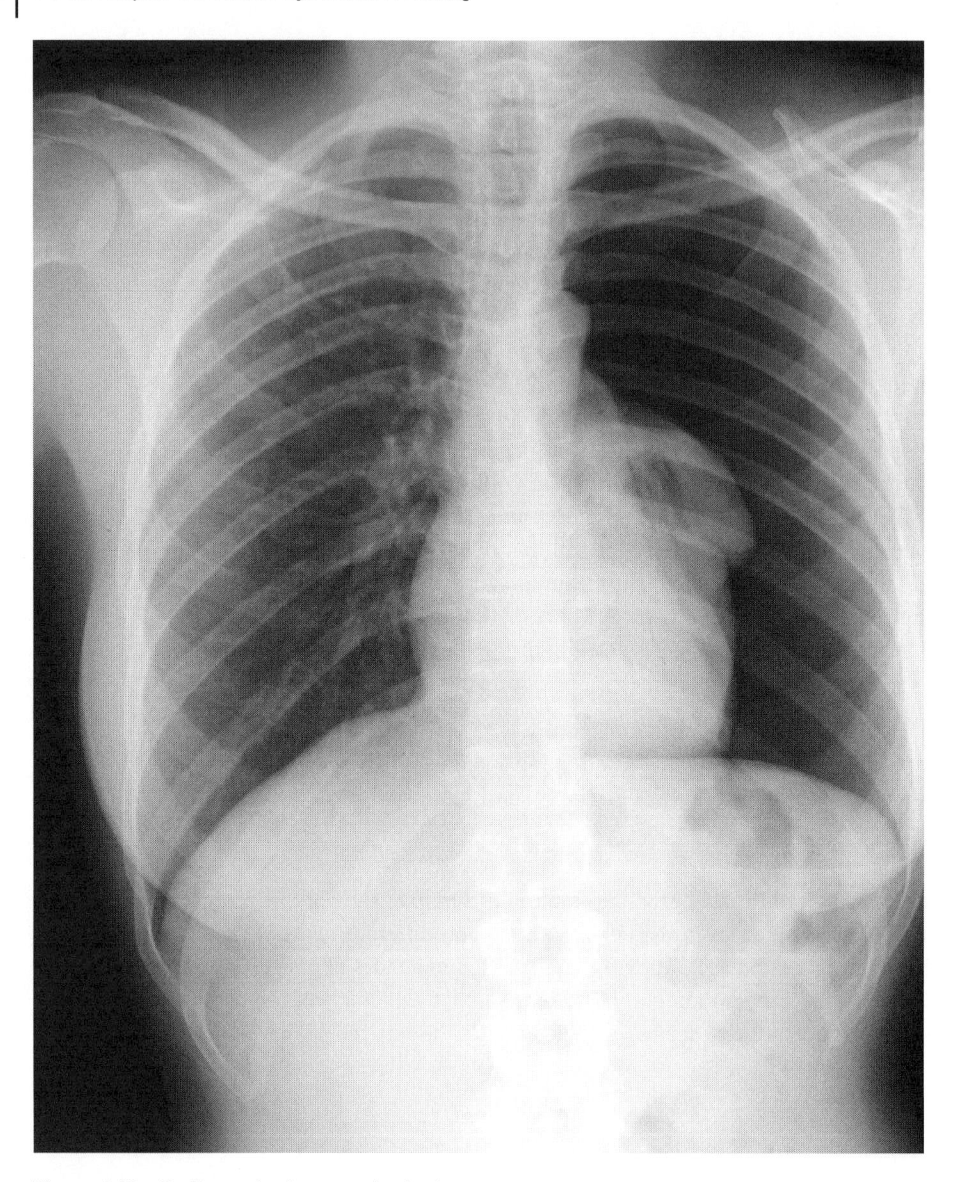

Figure 4.20 Radiograph of a tumor in the lung.

reveals spontaneous pneumothorax caused by pneumonia; diagnosis is only possible given the clarity of contrast exhibited. This can be compared with the right lung, which is perfectly normal and appears much lighter on the image. Successful diagnosis requires an image to be received with relevant details intact; image analysis will become meaningless if the details are lost during any stage of image transmission and processing.

The visual world is composed of analog images. This sentence makes good sense since the images we see in the real world are collections of a continuous spectrum of colors with an infinite amount of details. It is practically impossible to send any image with infinite details. So, the process of digitizing an image would reduce it to a finite size so that sending or storing becomes possible. The transmission of images requires the efficient use of the available channel bandwidth as a vast amount of data is involved. For example, a simple "bitmap" (matrix of gray or color dots called "pixels") image of 3000 × 2000 pixels – i.e. 6 megapixels (MP) resolution – with 256 shades of gray between deep black and pure white, when "uncompressed," has a file size of:

$$\text{Uncompressed bitmap file size} = \text{H} \times \text{W} \times 2^{\text{b}} \tag{4.11}$$

excluding any redundancies such as error checking and additional information about the image including the image type and date taken embedded into the file. Here, b is the number of bits per pixel that gives the levels of shades or color depth, while H and W are the height and width of the image, respectively. In this example, substituting the numbers into Eq. (4.11) yields: $3000 \times 2000 \times 8$ ($b = 8$ because $2^8 = 256$ that gives the number of shades) = 5.72 MB. The calculation is very simple: we multiply H and W together to get the total number of pixels in the image. After multiplying b for the number of bits per pixel, we convert the number into units in "bytes" by dividing the product by eight because each byte contains eight binary bits. From there, since each kilobyte (KB) contains 1024 bytes, we divide the number of bytes by 1024 to express the size in KB. Similarly, we further divide this number of KB by another 1024 to express the file size in megabytes (MB), since one megabyte consists of 1024 KB (not 1000 KB). This gives us some ideas about how much data is involved when handling a digital image. It is therefore desirable to make the image smaller for easier transmission and storage.

There is always a compromise between image quality and file size in digital imaging, with file size being an important consideration in terms of storage requirement and image transfer time, especially when there is a large number of images to deal with. In the context of medical imaging, an image of 8 MP is generally sufficient for viewing on a UHD 4K display (4096 × 2160 pixels), because any excess pixels can be subject to removal by bilinear interpolation or bicubic convolution resulting from the limitation of the display (Sano et al. 2017). This is simply to say that extra pixels that cannot be displayed are simply discarded by resampling from nearest-neighbor, i.e. adjacent pixels. However, there are times when subtle fine details are needed for diagnosis so that a certain portion of an image is zoomed in on, where resolutions higher than that of the display unit's capability become desirable.

4.2.3 Image Compression

Compression sets in when shrinking an image for transmission or storage in order to improve transmission efficiency or save space. The problem is that many data compression algorithms are "lossy." This means some details of the original image are not preserved so that the processed image recovered by decompression is not exactly the same as the original image before compression. Whereas with "lossless" compression, the original image can be converted back into its exact form after decompression without any loss of detail or clarity. This is to say that no difference should be detectable when comparing the two images before and after compression and subsequent decompression. Before going further into this topic, we should remind ourselves that a digital image (one that has been digitalized) consists of an array of pixels that is represented by a long string of 0s and 1s. We begin our discussion by summarizing the pros and cons of lossless and lossy compression methods as reviewed by the tutorial of Lin and Chen (2018). Medical image compression is important for improving the efficiency of transmission over telemedicine networks and to reduce the cost of storage for mass electronic patient records.

Color is represented in digital images by using varying amounts of red, green, and blue light; the three primary colors (not to be confused with the definition of primary colors as: yellow, magenta, and cyan; these are classified as "subtractive colors"). This color representation of each pixel in a digital image is the same as almost all consumer electronics appliances, such as TVs, computers, and cameras. Any color can be reproduced by adding together percentages of red, green, and blue of varying proportions. "Additive color" is the process of mixing red, green, and blue light to achieve a wide range of colors. In a simple color bitmap, each pixel is represented by three numbers to store the amounts of red, green, and blue light that define the color of that particular pixel. In such a simple bitmap, each pixel requires 1 byte for each primary color, giving a total of 3 bytes/pixel. Since 1 byte contains 8 bits, each pixel requires 24 bits to store all the color information. So, the total number of possible discrete colors this bitmap contains is $2^{24} = 16\,777\,216$, approximately 16 million possible colors. Images of 24-bit color are known as "true color" images in computing terms. The total number of possible colors is given by:

$$\text{Number of Colors} = 2^{\text{b}} \tag{4.12}$$

where b is the number of bits per pixel or "bit-depth." For more faithful and vivid color reproduction, may consumer digital cameras have 12–14 bits per color.

Compression works by finding areas in an image which are all the same color. These are then marked as, "This area is all the same color." Compression is essentially a process of eliminating gaps, empty fields, and redundancies within an image. The main problem with compressing medical images is that they usually contain a vast amount of subtle and important details, making lossy compression generally not suitable. These details are what stop areas from being all the same color or shade of gray, and as such the details can easily be lost due to compression. In many medical images, the details represent very subtle color and gray shade variations that may be too subtle to be discernable by a human eye while containing vital information about the health status of a patient. This includes situations such as the early development of a cancer tumor or an abnormal fetus. Lossy image compression algorithms may involve discarding faint details, the term "quality factor" is commonly used to describe the extent of image quality degradation. In Figure 4.21a, we compare the effect of varying "compression ratio" with reference to an uncompressed MRI scan. Figure 4.21b has undergone a moderate compression of ratio 1 : 20. Figure 4.21c has been compressed to 1 : 100. Is there any noticeable difference? Under close investigation, b is a bit coarser than a, and c is fairly abrasive and blurry.

Unlike lossy compression, lossless image compression maps the original information sequence into a string of data bits to reduce the file such that the original image can be recovered exactly from an encoded bit stream. Lossless compression does not achieve as high a compression ratio as lossy methods so the compressed file size of the same image will be larger.

4.2.4 Biopotential Electrode Sensing

Electrical activities such as ECG, EEG, electromyography (EMG), and graphic hypnograms incur the measurement of heart, brain, muscle, and sleep behavior over time. These are usually measured by the electric potentials on the surface of relevant tissue that correspond to nervous stimuli and muscle contraction over the duration of the measurement. These are graphical representations of biomedical waveforms generated by plotting electrical current amplitude over time. For the purpose of illustration, we shall concentrate our discussion on ECG data processing as other parameters exhibit very similar properties. Figure 4.22 shows examples of each of these four measurements. One important attribute in common is that all plots are irregular variations of amplitudes over a long measurement time.

ECG records the electrical activity of the heart as it beats. It should be noted that no electricity is sent through the body in the entire measurement process. The electrical impulses made while the heart is beating are plotted so that any abnormal activity with the heartbeat rhythm can be

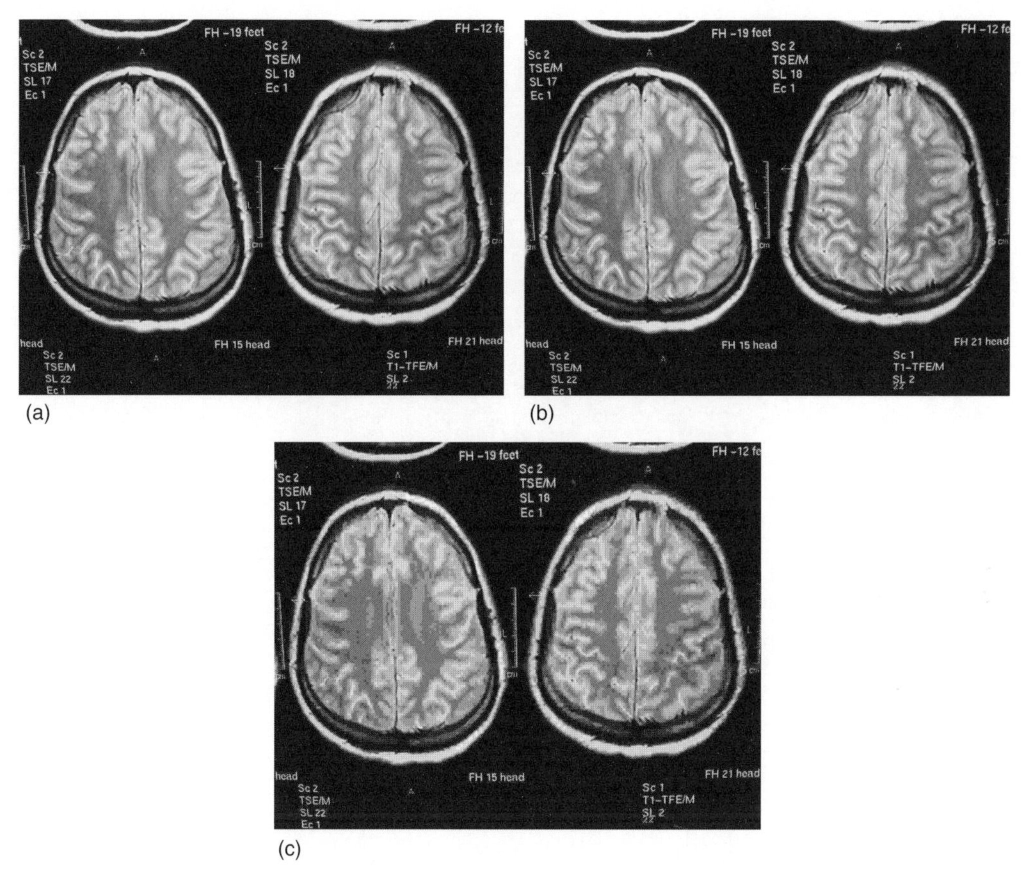

Figure 4.21 MRI scanned image: (a) without data compression; (b) moderate compression of 1 : 20; (c) compressed to 1 : 100.

identified. A range of possible causes can also be deduced from the plot. ECG is extremely useful in detecting and monitoring problems such as heart attack, coronary artery disease, prevalence of left ventricular hypertrophy, and carotid thickening. Numerous sources of noise can impair the measured signal. These include ablation, electric cautery, defibrillation, and pacing. Any impulse noise, of excessive amplitude and short duration, can severely affect the detection of abnormalities in the signal. Certain measurement procedures may also affect the effectiveness of ECG measurement. For example, patients suspected of having narrowing of the arteries to the heart may need to undertake ECG measurement while exercising on a treadmill since the plot can appear misleadingly normal if the measurement is performed while the patient remains stationary under such medical condition. Since measurement is taken with electrodes adhered to the chest, movement and shock may affect the accuracy of the measurement. Depending on its specific application, a measurement session can last as short as one minute, or it can be much longer. A short measurement is likely to be less tolerant to additive noise and interference.

The study of ECG graphs is usually performed manually by a physician. In case the graph is transmitted or stored electronically, there is bound to be a loss of quality – as in the case of the image processing discussed in the previous section. The electrocardiographic patterns need to be reproduced with such clarity to preserve all the useful features. Scanning can be a little difficult because the signal must be clearly separable from the background grid. For this reason, pure black

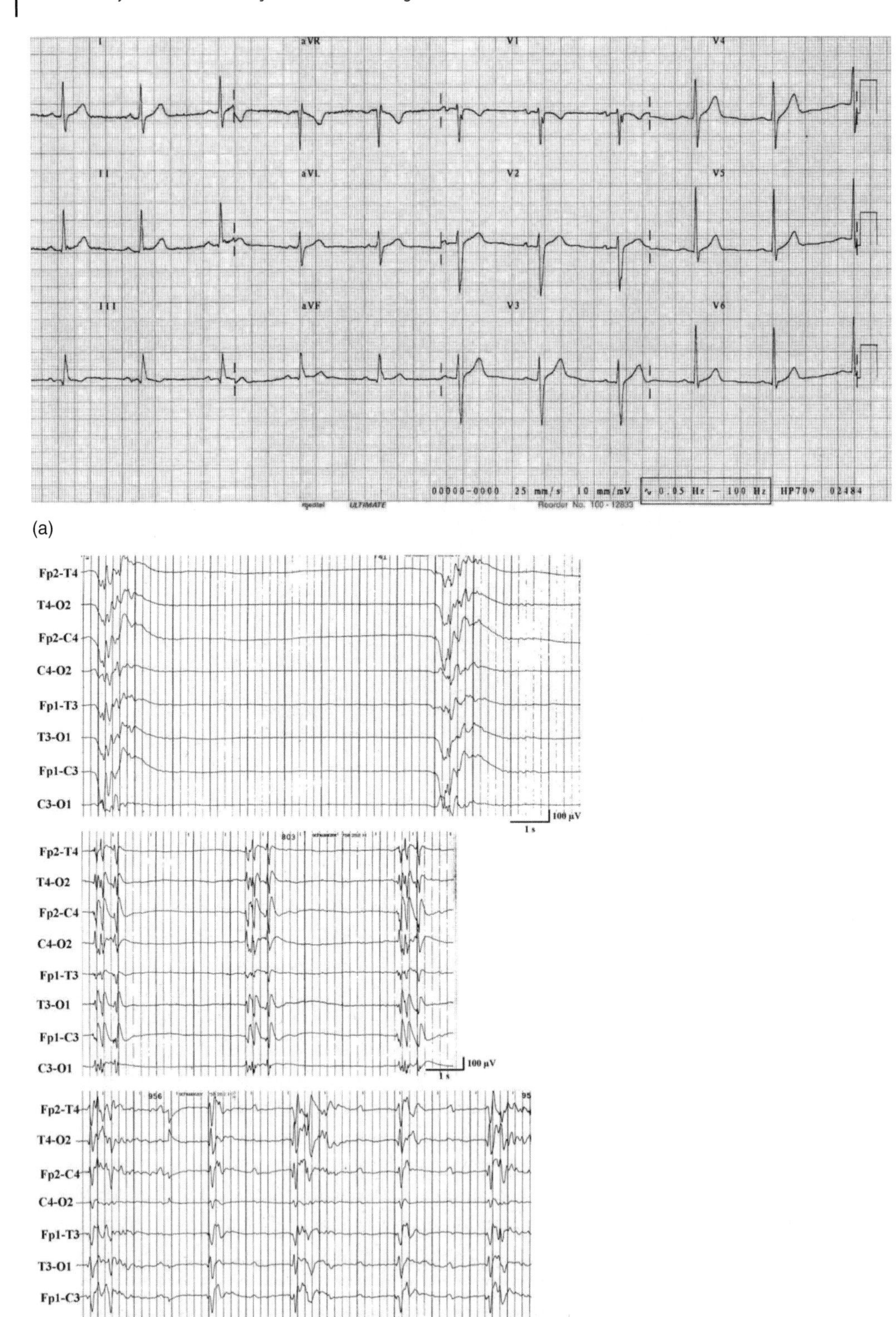

Figure 4.22 Electrical activities: (a) electrocardiogram (ECG); (b) electroencephalography (EEG); (c) electromyography (EMG); (d) graphic hypnogram.

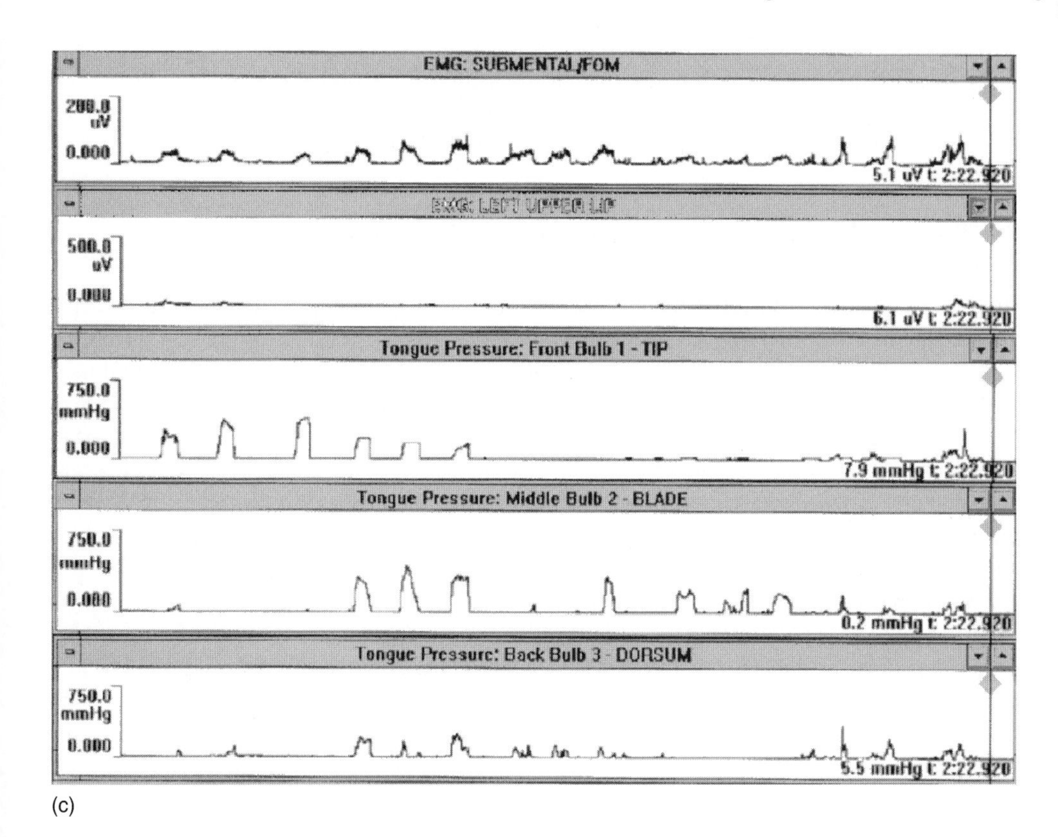

(c)

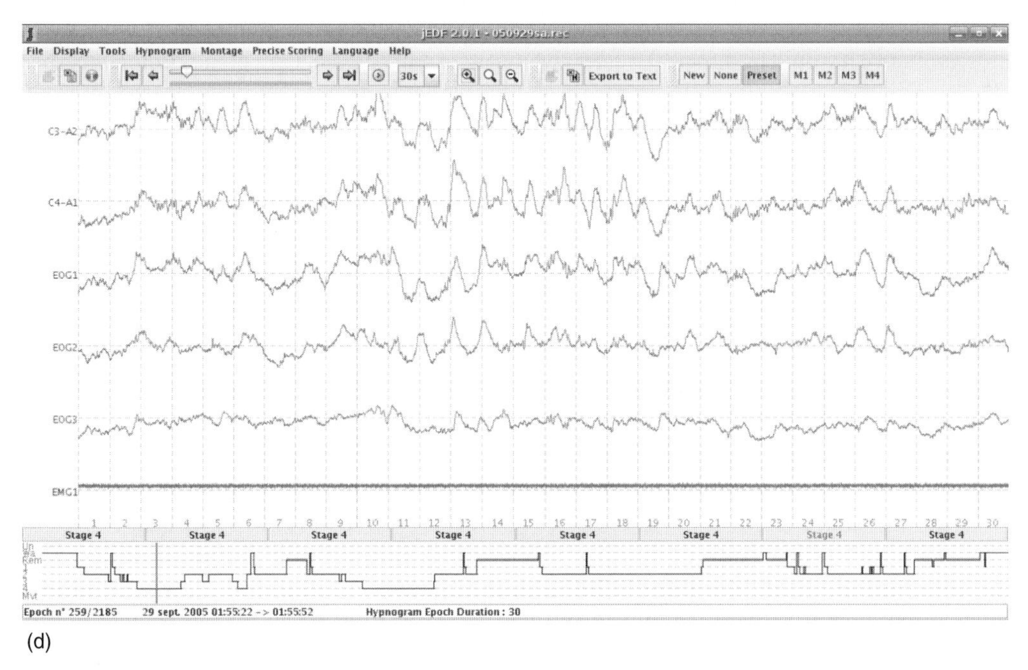

(d)

Figure 4.22 (*Continued*)

and white is not desirable, despite the plot representing the signal being a monochrome line. Sometimes, separating the image into three separate primary color channels helps extract the signal from the background grid. The pink grid, which appears only in the red channel, can therefore be easily removed from the plot simply by eliminating the red component of the plot.

4.3 Patient Records and Data Mining Applications

History about a patient's doctor visits have been kept for almost as long as medical science began centuries ago. Legacy paper log cards are still widely seen in many clinics for the sole purpose of recording details of each visit by each patient. There must be a good reason for keeping all these records. First, the patient's conditions over time can be tracked. It can also alert the doctor on conditions such as allergies to certain substances or drugs. Also, the repetitive appearance of certain symptoms may indicate something serious. All these can clearly show up on the patient's log card. Private doctors, especially those who have been practicing for decades, are so reluctant to switch to electronic patient records since the migration may involve manual data entry of many records. Another deterrent is perhaps the time necessary to get used to a new electronic system, both for updating the records and for information retrieval. They may be so much used to systematically filing paper records by patients' names. A clinical assistant would manually dig out the record of a patient and make it available to the doctor before the visit. The doctor updates the record in writing at the end of the visit and the assistant puts it back onto the shelf. This process sounds simple but there are several major problems. First, the doctor or the patient may move, or when the doctor ceases practicing due to retirement or whatever reason the records will be left behind. One common question many may ask is whether the writing on those cards is legible. It would be meaningless if the new doctor comes in and is unable to read the information scribbled on the cards. Another major problem is that records just keep on appending. Some patients may have a thick block of log cards, and since it would be difficult to detect who has ceased to be a patient of the clinic concerned so some records may just sit on the shelf for ever. If the patient has emigrated, the record will be left there redundant, and so this medical history will not be available to the patient's new family doctor.

A rural clinic may have hundreds of patients, whereas a large hospital in a metropolis can serve over 100 000 patients. Consider storing medical records for each individual patient from the time of their birth, including all test and diagnosis results, prescriptions given, details of each visit – and all of these for decades. How much data is involved? Each patient may have megabytes of medical history. Those with a long history may even run into gigabytes each. It is not difficult to grasp how vast a medical data bank a single hospital might be, but what about a national medical database for every single citizen of a country? What kind of data backup facility would be needed and how could an individual entry within the massive database be retrieved quickly and reliably? These are fundamental questions that we need to ask. Although nothing is like the size of the Internet, the information stored is still vast. This is where data mining technology comes in.

Take Mexico City as an example, with a population of over 22 million. An outbreak of swine influenza in March 2009 drove over 10 000 of its citizens with flu-related symptoms to visit its hospitals in a single day. These consisted of thousands of unrelated cases, hundreds of suspected cases, and tens of confirmed cases of A(H1N1) infection. This certainly involved a large amount of data if any attempt to keep medical records of all cases was to be made. Retrieval of informative data for analysis of disease mutation and spread among the cluster of data collected on a daily basis is only made possible by data mining technology.

Data mining relies on statistical models for fast retrieval of information from a vast database. One similar application where we frequently make use of data mining is searching over the Internet, for example, using Google™ Web search. What we have is a "search engine" that is linked to millions of websites throughout the world. Once we enter the word or phrase to search for, it will grab all pages containing it in a fraction of a second. So, how does it work? To facilitate our discussion, we illustrate by searching for a single word for simplicity's. Of course, as far as the computer is concerned a phrase is just a very long word with "space" as a character such that it just treats it as a letter in a word. Computers understand characters (letters, symbols, spaces alike) in ASCII (American Standard Code for Information Interchange) codes, each character is given a unique seven-bit code for identification. For example, an "A" is known to a computer as "1000001," equivalent to number "65" in decimal representation. So, any word, or indeed phrase, is just a string of ASCII codes or sets of seven-bit codewords entered in a sequential manner.

Data mining involves pattern extraction by examining records in vast "relational databases" from various dimensions and categorizing them. As computational processing power and disk storage capacity increases, more effective statistical analysis software is made possible to search through very large amounts of information in a fraction of a second. To illustrate the power of modern search engines, the authors performed an Internet search for the phrase "data mining." Over 21 million results were displayed in a mere 0.18 seconds. The process analyzes relationships and patterns in records based on open-ended user queries as a search is initiated. Generally, there are four distinct steps necessary for information retrieval, as illustrated in the flow chart of Figure 4.23. Although

Figure 4.23 The information retrieval process.

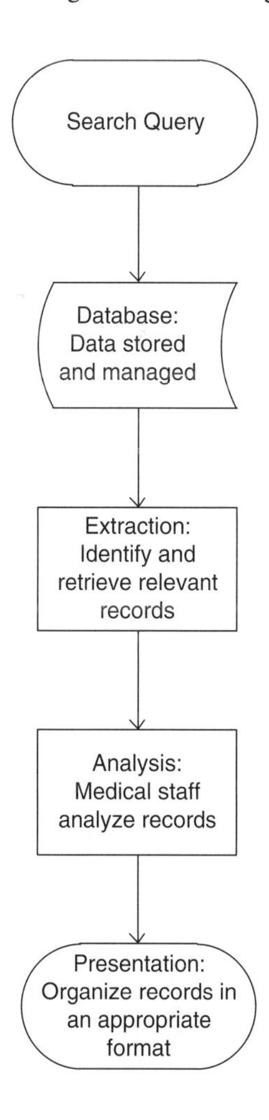

searching through electronic patient records requires very similar technologies, medical records may contain more than simple text and numbers. As we have learned above, a number of medical image types are also related to individual patients. And even real life image searches are often frustrating as often few of the displayed images are related to what we are actually looking for. Essentially, data mining extracts items based on four types of relationship:

- Associations: data is extracted to identify associations or links. For example, patients may be linked between diabetes and obesity as many diagnosed with diabetes are obese. However, it is not necessarily true that someone with diabetes must be obese.
- Classes: data is grouped according to set classes. For example, patients with diabetes can be grouped together in one class.
- Clusters: data is grouped according to logical relationships. For example, patients can be grouped by geographic or demographic criteria. This is particularly useful in studying statistical disease patterns.
- Sequential patterns: data is extracted to predict behavioral patterns and trends. For example, obesity can be linked to chronic disease so that a diabetes patient may be obese, and is more likely to suffer from chronic problems than a patient not classified as diabetic.

We shall go further by looking at a case study of an electronic patient record for a diabetes suffer. Vital but sensitive information (including name, gender, date of birth (that also tells the age), and contact details) is stored. The age alone can give an indication as to classify whether this patient is of diabetes type 1 or 2, since type 1 diabetes is usually developed during childhood, whereas type 2 diabetes mainly affects adults of over 40 years old. A large part of an electronic patient record contains information about each clinical visit, such as date of visit, nature of conditions, what remedies have been prescribed since diagnosis, and glucose result with respective measurement unit: milligrams per deciliter (mg/dl) or millimoles per liter (mmol/l). Table 4.1 shows a list of countries

Table 4.1 The two main blood glucose units used worldwide.

mg/dl	mmol/l
Argentina	Australia
Brazil	Canada
Caribbean countries	China
Chile	Ireland
Israel	The Netherlands
Japan	New Zealand
Korea	Russia
Mexico	Scandinavia
Most EU states	Singapore
Most Middle Eastern countries	Slovakia
Peru	South Africa
Taiwan	Switzerland
Thailand	Ukraine
USA	UK
Venezuela	Vietnam

and units of measurement. Any follow-up and known effectiveness of remedies is also recorded. In addition to text describing the above, digital images and audio recordings may also be included for different situations such as X-ray radiographs and heartbeat rhythms for a complete record.

There are methods in searching by pattern recognition, as described in Elmaghraby et al. (2006), where rule-based techniques are illustrated. In fact, a number of possibilities exist, and their effectiveness depends primarily on the database size and query complexity, as these would demand more computational processing power. The commonly used analysis methods include:

- Neural networks: predictive computational model that replicates the biological nervous system consisting of many interconnected processing elements as "neurons." The model has to be "trained" through its learning process. So, its performance increases over time given "adequate training."
- Data visualization: visual analysis of complex relationships in the data, involving the schematic abstraction of data graphically.
- Decision trees: branch structure that leads to sets of decisions, which derive rules for classification of data records. Two commonly used methods are classification and regression trees (CART) and χ-square automatic interaction detection (CHAID); in CHAID, the "CH" is taken from the Greek alphabet "χ," equivalent to *chi*. These rules are applied to new and unclassified data for data extraction.
- Genetic algorithms: adaptive heuristic search algorithm that replicates the natural evolution process. It relies on a combination of selection, recombination, and mutation to evolve a set of rules. Although everything is based on Charles Darwin's work of the nineteenth century, it was first applied to data mining by Holland (1962).
- Nearest neighbor: classification of each record is accomplished by a combination of the classes of the k number of records with most similarity historically. Also known as the "k-NN method" and is often used in ECG pattern recognition since its operation relies on statistical pattern recognition. It is a "supervised" learning algorithm whose results of new instance query are classified according to the majority of k-nearest neighbor categories. Its operation is fairly simple: with a query point, it finds k number of objects as training points that are closest to the query point. The classification uses the majority vote among the classification of these k objects. Neighborhood classification is used as the prediction value of the new query instance.
- Rule induction: the simplest method to implement as it only relies on a set of "if and then" rules derived through observation.

Although the importance of data mining in an electronic patient record system is well understood, we have not addressed solutions for supporting the extraction of medical images and structural information. Most image searches are currently done by associated text, for example, by adding text markers to accompany an image. Any search for a medical image then requires prior input of accompanying text, and a systematic label system thus becomes necessary. Current technologies in image feature extraction are still at a primitive stage. Algorithms used in video processing for consumer electronics mainly rely on certain image attributes, such as color, contrast, and texture. None of these provides adequate solutions for medical images.

4.4 Knowledge Management for Clinical Applications

Electronic patient records are kept in many countries for purposes ranging from patient care to statistical analysis of health risks as well as insurance claims. The behavior of how data is sought,

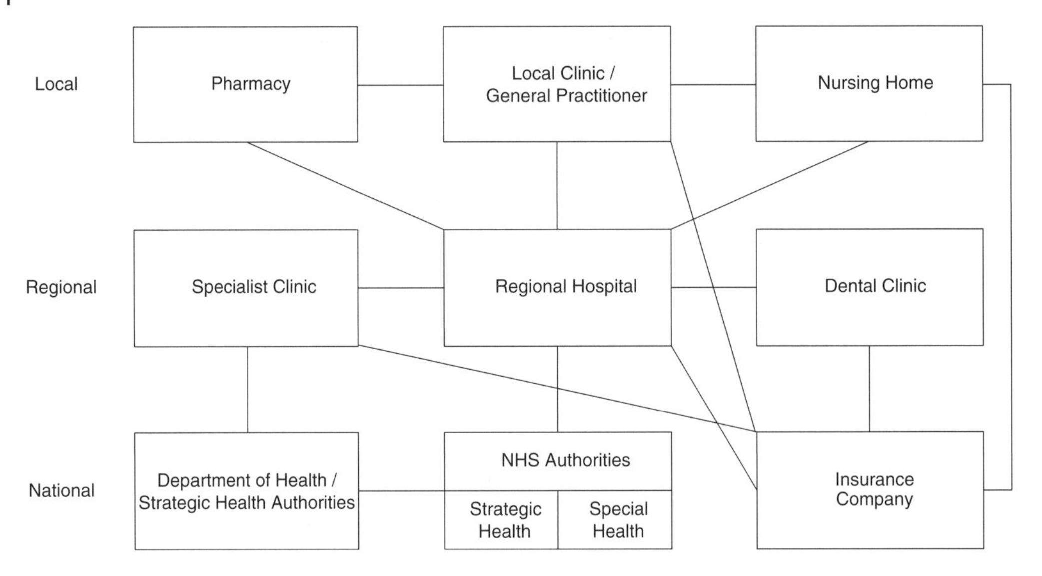

Figure 4.24 Clinical knowledge system.

outlined in Dawes and Sampson (2003), suggests that text searches through a vast amount of material remain a popular practice among physicians. From this observation, we need to find a way to efficiently handle the filing and storage of medical information since it would involve far more data than patients' information alone. Knowledge-based clinical applications span across areas from administration to medical practicing and dispensary. The block diagram in Figure 4.24 shows the complexity behind an electronic clinical knowledge system where many entities have their own information to process and share. As we can see, a lot of information is exchanged in this system, both in terms of the types of data and amounts of data. Let us take a closer look at the role of a general practitioner in this context by referring to Figure 4.25, which shows the information or knowledge shared with the outside world. There are a lot of interactions between the local doctor and related entities. So, Figure 4.25 is external to the GP's clinic. We then go deeper and expand within the doctor's clinic, where we assume that it is a small rural clinic with only one physician. The doctor obtains and shares information in many ways, as shown in Figure 4.26, which shows the information dealt with internally inside the clinic. Even within a small local clinic, there are many sources of information.

Knowledge management is all about creation, transfer, and the identification of useful information. The simple process describing it can be summarized in the knowledge management model shown in Figure 4.27. The knowledge conversion process is a continually changing and improving process that consists of the preservation and enhancement of knowledge. The knowledge conversion process can also be viewed as knowledge creation, transferring, and sharing; with the objective of improving knowledge access. The output of the process can be fed back to the input for the next round process for the purpose of continual improvement. In the clinical environment, knowledge management activities are mainly for the creation and maintenance of processes for improving healthcare services so that the general public will be healthier and live longer with less demand for medical services. So, the diagnostic process for providing the best possible treatment would vastly depend on the effectiveness of knowledge management. Following the process in Figure 4.28, constant monitoring of results from previous treatments from electronic patient records would result in more optimized treatment to be developed through prior experience. Diagnosis given the symptoms is usually completed according to clinical investigation and laboratory test results. Sometimes, diagnostic tests can be time consuming given the urgency of deriving a treatment plan.

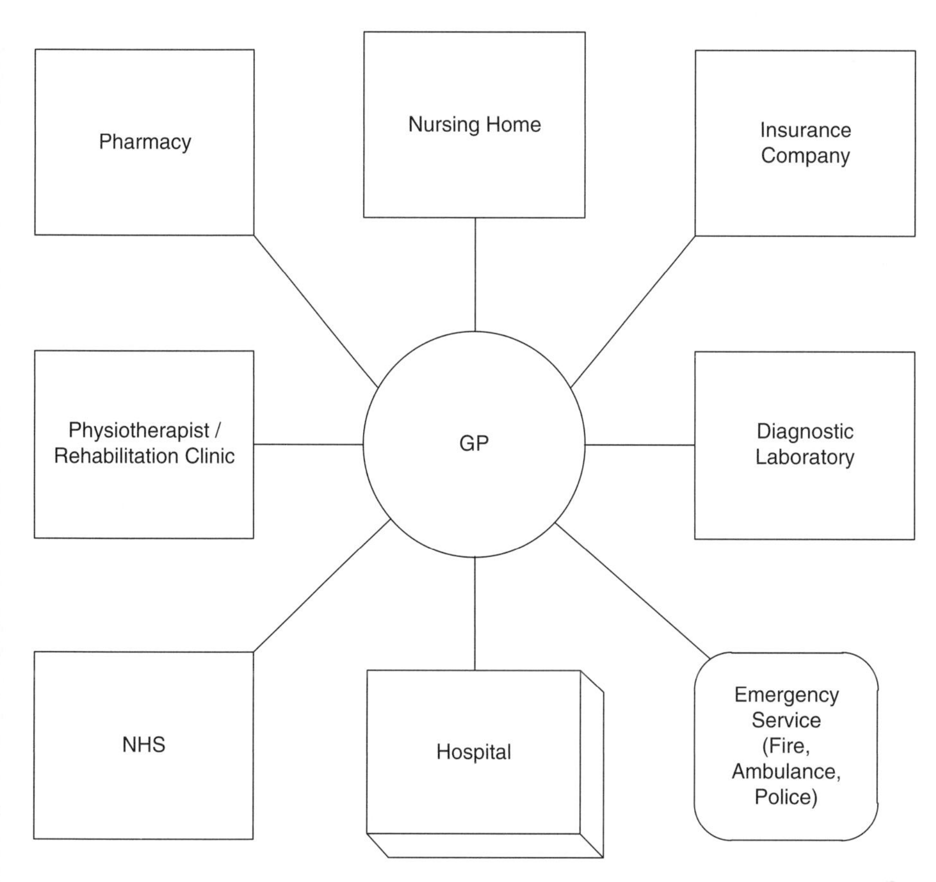

Figure 4.25 System linking a physician to the outside world.

So, knowledge from previous cases can be of significant assistance in drafting an action plan to provide the necessary treatment with minimal delay.

To illustrate this, we take a quick look at the case of using ultrasound to burn off a cancer tumor. A beam of ultrasound when focused on the tumor can rapidly heat it up to temperatures in excess of 70 °C and this process is very effective in damaging cancerous cells as it causes hypoxia, which cuts off its oxygen supply. However, there are many restrictions that prohibit its being effectual to many areas of a human body due to the risk of burning skin and fat. By keeping a record of the effectiveness and result of each treatment, it is possible to compile a list of tumor type and size that such treatment can be used on. This example shows us the importance of maintaining a knowledge database for information sharing.

The electronic patient record contains different types of information about a patient in various areas, including diagnosis, prescription, appointment record, description of symptoms, and treatment provided. It is also useful for treating similar cases in the future. There are risks and challenges that encumber the development of comprehensive electronic patient record systems. First, we refer to the famous "SOAP note" (Schimelpfenig 2006), which refers to:

- Subjective: condition of the patient, what symptoms have been described.
- Objective: collection of vital signs, visual inspection to look for signs of abnormalities and to conduct appropriate laboratory diagnostic tests.
- Assessment: summarize the above based on symptoms and diagnosis.
- Plan: derive an action plan for treatment, e.g. prescription and any follow-up action.

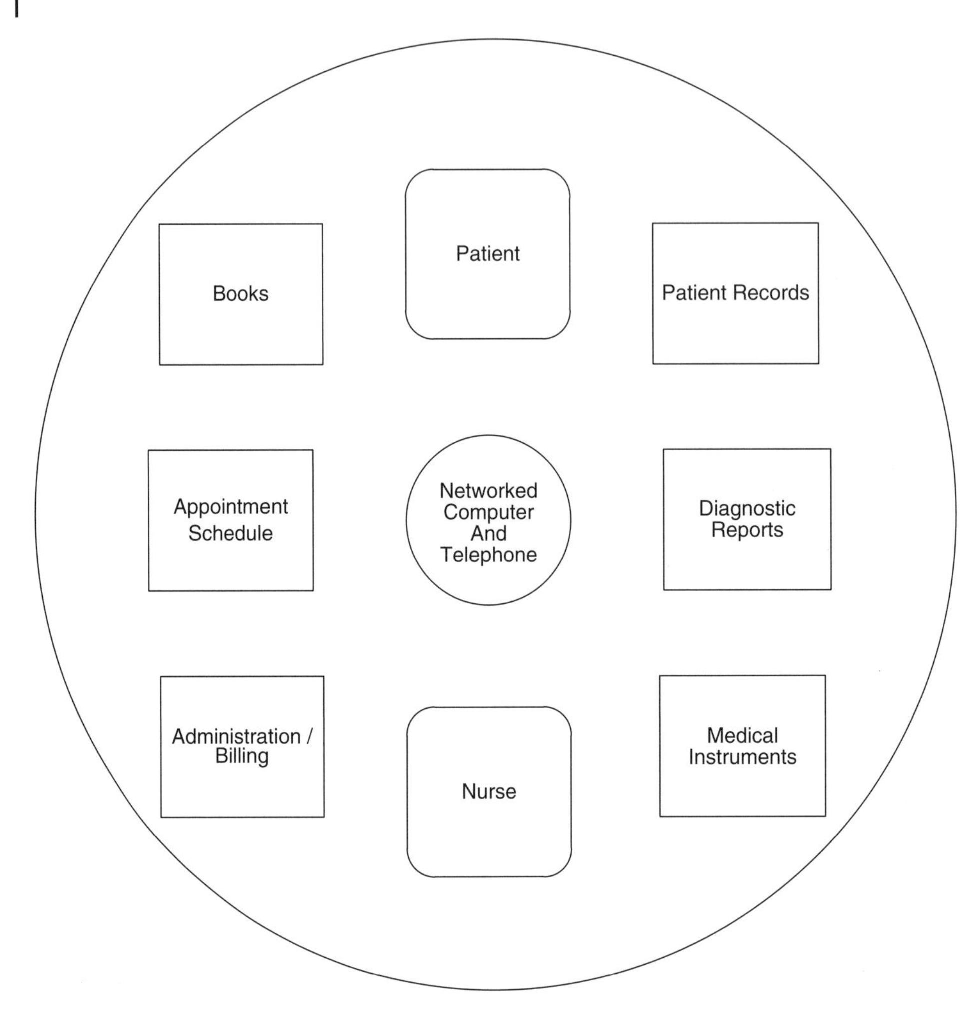

Figure 4.26 Inside the clinic.

Ultimately, this is to facilitate effective patient assessment through knowledge management, thereby providing a basis for communication between the patient and healthcare providers. The SOAP note format is often used to standardize healthcare evaluation entries made in clinical records for consistency. As a final note, medical records are legal documents. Therefore, data entry must be done in an accurate and responsible way, and access to information must be strictly controlled.

4.5 Artificial Intelligence (AI) in Digital Health

AI can be perceived as some kind of intelligence that is artificially made to *think* like an animal brain using neural networks, which fundamentally differ from biological brains in both size (there are far more neurons and synapses in a brain) and organization (brains are self-organizing and adaptive). Neural networks in AI are merely artificially organized based on a certain architecture that essentially resembles a graph that often relates to types of statistical models (e.g. curve fitting or regression analysis) rather than an ordered network (Fayyad et al. 1996).

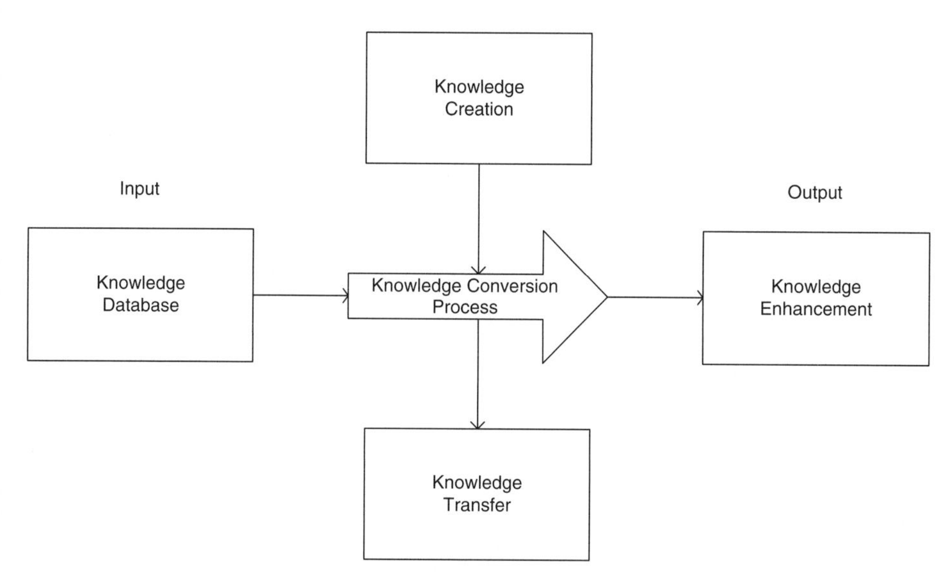

Figure 4.27 Knowledge management for electronic patient records.

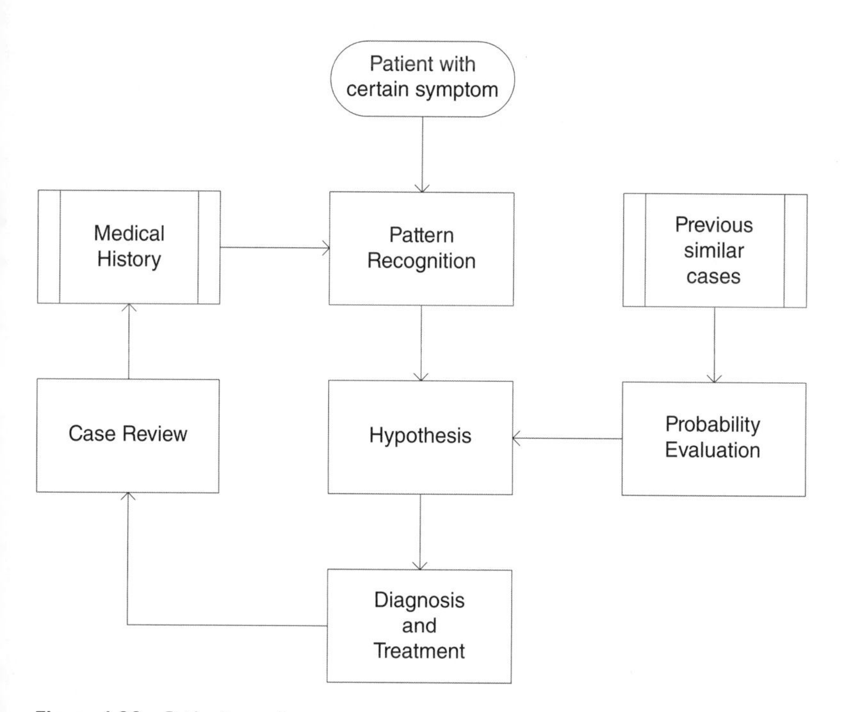

Figure 4.28 Patient monitor.

AI allows machines to learn from people so that various human-like tasks can be automated. We have seen AI being put into action to assist us with carrying out various tasks such as driving autonomous vehicles (Wong et al. 2015) and talking to people through natural language processing (Green et al. 2015). Very briefly, machines are trained to accomplish certain predetermined tasks through AI by processing input data as training sets and recognizing patterns within the data to produce an output. AI can learn about a patient autonomously. Machine learning is one of the most prominent methodologies for personalized care, as it is designed to analyze the needs of the individual patient.

When we pick up our digital camera and half-press the shutter button, the camera will automatically determine which subject it should focus on, thanks to AI (Choi et al. 1999). The onboard computer of the camera is trained to recognize objects within the image frame. Similarly, AI can be applied to skin cancer diagnosis through machine training on how to recognize features of affected areas (Pandey et al. 2019). They both work by training the machine to recognize objects of interest. AI not only works on imaging but is also widely used in analyzing many different types of data learning (Fatima and Pasha 2017).

AI differs from robotic automation in that it does not merely support repetitive automated tasks such as those commonly found in factory production lines but will also add intelligence to devices and systems so that they become "smart" in order to "think" and "analyze" in order to carry out prognosis as well as diagnostic tasks.

4.5.1 Deep Learning

Neural networks consist of layers of interconnected nodes. Each node is known as a "perceptron" (as in perception) that resembles a multiple linear regression (more than one explanatory variable). Its structure consists of one input layer, a hidden layer(s), and the output layer. The input layer receives input patterns. Hidden layers adjust the weightings on inputs to minimize the neural network's computational error. Finally, the output layer maintains a list of classifications for mapping those input patterns. It can be viewed as the hidden layers are used to extract salient features within the input data in order to carry out feature extraction. Neural networks that contain a large number of hidden layers are referred to as being "deep," in that they have the ability to extract much deeper features from the data than those with a few hidden layers.

The learning process of a neural network attempts to optimize the weights of the neural network of until some user-defined stopping condition is met. The process is repeated until a specific user-defined stopping condition is met. This condition is usually set as either a threshold of network errors that meets an acceptable level of accuracy on the training set or when an allocated amount of computational resources has been used up. Such exhaustion of computational resources is particularly important in many telehealth applications since health monitoring is often carried out using wearable devices, where data processing time should be minimized to allow a sufficient number of readings to be taken.

Deep learning describes the process of training a computer to perform human-like tasks, such as understanding medical images (Giger 2018). Deep learning involves training through establishing elementary parameters about the data type of interest (e.g. vital signs, images, infection cases, etc.) and teaches a machine to learn by itself through recognizing patterns using *multiple* layers of processing. This is in contrast to simple machine learning through organizing a set of training (reference) data to run through predefined equations (Wu et al. 2010). The main purpose of implementing deep learning as in AI for healthcare is to improve the classification, recognition, detection, as well as the description of medical information. Systems such as voice recognition in smartphones and autonomous driving in driverless vehicles all involve deep learning in classifying images, recognizing speech, detecting objects, and so on. Deep learning has the ability to provide highly personalized healthcare services by learning the individual needs of a patient. Additionally, the ability to analyze images can also be extended for integration with sports and exercise in improving performance and recommending a healthier diet. Machine learning not only facilitates the learning of patients' specific needs, it also helps in improving health monitoring through continuously learning about the usage and operational environment. The objective is to improve prognostic and diagnostic capabilities. Here, we take a look at an example of machine learning implementation in an m-health environment.

4.5.2 AI in Mobile Health

Accurate prognosis and diagnosis relies heavily on the collection of reliable health data for subsequent analysis (Stewart et al. 2015). Mobile health (m-health) provides a convenient means of collecting a patient's data for prognosis and diagnosis through appropriate sensors, yet the user's motion can significantly affect the data's integrity (Kim et al. 2016). One of the greatest challenges between controlled laboratory testing and the actual usage environment is that there is no way of anticipating *how* a device is actually used by an individual user (Fong and Li 2012). For example, a subject can be requested to sit still while testing a wearable monitor in a laboratory so that measurement is taken when the subject remains stationary, whereas the same monitor can be worn by an athlete so that measurement is taken while the monitor moves around. To compensate for the difference conditions of use, which can have a significant impact on monitoring performance, we investigate the use of AI in heart rate measurement using a wearable cardiac monitor, as discussed in Section 4.1.2.

A photoplethysmogram (PPG) is often used for detecting blood volume change by counting optical pulses through measuring each pulse's light absorption (Spierer et al. 2015). A PPG is a preferable option in wearable monitors over ECG as a tradeoff between measurement accuracy and ease of implementation in a mobile environment (Jo et al. 2016). In theory, the PPG measurement process can be as simple as using a pair of LED and photosensors measured across the user's finger, as shown in Figure 4.29. The light from an LED is focused with a fixed lens before shining through the user's finger to be picked up by a photosensor. The PPG signal represents the flow of blood and is measured by a pulse oximeter at the sensor end. In this illustration, the photosensor is misaligned due to movement so that it no longer picks up the light signal. To ensure that measurement can be taken without the subject being stationary, the monitor can be either fastened with athletic taping or motion compensation can be carried out. The latter is certainly a preferable option as it allows the subject to move around freely while undertaking daily activities. Motion artifacts can be eliminated by subtracting body acceleration (ACC) signals (Mashhadi et al. 2016), which is represented

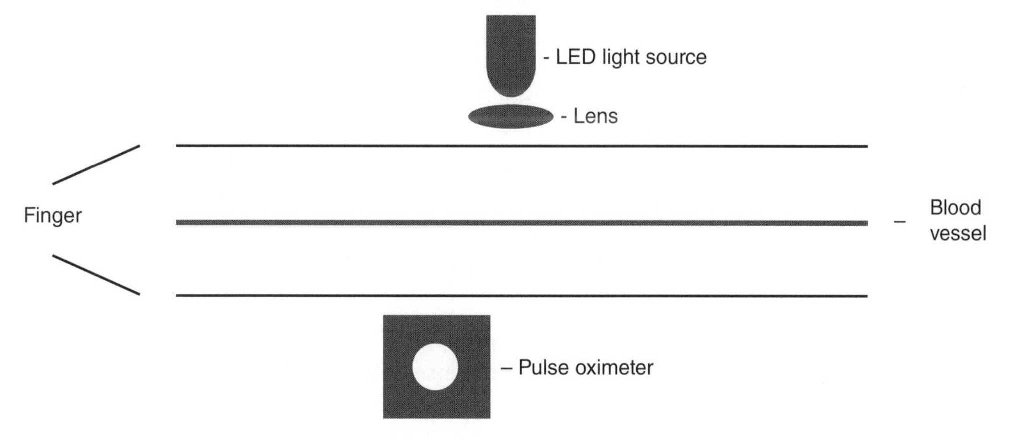

Figure 4.29 Photoplethysmogram (PPG) measurement through counting the light pulses transmitted from an infrared LED. The operating principle is similar to the oxygen meter featured in Figure 2.8.

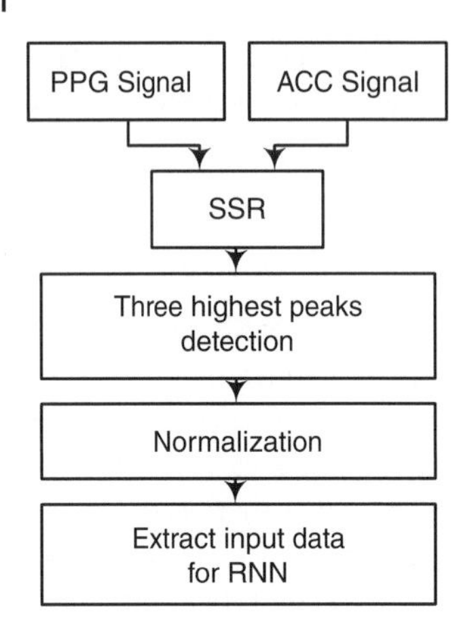

Figure 4.30 Extraction of feature frequencies using SSR.

by three signals of the x, y, and z axes. A recurrent neural network (RNN) structure that carries out time-series data pattern analysis can be used for counting heart rate from PPG signals.

RNNs have been widely used for learning time-series data such as speech disease prognosis (Choi et al. 2016). The main problem of running an RNN in a wearable monitor is that it needs a vast amount of training data, which results in a lengthy learning process (Bekhet et al. 2018). Feature extraction, as shown in Figure 4.30, uses sparse spectrum reconstruction (SSR) to generate the input values of the RNN such that the spectra derived from SSR can be shortened by extracting only a few peaks (Reiss et al. 2018). SSR is used to represent spectral characteristics using only the minimum number of frequencies by simply using a single measurement vector (SMV) (Karim et al. 2019).

To remove the ACC spectral signal at the exact frequency location of the PPG spectrum signal using an SMV, SSR is computed on several incoming signals sequentially. The spectrum of each signal is computed from the characteristics of all signals. Obtaining the spectrum for each signal ensures that the characteristics of each signal are preserved, since the PPG and ACC signals are measured independently of each other as different sources.

From the spectra, the largest three nonconsecutive values are extracted from each spectrum. Figure 4.31 shows a RNN structure to process the three frequency values of the PPG and ACC signals detected through the SSR (3 PPG + 3 ACC = total of 6 signal components with their respective magnitudes). Each of these six signals is combined with one output value from the previous output layer (being used as input values for the RNN), making a total of 12 input values. Computational time is primarily determined by the number of hidden layers within the RNN that is determined by the Occam's razor principle (Pereira and Borysov 2019).

In addition to mobile health monitoring, AI is also widely used in diagnosis and prognosis, such as in cancer prediction (Kudo 2018). Automated diagnosis and prescriptions enable fast prognosis, which leads to more efficient remedial actions. Machine learning facilitates the assessment of a disease as well as its progression. The ability to learn about an individual patient also allows optimal treatments to be determined for individual patients (Lu et al. 2019). The assessment process uses

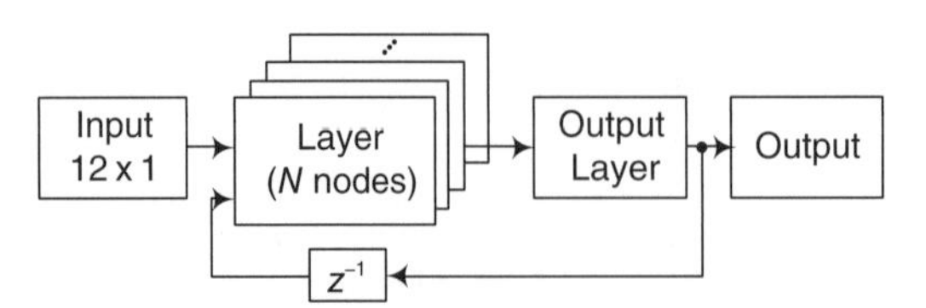

Figure 4.31 Recurrent neural network (RNN) that processes PPG and ACC signals from the SSR.

tumor imaging to identify disease progression and spread so that the most effective therapeutic options can be identified.

4.5.3 Virtual Reality (VR) and Augmented Reality (AR)

Before jumping into VR and AR technologies in healthcare applications we should first take a look at the fundamental differences between the two. Very briefly, VR often uses a headset to provide a *simulated* environment using computer programs. Whereas AR adds a 3D virtual object (one that does not physically exist but generated by a computer program) to an actual scene. This is essentially to say that the entire VR image is *animated*, whereas in AR only certain object(s) within an image is virtual but the surrounding environment is real. For example, surgical practice can be carried out using VR where everything from the operating theater to the patient and the tools are all computer-generated graphics, whereas using AR one can practice an operation using real surgical tools on a virtual patient in a laboratory that resembles an operating theater. In this sense, AR usually requires a camera that captures the surrounding environment and a virtual object is added to this captured scene.

Both VR and AR provide numerous opportunities for digital health applications. VR in surgical simulation has been used for teaching and assessment of surgical skills for well over two decades (Satava 1993). Display resolution is often a significant issue when providing a realistic surgical experience using VR (Traub et al. 2008). The introduction of AR makes it possible for practicing surgical skills in a simulation environment that very closely resembles an actual surgery (Yamamoto et al. 2012). Figure 4.32 shows a laboratory that appears and feels like an actual operating theater. The surgeons can practice on a simulated patient using either physical or virtual tools, depending on preference. In this particular example, a patient's head is opened without any safety risk, because

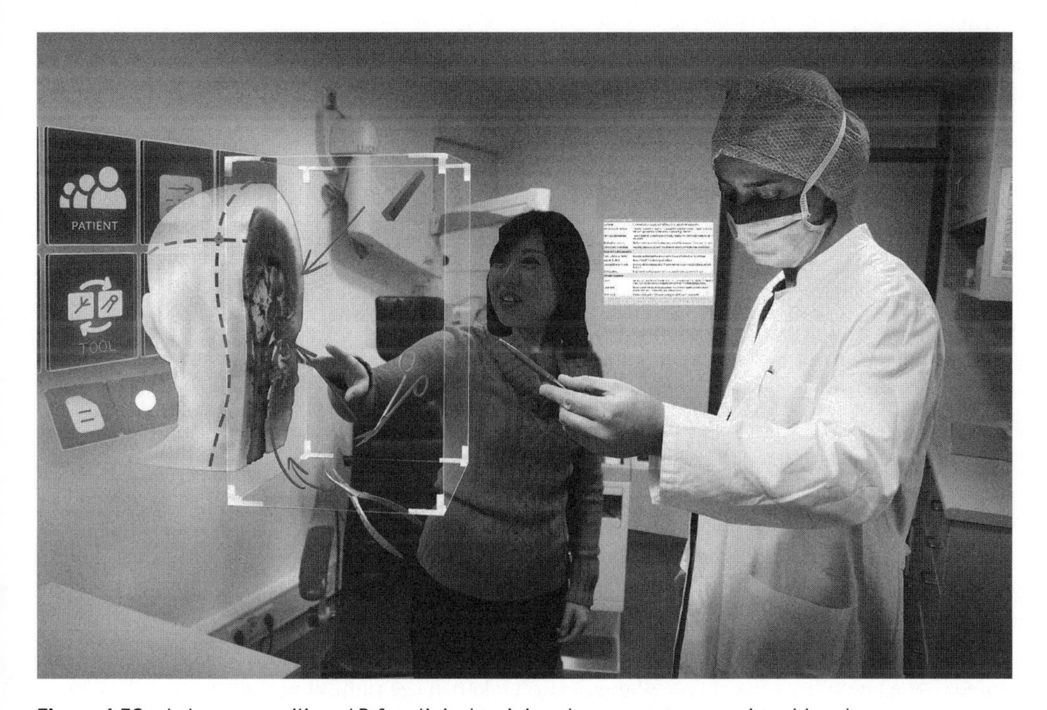

Figure 4.32 Laboratory utilizes AR for clinical training that operates on a virtual head.

the patient is not real. A wide range of information can be projected onto the patient, and the surgeons can even switch between patients with different medical conditions simply by gesture control. The patient can also be programmed with any conditions and complications.

4.5.4 Electronic Drug Store

We conclude this chapter by looking at medical information sharing using an electronic drug store for healthcare professionals and end users. The word "telemedicine" may be closely linked to the remote dispensary of medicine. Electronic drug stores make it much easier to organize and dispense medication for people in isolated areas or those with mobility problems. Although medicine must be physically delivered by some means of transportation, it does provide rural areas better access to medications and related information. One of the features of the electronic drug store is to assist with dispensing medication securely with automated auditing procedures for quality assurance and to reduce administrative costs. It is not simply a vending machine selling nonprescription medicine. Note, incidentally, that the term "dispensary" is not used here, because a dispensary in the United States is an agency that provides substances, such as alcohol and legalized cannabis for herbal therapy (Martinez and Podrebarar 2000). Another major application is the analysis of drugs so that effectiveness and any side effects can be duly recorded in an efficient and organized manner in the process of developing new medication. Most importantly, an electronic drug store serves as a library for patients to learn about the proper use of medications and their side effects, and automatically generates reminders for the replenishment and disposal of expired/outdated medications. It keeps patients connected to their local pharmacies. Pharmacists can therefore provide a healthcare service that does more than just dispense medication. Both patients and pharmacists can obtain information about possible adverse drug reactions and allergies. Also, any product recall exercise and expiration of drugs or their respective license/registration can keep pharmacists up to date.

To the general public, e-prescribing ensures they get the best medicine while the risk of drug mix-up is kept to an absolute minimal as technology is available at every step to ensure proper procedures are followed. Electronic patient records are also integrated to ensure that what they have taken is recorded. In addition, there is an electronic link between physicians, pharmacist, and the patient. Patients can collect their medicine after a doctor visit without having to bring a prescription form from the clinic, since the pharmacy can retrieve this electronically. The idea of an electronic drug store is to employ a remote drug ordering system such that licensed pharmacists can receive e-dispensing orders and patient records irrespective of the time of the day. Pharmacists then check and profile the accuracy of each order and authorize the hospital pharmacy system to dispense the medicine. The pharmacists also monitor allergies, drug interactions, dosage, and each patient's pharmaceutical history before issuing an authorization. Also, the system can check whether the medication is covered by the patient's insurance so billing can be made accordingly. The generation and storage process of prescription records are all done automatically. Last but not least, patients can be reminded to take medicine at the right time and to replenish their medicine. For older patients and those with visual impairments who require long-term medication, a small device can be installed at the patient's home as a "personal medication assistant." This device, shown earlier in Figure 4.29, can also be implemented as a software application installed on any home or laptop computer with a radio frequency identification (RFID) reader. It reads the RFID tag attached to each bag containing the drug, and information about dosage and time for medication is stored in the tag so the correct type of medication taken at the right time can be assured. A log of which drug was taken at what time is kept.

Telemedicine and electronic drug stores help not only patients to get their medications easily and virtually risk free but also inventory to be up to date at all times. This is particularly important when keeping stock for defending against spreading pandemics. A case study to illustrate this is the March 2009 spread of a new strain of A(H1N1) influenza virus. The stock levels of seasonal flu vaccine and related drugs must be closely monitored to ensure that at least the most high risk population gets an adequate supply. Telemedicine provides a communication link between manufacturers and pharmacies that makes the drug ordering process more efficient and more accurate while also allowing sudden, abnormally high demand for certain drugs to be met swiftly and effectively.

References

Bekhet, L.R., Wu, Y., Wang, N. et al. (2018). A study of generalizability of recurrent neural network-based predictive models for heart failure onset risk using a large and heterogeneous EHR data set. *Journal of Biomedical Informatics.* 84: 11–16.

Bousso, R. (2017). Universal limit on communication. *Physical Review Letters* 119 (14): 140501.

Choi, K.S., Lee, J.S., and Ko, S.J. (1999). New autofocusing technique using the frequency selective weighted median filter for video cameras. *IEEE Transactions on Consumer Electronics* 45 (3): 820–827.

Choi, E., Schuetz, A., Stewart, W.F., and Sun, J. (2016). Using recurrent neural network models for early detection of heart failure onset. *Journal of the American Medical Informatics Association* 24 (2): 361–370.

Cover, T.M. and Thomas, J.A. (2006). *Elements of Information Theory*, 2e. Hoboken, NJ: Wiley-Interscience.

Cranston, W.I., Hellon, R.F., and Mitchell, D. (1975). Proceedings: fever and brain prostaglandin release. *Journal of Physiology* 248 (1): 27P.

Dawes, M. and Sampson, U. (2003). Knowledge management in clinical practice: a systematic review of information seeking behavior in physicians. *International Journal of Medical Informatics* 71 (1): 9–15.

Delbeck, S., Vahlsing, T., Leonhardt, S. et al. (2018). Non-invasive monitoring of blood glucose using optical methods for skin spectroscopy: opportunities and recent advances. *Analytical and Bioanalytical Chemistry* 411: 63–77.

Elmaghraby, A.S., Kantardzic, M.M., and Wachowiak, M.P. (2006). Data mining from multimedia patient records. In: *Data Mining and Knowledge Discovery Approaches Based on Rule Induction Techniques* (eds. E. Triantaphyllou and G. Felici), 551–595. Berlin: Springer.

Fatima, M. and Pasha, M. (2017). Survey of machine learning algorithms for disease diagnostic. *Journal of Intelligent Learning Systems and Applications* 9 (1): 1.

Fayyad, U., Piatetsky-Shapiro, G., and Smyth, P. (1996). From data mining to knowledge discovery in databases. *AI Magazine* 17 (3): 37.

Fong, B. and Li, C.K. (2012). Methods for assessing product reliability: looking for enhancements by adopting condition-based monitoring. *IEEE Consumer Electronics Magazine* 1 (1): 43–48.

Giger, M.L. (2018). Machine learning in medical imaging. *Journal of the American College of Radiology* 15 (3): 512–520.

Green, S., Heer, J., and Manning, C.D. (2015). Natural language translation at the intersection of AI and HCI. *Communications of the ACM* 58 (9): 46–53.

Holland, J.H. (1962). Outline for a logical theory of adaptive systems. *Journal of the ACM* 9 (3): 279–314.

Jo, E., Lewis, K., Directo, D. et al. (2016). Validation of biofeedback wearables for photoplethysmographic heart rate tracking. *Journal of Sports Science & Medicine* 15 (3): 540.

Karim, A.M., Güzel, M.S., Tolun, M.R. et al. (2019). A new framework using deep auto-encoder and energy spectral density for medical waveform data classification and processing. *Biocybernetics and Biomedical Engineering* 39 (1): 148–159.

Kim, Y.K., Wang, H., and Mahmud, M.S. (2016). Wearable body sensor network for health care applications. In: *Smart Textiles and Their Applications* (ed. V. Koncar), 161–184. Woodhead Publishing.

Koeningsberger, D.C. and Prins, R. (1988). *X-ray Absorption: Principles, Application, Technique of EXAFS, SEXAFS and XANES*. New York: Wiley.

Kudo, Y. (2018). Predicting cancer outcome: artificial intelligence vs. pathologists. *Oral Diseases* 25 (3): 643–645.

Lin, L.H. and Chen, T.J. (2018). Mutual information correlation with human vision in medical image compression. *Current Medical Imaging* 14 (1): 64–70.

Londeree, B.R. and Moeschberger, M.L. (1982). Effects of age and other factors on maximal heart rate. *Research Quarterly for Exercise and Sport* 53 (4): 297–304.

Lu, H., Arshad, M., Thornton, A. et al. (2019). A mathematical-descriptor of tumor-mesoscopic-structure from computed-tomography images annotates prognostic – and molecular – phenotypes of epithelial ovarian cancer. *Nature Communications* 10 (1): 764.

Mackowiak, P.A., Wasserman, S.S., and Levine, M.M. (1992). A critical appraisal of 98.6 F, the upper limit of the normal body temperature, and other legacies of Carl Reinhold August Wunderlich. *JAMA* 268 (12): 1578–1580.

Maintz, J.A. and Viergever, M.A. (1998). A survey of medical image registration. *Medical Image Analysis* 2 (1): 1–36.

Malik, M. (1996). Standards of measurement, physiological interpretation, and clinical use. *Circulation* 93: 1043–1065.

Marchiando, R.J. and Elston, M.P. (2003). Automated ambulatory blood pressure monitoring: clinical utility in the family practice setting. *American Family Physician* 67 (11): 2343–2350.

Martinez, M. and Podrebarar, F. (2000). *The New Prescription: Marijuana as Medicine*, 2e. Quick American Archives.

Maruo, K. and Yamada, Y. (2015). Near-infrared noninvasive blood glucose prediction without using multivariate analyses: introduction of imaginary spectra due to scattering change in the skin. *Journal of Biomedical Optics* 20 (4): 047003.

Mashhadi, M.B., Asadi, E., Eskandari, M. et al. (2016). Heart rate tracking using wrist-type photoplethysmographic (PPG) signals during physical exercise with simultaneous accelerometry. *IEEE Signal Processing Letters* 23 (2): 227–231.

Mosco, V. (2017). *Becoming Digital: Toward a Post-Internet Society*. Emerald Group Publishing.

Ozana, N., Margalith, I., Beiderman, Y. et al. (2015). Demonstration of a remote optical measurement configuration that correlates with breathing, heart rate, pulse pressure, blood coagulation, and blood oxygenation. *Proceedings of the IEEE* 103 (2): 248–262.

Pandey, P., Saurabh, P., Verma, B., and Tiwari, B. (2019). A multi-scale retinex with color restoration (MSR-CR) technique for skin cancer detection. In: *Soft Computing for Problem Solving* (eds. A.K. Nagar, K. Deep, J.C. Bansal and K.N. Das), 465–473. Springer.

Pereira, F.C. and Borysov, S.S. (2019). Machine learning fundamentals. In: *Mobility Patterns, Big Data and Transport Analytics* (eds. C. Antoniou, L. Dimitriou and F. Pereira), 9–29. Elsevier.

Pickering, T.G. (1999). 24 hour ambulatory blood pressure monitoring: is it necessary to establish a diagnosis before instituting treatment of hypertension? *Journal of Clinical Hypertension (Greenwich, Conn.)* 1 (1): 33–40.

Reiss, A., Schmidt, P., Indlekofer, I., and Van Laerhoven, K. (2018). PPG-based heart rate estimation with time-frequency spectra: a deep learning approach. In: *Proceedings of the 2018 ACM International Joint Conference and 2018 International Symposium on Pervasive and Ubiquitous Computing and Wearable Computers*, 1283–1292. ACM.

RSNA (2009). Radiation exposure in X-ray examinations. http://www.radiologyinfo.org/en/pdf/sfty_xray.pdf (accessed 16 January 2009).

Sandsunda, M., Gevinga, I.H., and Reinertsena, R.E. (2004). Body temperature measurements in the clinic; evaluation of practice in a Norwegian hospital. *Journal of Thermal Biology* 29 (7): 877–880.

Sano, Y., Mori, T., Goto, T. et al. (2017). Super-resolution method and its application to medical image processing. In: *2017 IEEE 6th Global Conference on Consumer Electronics (GCCE)*, 1–2. IEEE.

Satava, R.M. (1993). Virtual reality surgical simulator. *Surgical Endoscopy* 7 (3): 203–205.

Schimelpfenig, T. (2006). *NOLS Wilderness Medicine*. Mechanicsburg, PA: Stackpole Books.

Shannon, C.E. (1948). A mathematical theory of communication. *Bell System Technical Journal* 27 (3): 379–423.

Spierer, D.K., Rosen, Z., Litman, L.L., and Fujii, K. (2015). Validation of photoplethysmography as a method to detect heart rate during rest and exercise. *Journal of Medical Engineering & Technology* 39 (5): 264–271.

Stewart, L.A., Clarke, M., Rovers, M. et al. (2015). Preferred reporting items for a systematic review and meta-analysis of individual participant data: the PRISMA-IPD statement. *JAMA* 313 (16): 1657–1665.

Tempkin, B.B. (2009). *Ultrasound Scanning: Principles and Protocols*, 3e. Saunders.

Traub, J., Sielhorst, T., Heining, S.M., and Navab, N. (2008). Advanced display and visualization concepts for image guided surgery. *Journal of Display Technology* 4 (4): 483–490.

Wong, C.C., Siu, W.C., Jennings, P. et al. (2015). A smart moving vehicle detection system using motion vectors and generic line features. *IEEE Transactions on Consumer Electronics* 61 (3): 384–392.

Wu, J., Roy, J., and Stewart, W.F. (2010). Prediction modeling using EHR data: challenges, strategies, and a comparison of machine learning approaches. *Medical Care* 48 (6 Suppl): S106–S113.

Yamamoto, T., Abolhassani, N., Jung, S. et al. (2012). Augmented reality and haptic interfaces for robot-assisted surgery. *International Journal of Medical Robotics and Computer Assisted Surgery* 8 (1): 45–56.

5

Wireless Telemedicine System Deployment

As we have seen in the previous chapter, data conveying information about a person's health can be obtained from many sources. Different types of data have different ways of acquiring and many have different requirements on transmission and subsequent processing. We have learned how various types of medical data can be captured and what to consider when making the data suitable for transmission through telemedicine networks. Vital signs and medical images are different in many ways. Some have more stringent requirement than others. The diversity of data acquisition makes both instantaneous and long-term measurement necessary in order to cater for different health monitoring situations. One key requirement in common is an efficient and reliable communications network to support patient caring. Network implementations are determined by the specific application supported, and so they are designed to satisfy the specific requirements imposed by the type of data sent, for example an X-ray radiograph has very different requirements from a prescription form that contains plain text information in terms of bandwidth.

Any communication channel, wired or wireless, has a specific theoretical limit to the amount of data that can be conveyed, the channel bandwidth governs how many data bits can "pass through" the given channel in one second. A network must therefore make use of communication channels that are capable of delivering all the data for an application without overflow (flooding if too many data bits attempt to get into the channel at a rate too fast for the channel). To understand more about its importance, we take a look at an example of attempting to send a high-definition (HD) video clip over an analog telephone channel, whose channel bandwidth is 3100 Hz. Obviously, we can tell straight away that far too many data bits are there for the available bandwidth without even doing any calculation. Even with "data compression" we still need bandwidth in the magnitude of MHz for HD video transmission.

In digital communications, information is acquired either as a "block" or as a "stream." The bursty nature of information, in the case of randomly taking a one-off measurement, does not usually have any statistical pattern of occurrence. So, a discrete block of data is collected when each reading is taken. No more data will follow until the next set of reading comes. An example of this kind of random event is the hospital ER admission where sometimes there is no patient, while at other times it may be treating several patients at the same time. The discrete probability distribution of data acquisition means statistical analysis of information flow is best dealt with using Poisson distribution modeling (Shmueli et al. 2005). In contrast, continuous monitoring, such as in the case of collecting data from a wearable device for health monitoring, will generate a stream of data as information will come in incessantly at a certain rate. We therefore handle sequential data of infinite duration, that is until monitoring is interrupted. Audio and video information is usually of such a nature. To understand more about how a communication system handles data of discrete blocks and continuous stream nature, we go back to the earlier example of attempting to send a

Telemedicine Technologies: Information Technologies in Medicine and Digital Health, Second Edition.
Bernard Fong, A.C.M. Fong, and C.K. Li.
© 2020 John Wiley & Sons Ltd. Published 2020 by John Wiley & Sons Ltd.

video clip over a telephone channel of bandwidth 3100 Hz originally designed only to carry mono voice signals. If the video comes as a burst of, say, a five-second clip, and nothing follows for the next few minutes, the entire clip can still get through with lengthy delay. The channel has sufficient time to "swallow" the large amount of data, just like pouring water into a narrow pipe through a funnel. If the funnel is large enough to act as a "buffer," and the water stops coming in before the funnel overflows, it is still possible to get the water through without spilling over. However, a continuous flow of water will not get through, as in the case of a stream. Imagine what would happen if we left the water tap fully open and let it run down a funnel continuously into a narrow pipe that does not have sufficient capacity to carry the water flow. The result is obviously overflowing from the funnel and we spill some water. Exactly the same would happen in data communications if we attempted to throw too many data bits into a channel that does not have adequate bandwidth to carry the data.

Communication networks are essential parts of modern healthcare systems that play a vital role for information exchange. With the capability of supporting a vast range of medical services, as seen in previous chapters, networks are inherently developed to support a multiplicity of innovative services as technology advances. In this chapter, we shall learn how to beat the challenges of developing and maintaining a future-proof network that will incorporate new features as and when they become available. From the fundamental knowledge on digital communications gained in Chapter 2, we proceed by first looking at some theories behind network planning and exploitation followed by necessary measures to ensure the network can be expanded for the future. As many networks are built on existing frameworks, what needs to be considered will be discussed. We then explore the pros and cons of outsourcing, and conclude the chapter by exploring network quality assurance.

5.1 Planning and Deployment Considerations

To thoroughly understand what lies behind network planning for telemedicine, we must first get a good understanding about what goes on behind the scenes. A good starting point is by referring to a primitive computer network that consists of two personal computers (PCs) linked together in a peer-to-peer (P2P) structure, as in Figure 5.1. Each PC has a network interface card (NIC, which is

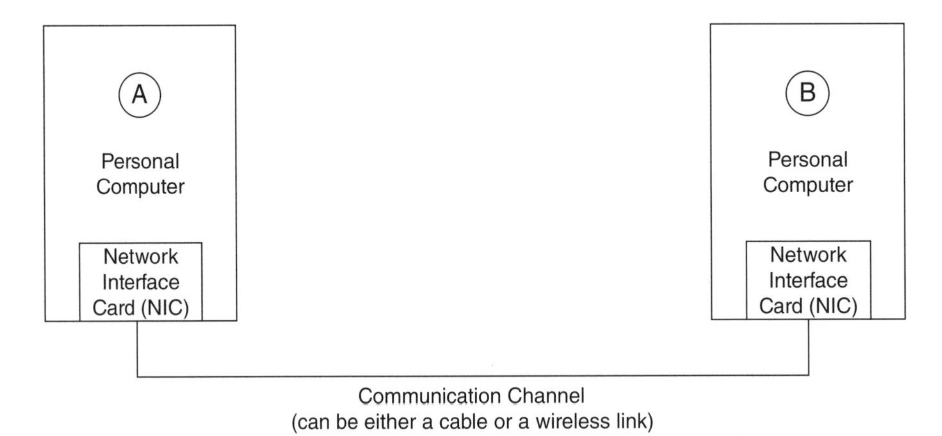

Figure 5.1 A simple peer-to-peer (P2P) network. This is the most basic form of a network with only two parties communicating with each other.

usually implemented as an integral part of a modern computer's motherboard), that connects it to the outside world, which is another PC in this example. As far as the PC is concerned, whether it is connected using a cable or via a wireless link makes no difference as long as data can be reliably exchanged between the PCs. The key feature of a P2P network is that it does not have a centralized location, and so all nodes (members of the network) are of equal status. Before advancing into the technical details of communication networks, let us first look at what happens inside the PC in the context of data communications by introducing the Open Systems Interconnection (OSI) reference model (ITU-R X.200 1994).

5.1.1 The OSI Model

The OSI model provides an outline for network communications. Its main purpose is to serve as a guide for network design. It is essentially a descriptive model for *layered* communication, that is network architecture is split into different layers each specifying a set of functions in data processing and formatting. The standard model in Figure 5.2 consists of seven distinctive layers. Each is responsible for a set of tasks. Communication between any two adjacent layers is said to be "direct." This involves exchange of data blocks through a port of a layer known as a service access port (SAP). Of the seven layers, they are broadly classified into two groups, namely the upper three layers, known as "host layers," and the bottom three layers grouped as "transport layers." The middle layer lies in a gray area that some in the literature group into host layers while others group in the transport layer (e.g. the definition given in *wiki*). As the middle layer itself is named "transport," it would logically be more appropriate to consider it as belonging to the lower transport layers.

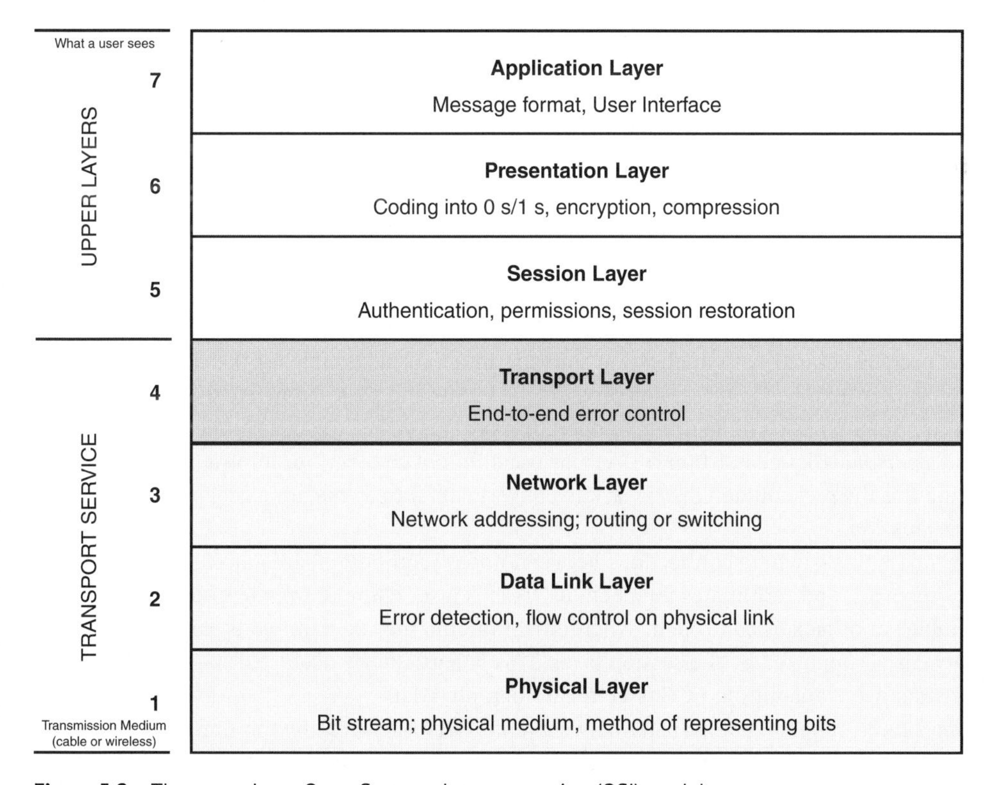

Figure 5.2 The seven-layer Open Systems Interconnection (OSI) model.

The OSI model basically classifies the entire communication process into functional layers. The communication process between each layer of a given host is maintained with a peer process on the corresponding layer, e.g. PC "A" in Figure 5.1 communicates with PC "B." Simply, any given layer n of A talks to B's layer n, where n can be any of the seven layers of the OSI model. The processes carried out at layer n are collectively known as "layer n entities." Since no direct connection link is established between layer n of the two PCs, communication between the two layers n is said to be "virtual." Communication is accomplished by the exchange of protocol data units (PDUs). A PDU is essentially an envelope that carries the data inside it. The user data carried inside a PDU is known as a service data unit (SDU). So, an SDU is part of a PDU that also contains a "header" of information about the data but is not part of the user data. The function of each layer n entity is managed by a set of rules called "layer n protocol." Each layer n entity functions by using control information to construct a header, along with the SDU to produce a PDU that is sent down to the layer $(n - 1)$ below for further processing. The function of layer n usually entails reception of PDUs from the layer above $(n + 1)$, delivering it to its peer processes then passing it down the stack (i.e. to the next layer below) and eventually to manipulate the data into a suitable form for transmission through the communication channel.

In case the data block exceeds the maximum size that can be handled by the next layer $(n - 1)$, the "segmentation and assembly process" becomes necessary to break the data block down into smaller units that can be "swallowed" by layer $(n - 1)$, and subsequently putting the segmented data back on the receiving end. Here, the SDU is segmented into multiple numbers of layer n PDUs. These smaller units are then passed further down the stack until it is sent out from PC A and reaches PC B. Reassembly (joining smaller pieces back together, as the reverse process of segmentation) is then performed at PC B's layer n.

Next, we take a brief look at the functions of each layer shown in Figure 5.2, starting from the top:

- *Application layer*: the top or seventh layer, where the human–computer interface (HCI) is supported. This layer makes direct interaction between a user and the application software possible. For example, a doctor generates a prescription by entering the information about the patient and the drugs, and sends it off to the pharmacy nominated by the patient. This entire process, as seen by the doctor, is handled by the application layer. Applications such as database entry, word processing, web browsing, email, etc., are all handled by the application layer.
- *Presentation layer*: the context between application layer entities is established here. It supports the application layer on data representation. Its main function is to convert information from human users generated with application software into a form suitable for the computer to process. So, data is "mapped" into session protocol data units and passed down to the next layer. Different computer types running different operating systems (OSs), e.g. communication between a PC and a personal digital assistant (PDA) may use different code sets for information representation, and it is the presentation layer's task to convert data into a "machine-independent" format for transmission.
- *Session layer*: this is where connections between nodes (computers) are administered. It establishes, manages, and terminates connections. This is also where the mode of communication, e.g. simplex or duplex, is controlled. Synchronization that may be required by some error recovery services is supported by the session layer. This is particularly useful for handling long data streams, such as transmitting ECG (electrocardiography) data.
- *Transport layer*: reliable data transfer between end users is assured by the transport layer. The link reliability is managed through flow control, segmentation, and reassembly, as well as error control. It uses the services provided by the underlying network for imparting the session layer

with the necessary quality of service (QoS) for data transfer, and keeps track of the segments and initiates retransmission when necessary.

- *Network layer*: here, data is converted into "packets" for transmission over the network. The process of identifying the optimal path for each packet to reach its destination, called "routing," is carried out by passing through a number of transmission links (in almost all telemedicine networks, far more network nodes and transmission links are present than the simple example shown in Figure 5.1). Effective routing requires information about the links' conditions from other network nodes. So, congestion in certain parts of the network can be isolated when an excessive amount of data exists.
- *Data link layer*: data transfers between network entities and error detection/correction are all taken care of here. Data blocks are converted into "frames." The header that contains control, address information, check bits for error detection, and framing information to mark the boundaries of each frame is inserted. Also in the data link layer, medium access control (MAC), manages data transmission from nodes in the communication channel (the "medium"), hence the process of controlling the access to the medium.
- *Physical layer*: finally, right at the bottom of the stack, we reach the first layer. The electrical and physical specifications between the node and the medium are defined here. These are attributes like how the 0s and 1s are represented; what the data rate, hence signal duration, is; and the pin configurations of plugs and sockets used are also defined here so that what the data from a specific pin represents can be known. Since the physical layer only deals with sending binary bits through a communication medium, it is not concerned about the actual meaning of the stream of data bits. So, whether the data contains medical images or body vital signs, it would make absolutely no difference to how the physical layer handles the data. Modulation is also performed here. Establishment and release of physical connection is another main task of the physical layer.

5.1.2 Site Survey

Surveying is a very important step at the early stages of network planning. This is vital to establish the correct number and placement of radio stations or access points (APs) in any wireless networks. Establishing the impact on signal reception under different scenarios cannot be taken lightly. Site surveys and simulations allow "stress tests" to be carried out to find out problem areas, such as poor reception, interference, susceptibility to hacking, etc. (Hummel 2007). Unlike the stress test performed by banks in May 2009 (which attempted to simulate the impacts on business operations of various financial and economic changes but ended up giving virtually no hints on their true state, thus giving the perception that stress tests are useless games), stress tests on wireless network sites reveal a great deal of useful information about "what if". There may be questions like "What if moving objects randomly obstruct a radio link that requires direct line-of-sight?", "What if frequent heavy downpour degrades a radio path?", or "What if coverage needs to be expanded to serve the new building across the road next year?" As we can see, there are all kinds of questions that we may need to address when we start planning a new network. Quite simply, we need to seek information about the location of each radio station and its coverage area.

Site surveys entail measuring wireless signals and using the measured data for network planning and optimization. The major factors to study are coverage, capacity, interference, and physical obstacles. Coverage area is directly related to the signal strength. To illustrate this, we recall the procedures of connecting our laptop computer to a wireless network in a café. When the network is found we see a number of bars of ascending heights. The more of these bars we see, the stronger the signal strength we can pick up from the AP. As we move further away from the AP, these bars vanish one after another until we move to a certain location where all bars eventually disappear and the computer is disconnected from the network. This is where network coverage is no longer available. The loss of signal strength (attenuation) as we move away from the AP follows the inverse-square law, such that the signal strength *S(d)* is proportional to the distance between the AP and the receiver as:

$$S(d) \propto \frac{1}{d^2} \tag{5.1}$$

This is under the assumption that we maintain a direct line-of-sight (LOS) path from the AP without any obstacle in between.

Network capacity concerns the maximum number of users that can be connected to the network at any given time. It also governs how much data can be transferred simultaneously by all users. Therefore, wireless networks can be saturated when at maximum capacity. When this happens, either any other users that attempt to make a connection will be blocked (denied from access) or all users will experience a degradation in QoS (data transfer becomes slow and intermittent network outage becomes more frequent).

"Interference" describes the strength of all distracting signals that come within coverage range. This can be a major issue when the network operates in a frequency band that is shared by other systems. For example, an IEEE 802.11n wireless local area network (WLAN) operating at 2.4 GHz may experience interference from a nearby cordless telephone that also uses the same carrier frequency. Also, WLANs of close proximity can interfere with each other, as in the case of an apartment block where several units have their wireless routers operating at the same time, at the same carrier frequency, and with the same channel. Connection reliability and data speed can be jeopardized by this kind of interference.

Physical obstacles can cause all kinds of problems to wireless signals, as discussed in Section 2.4. Absorption and reflection can be serious problems in many situations. Walls and partitions, especially thick concrete with steel beams, can shield off radio waves. Also, glass panels with film coatings or embedded wire mesh can degrade propagating signals.

During a site survey, measurement of signal strength and interference is conducted in various locations, and usually by a laptop computer installed with a network management system (NMS) that captures the measured data. Test APs are sometimes used and moved across various locations to test relative signal reception quality from a reference location in order to find out the optimal location to install an AP. Some modern surveying software allows prior input of the location map so that the surveyor can simply stroll along the site with a laptop computer while it measures and analyzes the received signal from each AP. For large sites, such as a hospital, it may be impracticable to scan the entire site so an estimation is normally performed based on extrapolation from certain sampled areas. Site surveys can give an indication of what to expect, but readers should understand that the results do not imply the completed network will behave as the test results suggest, since there are far too many uncontrollable variables involved. As a final note, certain operational and safety limitations may be imposed by local authorities, and any such applicable requirements must be met before network setup is completed. In addition, AP placement must be free from interfering with any nearby delicate medical instruments.

5.1.3 Standalone Ad Hoc Versus Centrally Coordinated Networks

Wireless networks can be deployed with either standalone APs or centrally coordinated. The former, such as that depicted in Figure 5.3, utilizes the integrated functionality of each AP to enable wireless network services. Each AP in the network operates independently of each other and is configured separately so that it does not respond to changing network conditions, such as data

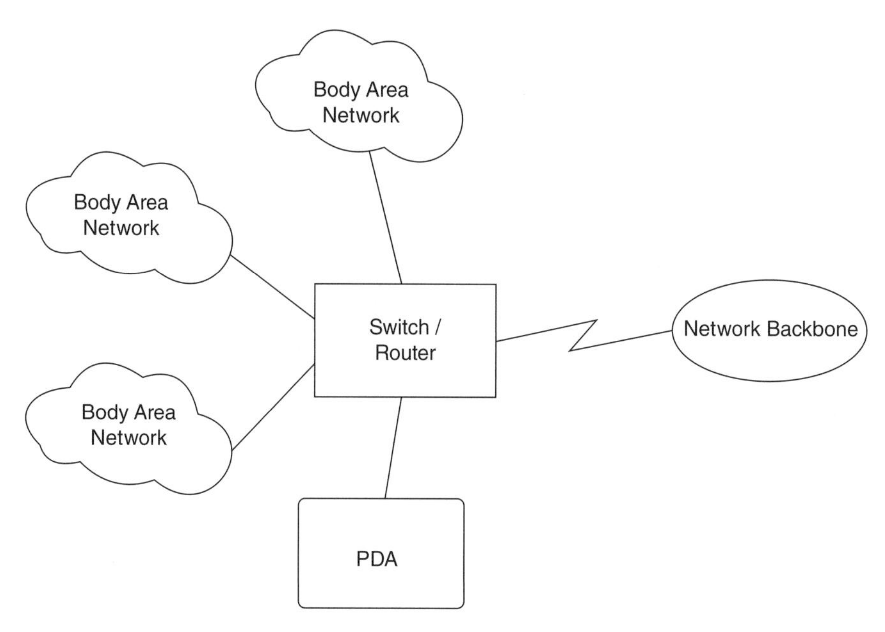

Figure 5.3 An ad hoc network that consists of individual devices communicating with each other through a direct connection.

traffic congestion or neighboring AP failure. In this network configuration, it does not have any centralized location that deals with user access or data flow control. Each body area network operates quite independently from each other. For patient monitoring applications that usually involve the continuous streaming of data, these require a great deal of energy to maintain adequate performance (Zou and Chakrabarty 2005). Such simple arrangement without centralized control fails to deal with issues such as power management, packet losses, security attacks, and similarities in applications that lead network performance degradation.

In a centrally coordinated wireless network, each AP is relieved from most of the data management tasks as they are taken care of by a central controller. Network performance monitoring throughout the entire network can be done centrally. Expanding coverage area with this kind of network arrangement can be easily accomplished by plugging more APs into the controller and letting it monitor traffic flow among all APs. The controller can be programmed to reconfigure each AP independently due to change in network conditions, for example, disabling failed APs or rerouting traffic for load balancing. This provides mechanisms for a "self-healing" network. Its configuration is very similar to that of Figure 5.3, except that the switch or router that connects different APs or body area networks together needs to be replaced by a controller.

Standalone configuration is usually more suited for a smaller isolated wireless coverage area with a very small number of APs, or in situations where temporary extended coverage is need to serve a nearby area. Otherwise, centrally coordinated wireless network configuration is more desirable, since it facilitates ease of deployment and rapidly responds to real-time changes in network conditions.

5.1.4 Link Budget Evaluation

This is an essential step to determine the link range or coverage by giving the system an operating margin in case any unforeseen events degrade the operating environment. For outdoor networks where rain can severely impact signal propagation, other factors such as modulation scheme and polarization also have a significant impact on the radio link (Fong et al. 2003a). The link budget describes the gains and losses incurred throughout the entire communication system of Figure 2.2 featured in Chapter 2. The concept is very simple: the received signal power P_R, after going through the entire communication system of antennas, channel (air, physical obstacles along the propagating path, etc.), and all cables/wires that connect components, such as between the receiving antenna and the demodulator, can be described as:

$$P_R = P_T + G–L \quad (5.2)$$

where P_T is the transmitted signal power in dBm and G and L are the sums of all gains and losses in dB throughout the entire communication system, respectively. Let us go back to Eq. (5.2) and expand on the calculation of link coverage. Obviously, link coverage decreases when the radio wave spreads out as it propagates along the path. Doubling the propagating distance will result in the received power reducing to a quarter so that:

$$L = 2 - \log_{10}\left(\frac{4\pi d}{\lambda}\right) \quad (5.3)$$

NIST (2006) provides a link budget calculator for a rough estimation of link budget in an outdoor environment. Since the transmitting and receiving properties of any pair of transceivers can be quite different, it is usually necessary to calculate the link budget for both signal directions.

In general, a telemedicine network requires at least 10 dB of link margin to tackle variations of signal strength due to reflection. Further, an additional 30 dB is necessary in case of a polarity mismatch between antennas in an orthogonally polarized configuration. So, the link margin is one important parameter that distinguishes telemedicine from general purpose wireless networks. We need to ensure that the network remains reliable even under persistent heavy downpours. The amount of extra link margin that we can afford to provide depends primarily on transmitter design and site environment. Maximizing the affordable link margin would ensure optimal system reliability.

5.1.5 Antenna Placement

Where to place an antenna is sometimes a moot point in practice, since the location offering optimal performance is not feasible for any number of reasons, including it is hard to reach or is on prohibited land. Also, impedance matching for the efficient transfer of energy between the antenna, radio, and cable connecting the antenna and the radio has to be identical in order to avoid loss due to impedance mismatch. Since most antennas' inherent impedance differs from that of the connecting cable, impedance matching circuitry is usually necessary for transforming the antenna impedance to that of the cable. Impedance is measured by the voltage standing wave ratio (VSWR). The VSWR should be less than 2 : 1 in order to ensure that a vast proportion of the power is forwarded with minimal reflected power. A high VSWR indicates that the signal is reflected and lost.

The ratio of VSWR and reflected power defines an antenna's bandwidth. The "percent bandwidth" is constant relative to the carrier frequency f_c expressed as:

$$BW = \frac{f_H - f_L}{f_c} \Big/ \text{x } 100\% \quad (5.4)$$

$$f_c = \frac{f_H + f_L}{2} \quad (5.5)$$

where f_H and f_L are the highest and lowest frequency in the band, respectively.

The angle of coverage or directivity of an antenna needs to be considered in antenna placement since many antennas do not provide 360° omnidirectional coverage. It should be noted that "omnidirectional coverage" usually refers only to horizontal directions but not necessarily extends to the vertical or elevation plane. The directivity of an antenna describes its focus of energy in a particular direction when transmitting or receiving energy from a transmitting antenna pointing toward it. Essentially, the directivity is measured as the ratio of an antenna's efficiency versus gain. Although most wireless routers we use at home are equipped with a cylindrical monopole antenna for 360° degree coverage, there are many antennas that are highly focused with a beamwidth of only a few degrees. Narrow beamwidth antennas are used when longer coverage is necessary by focusing energy toward a certain direction. This situation is very similar to comparing an ordinary lightbulb with a spotlight of the same power rating, where a spotlight concentrates its illumination over a more confined area but is much brighter than a lightbulb of the same wattage covering the same area. The radiation pattern, showing the relative strength of the radiated field in different directions from the antenna, in the coverage area close to the antenna diverges from that of the pattern over large distances. This leads to the introduction of the terms "near-field" and "far-field," also known as "induction field" and "radiation field," respectively. Far-field is normally used for measuring an antenna's radiated power. The minimum distance d to conduct far-field measurement is given by:

$$d = \frac{2 \times l^2}{\lambda} \tag{5.6}$$

where l is the length of the longest side of any dimension of the antenna and λ is the wavelength of the carrier frequency, as in Eq. (2.6). Near-field measurement is far less important when considering an antenna's placement. The only situation where it is useful is when deducing the minimum safe distance when dealing with ultra-high-power antennas that may radiate sufficiently large amounts of energy to cause health concerns.

5.2 Scalability to Support Future Growth

"Network scalability" refers to the ability of a network to scale up for increase in capability, in terms of performance, capacity, and coverage. Any communication network should be designed to handle future expansion in terms of both data throughput for new services and number of subscribers. It should also be made possible for extending coverage area. A properly designed scalable network should at least maintain its performance, if not enhanced, when expanded.

Scalability often involves installation of new hardware to an existing network. While a network is fully operational any interruption to its normal operation should be restrained during the process of expansion. Maintaining network availability when works are carried out is particularly important with systems that support life-critical missions. It is unreasonable to assume that a network can be temporarily shut down for improvement work to be taken, since no mechanism exists for predicting when a telemedicine network is not used, because accidents won't wait till service resumption to happen. Scalability almost certainly involves laying new cables in the case of wired networks, whereas with wireless networks several parameters can be adjusted within the network infrastructure. It is therefore far easier to perform enhancement work on a wireless network than with cables.

We begin our technical coverage by referring to Fong et al. (2004) as we look at what relevant parameters exist within the network backbone. Obviously, it is impossible to alter anything along the path, namely the air as the signal propagates through it. We have to work on the transmitter side and make sure that each receiver in the network is capable of processing the incoming stream of data in the case of increasing data throughput. Remember, the main objective is to efficiently utilize available system resources and minimize errors so to keep the need of retransmission to an absolute minimum. Before we look at how to fiddle around with various system parameters, let us recap on what within a wireless network can be altered to improve network capacity.

5.2.1 Modulation

Modulation, being the process of “putting” data bits into the signal, can be changed to “squeeze” more data bits into the signal. As we can see from the “constellation diagram” of Figure 5.4, which shows a representation of the signal quality as well as any presence of distortion. As seen from the constellation diagram whose axes represent the signal’s amplitude and phase, higher order modulation (e.g. 64 or above) have more “dots” so that more data is carried. At the same time, an increase in order packs the dots closer together while squeezing more bits into the signal. The nearer these dots get to each other, the harder it becomes to identify each individual dot. The net result is a tradeoff between spectral utilization efficiency and receiver structure complexity, since more complicated receivers are required to resolve each dot on a closely packed constellation diagram. With something as high as, say, 1024, the amplitude and phase difference between adjacent signal points may make them indistinguishable from each other.

In general, signals transmitted with lower modulation order, e.g. QPSK, can be properly received from a greater distance with an identical transmission power compared to higher order modulations, as signal loss is a less significant issue. It is therefore possible to combine the coverage

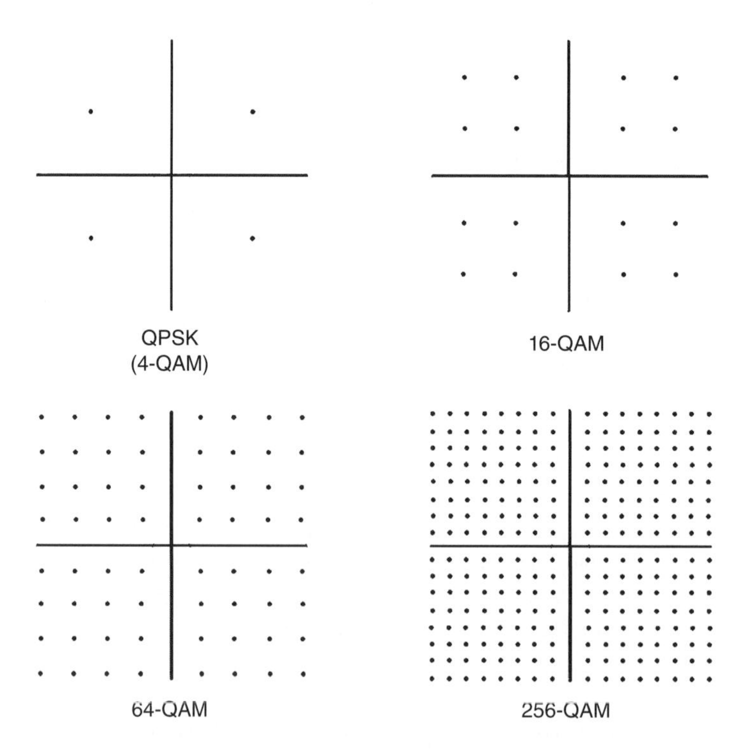

Figure 5.4 Constellation diagram that shows a signal modulated by any digital modulation scheme. In this illustration, four different types of modulation schemes from QPSK to 256-QAM are represented.

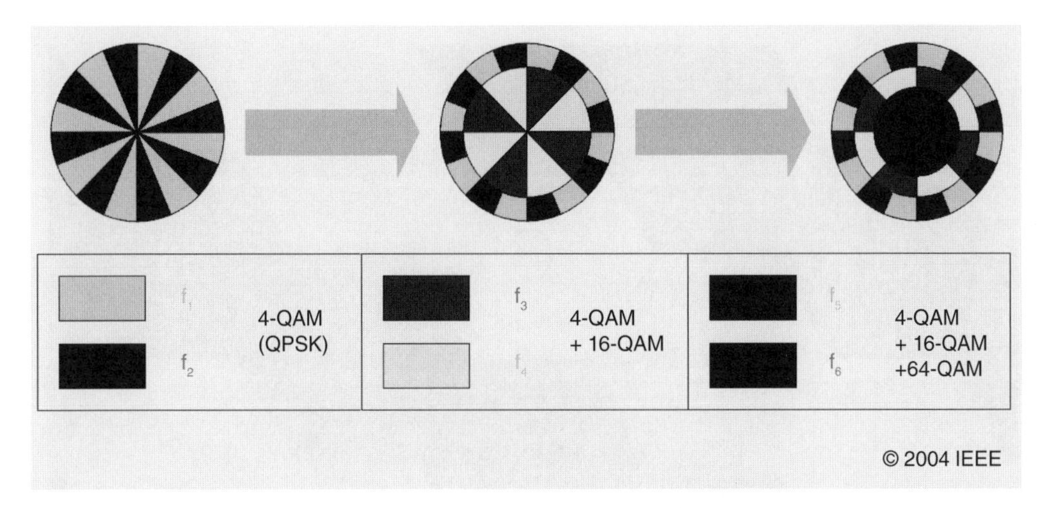

Figure 5.5 Coverage enhancement through augmentation. Source: Reproduced by permission of ©2004 IEEE.

distance of QPSK to serve only areas further away from the transmitter, and the higher bandwidth efficiency of 16-QAM for serving receivers closer to the transmitter. Further augmentation is possible to enhance overall coverage, as illustrated in Figure 5.5.

5.2.2 Cellular Configuration

A wireless network can be either single cell or macro cell (a cellular infrastructure with multiple cells to cover the overall coverage area), as illustrated in Figure 5.6. They can both utilize the "frequency reuse" technique but in different ways. Frequency reuse is a method of enhancing spectral efficiency and network capacity through reusing channels and frequencies of the same network by dividing an RF (radio frequency) radiating area (coverage area) into segments of a cell, as shown in Figure 5.7, where a simple example shows the reuse of two different frequencies. These frequencies, although in the same frequency band, need to be allocated with adequate separation from all neighboring segments to minimize any risk of interference. Any given frequency is reused at least two segments away from each other.

In a single-cell structure, broader geographical coverage is usually provided by high gain antennas with direct LOS to receiving antennas, and frequency reuse is possible by using different polarizations, whereas macro-cell systems use spatial frequency reuse that can usually provide acceptable signal reception properties without LOS.

So, spectral efficiency can be improved by a combination of higher order modulation and frequency reuse. There is, however, an inherent problem here as frequency reuse amplifies co-channel interference (CCIR), which means that two sufficiently nearby channels sharing the same frequency interfere with each other. This is usually caused by problems such as network congestion or bad frequency planning during the system design stage. This, contradictorily, reduces the modulation order. The spectral efficiency, measured in bits per second per hertz within a cell (quite simply, bit rate per Hz of bandwidth per cell = bps/Hz/cell, or BHC for short), is the data rate sent across each cell per Hz of the channel bandwidth.

CCIR in a single cell configuration is primarily caused by scattering from reuse sectors within the cell. A macro cell configuration employs frequency reuse in spatially separated segments, and the

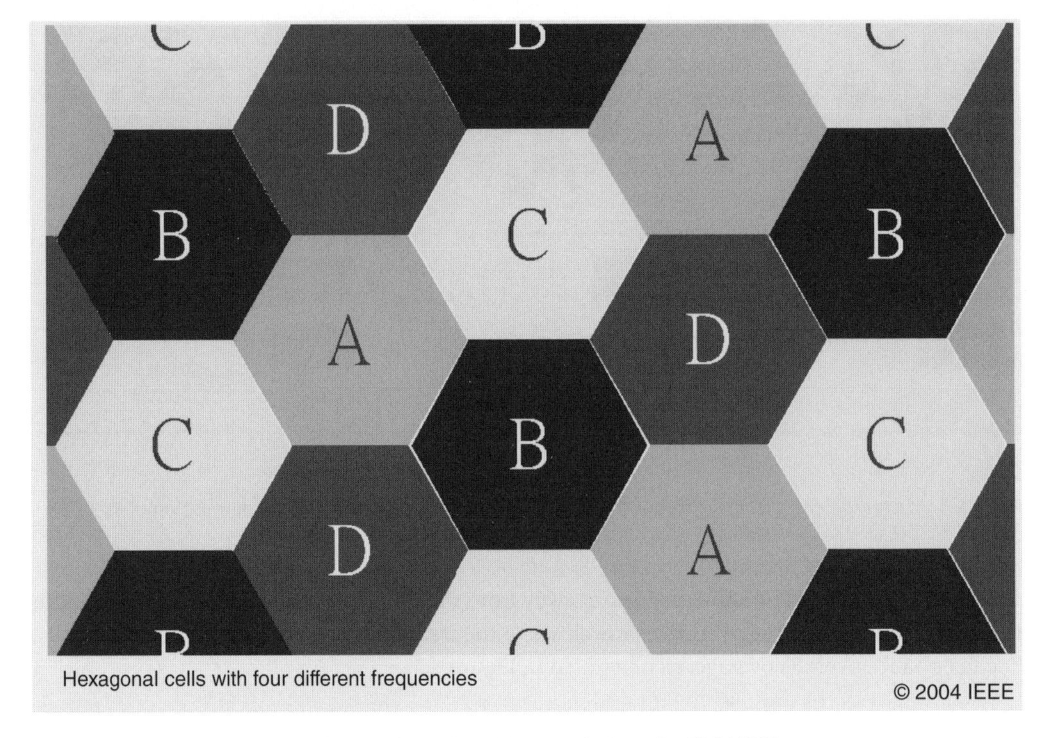

Figure 5.6 Cellular coverage. Source: Reproduced by permission of ©2004 IEEE.

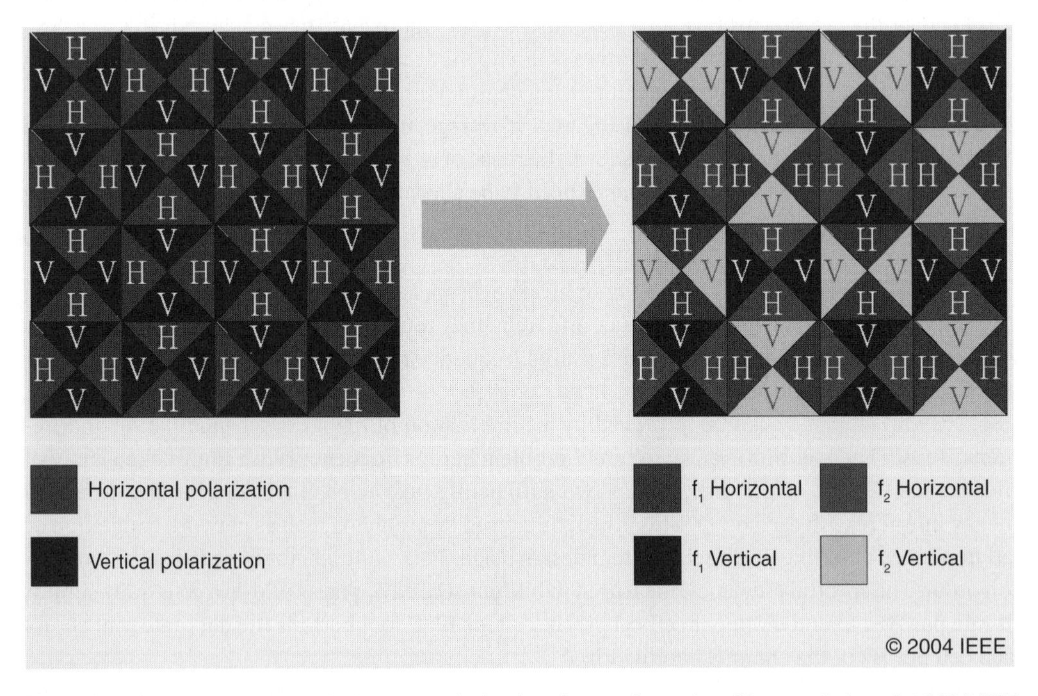

Figure 5.7 Frequency reuse with alternate polarization. Source: Reproduced by permission of ©2004 IEEE.

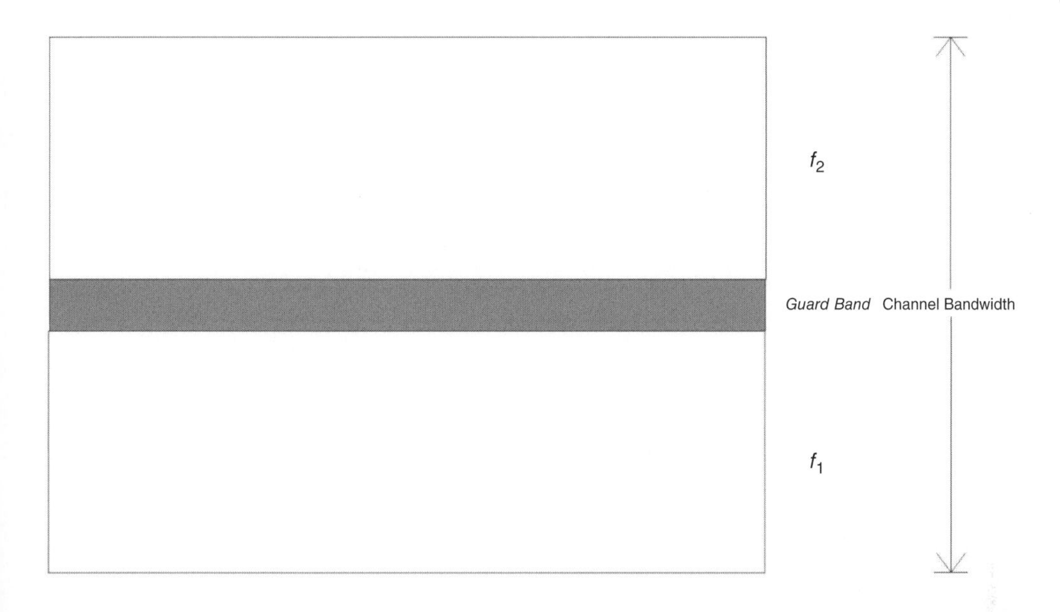

Figure 5.8 Subchannels separated by a guard band. This effectively provides a buffer between the two adjacent subchannels.

sharing of frequency among nearby segments induces CCIR. By applying Shannon's information theorem (see Chapter 2), frequency reuse is possible if:

$$BHC \leq \frac{\log(1 + C/I)L}{K \times m} \tag{5.7}$$

where C/I is the channel to interference ratio, L is the number of times a channel is reused, K is the spatial reuse factor ($K = 1$ in single-cell configuration), and m is the overhead assigned to "guard bands." A guard band is defined as a narrow portion of frequency band that is assigned to separate between two channels of similar frequencies without carrying any data, as illustrated in Figure 5.8.

Looking at Eq. (5.7) more closely, we can deduce that frequency reuse can be optimized by increasing L in a single-cell configuration or reducing K with a macro cell. In practice, C/I for single-cell and macro-cell structures can be approximated as Eqs. (5.8) and (5.9), respectively (Sheikh et al. 1999):

$$C/I = \frac{c}{L} \tag{5.8}$$

$$C/I = c \times K^2 \tag{5.9}$$

where c is an arbitrary constant specific to a given network deployment.

5.2.3 Multiple Access

As its name suggests, "multiple access" refers to multiple devices connected to a single wireless channel so that its available bandwidth is shared by all devices within the network. Rephrasing this sentence tells us that we are simply looking at "splitting" the channel into different portions using some kind of "multiplexing" technique. In its simplest form, we can split the channel into either time or frequency slots that leads to time division multiple access (TDMA) or frequency

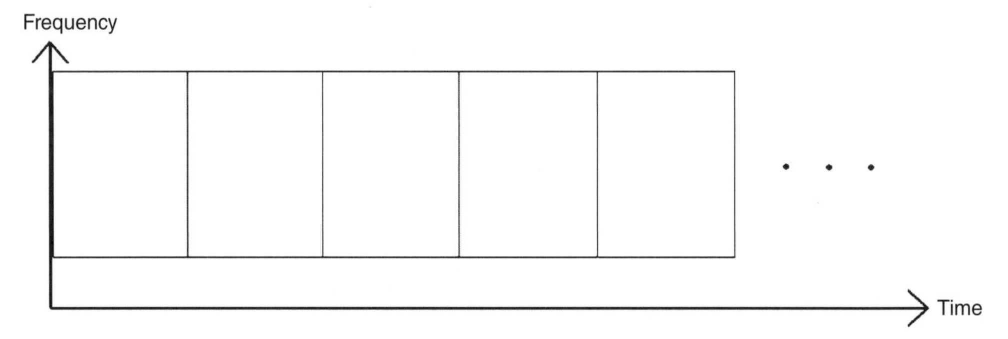

Figure 5.9 Time division multiplexing where the channel is divided into time slots.

Figure 5.10 Frequency division multiplexing where the channel is divided into smaller chunks of frequency bands.

division multiple access (FDMA). These names may sound familiar as we heard them mentioned many times when digital cellular phones were launched in the early 1990s. The operating principle is very easy to understand if we look at Figures 5.9 and 5.10.

In Figure 5.9, the channel is split into different time slots. The time slots are of equal duration in this particular example but this does not necessarily have to be the case. Each transmitting device is assigned with a time slot and the next device will use the next time slot. Taking an example of three devices, *A*, *B*, and *C*, the duration of each time slot is 10 ms. So, when the transmission process begins at time = 0, *A* will have exclusive access to the channel for the first 10 ms. At time = 10 ms *B* will take over and given exclusive use of the channel for the next 10 ms so that it stops transmitting at time = 20 ms. This then allows *C* to take over the channel between 20 and 30 ms. The entire channel sharing process then repeats itself when *C*'s turn is completed at time = 30 ms so it returns to *A* again for its next time slot of another 10 ms, and so on. This, of course, only briefly describes the theoretical principle. In practice, switching between transmitting devices takes a finite amount of time so that *A* stops transmitting and releases the channel before *B* starts transmitting. This effectively induces a short idle time during the switching process that effectively reduces transmission efficiency. A short amount of time as overhead for switching must therefore be taken into consideration. The process is shown in Figure 5.11, where a switch allows each transmitting device exclusive access to the channel.

Instead of splitting the channel into time slots, FDMA splits the channels into different frequency bands within the allocated bandwidth, as shown in Figure 5.12. For example, a channel assigned with the band 100–400 MHz, when equally split between three transmitting devices, will

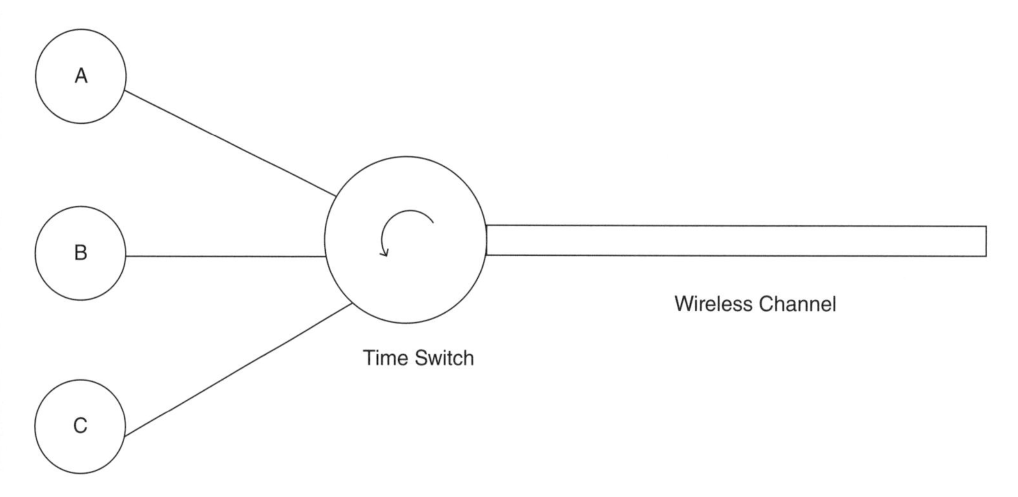

Figure 5.11 Switching between time slots.

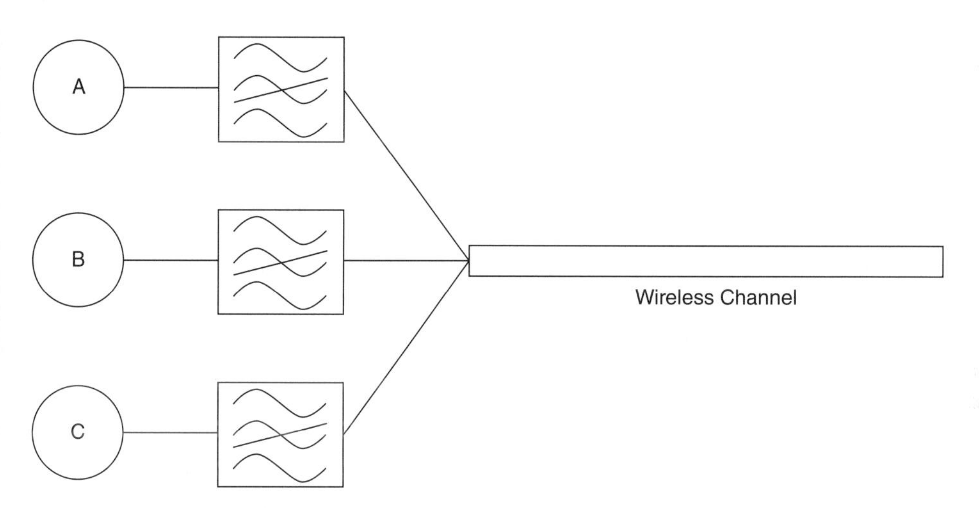

Figure 5.12 Filtering for different sub-bands.

be as follows: three transmitters require subdividing the 300 MHz frequency range (from 100 to 400 MHz) into three equal portions, which would mean we have 100–200 MHz, 200–300 MHz, and 300–400 MHz subchannels to be assigned to the transmitters so that each gets 100 MHz, or one-third of the total channel bandwidth. Again, in practice we'll never be able to allocate a full 100 MHz for each subchannel, simply because the cutoff bandpass filters can never have sharp cutoff. We need to spare a guard band for filter cutoff in order to leave an adequate margin for the filters. In Figure 5.13, we can see an "ideal filter" cuts off at exactly one frequency. Of course, an ideal filter does not exist in real life. Instead, all practical filters require a range of frequencies to cut off as the cutoff process is a gradual one. Better-performing filters will cut off with a steeper slope.

These access techniques have an impact on the way a wireless network is shared between different users and entities. While other multiplexing techniques such as code division multiple access (CDMA) are also used, TDMA and FDMA are the two most popular options. As we can see, the main tradeoff between the two is full bandwidth availability for periods of time versus constant availability but of a reduced bandwidth. It is therefore most appropriate to utilize TDMA

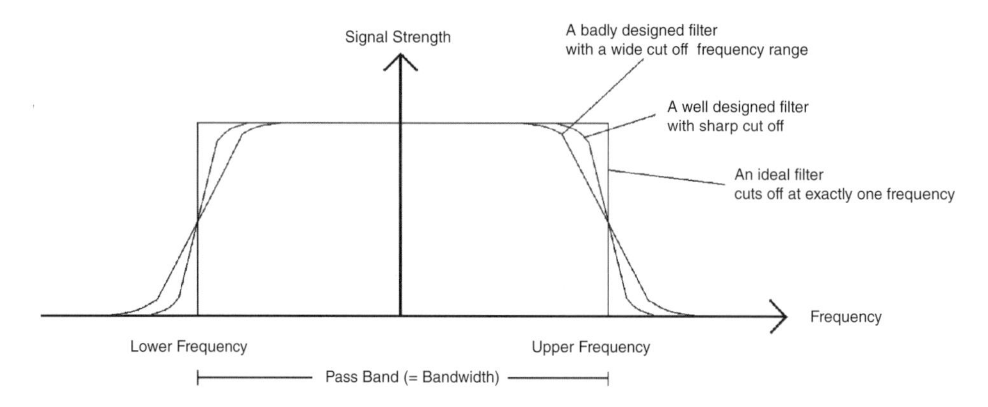

Figure 5.13 "Ideal" filter with sharp cutoff.

for downstream data traffic and FDMA for upstream, owing to the bursty nature that makes TDMA a better choice, whereas FDMA provides a constant pipe, which makes it more suitable for upstream data traffic. Dynamic bandwidth allocation improves channel efficiency by assigning more resources to that of higher demand.

5.2.4 Orthogonal Polarization

It is possible to expand data throughput by fiddling around with antennas. For example, with two signal paths with both vertical and horizon polarizations, we can essentially have two separate channels simply by mounting two antennas of orthogonal polarizations perpendicular to each other. Relative to the earth's surface, the electric field of a vertically polarized antenna is perpendicular, whereas a horizontally polarized antenna is parallel. These are both said to be linearly polarized antennas. A classic example would be the old-fashion TV antennas mounted on rooftops similar to that in Figure 5.14. As we can see, this type of TV antenna is parallel to the earth's surface.

Figure 5.14 Conventional outdoor TV antenna, used over several decades from VHF to UHF and digital broadcast over recent years with essentially the same design.

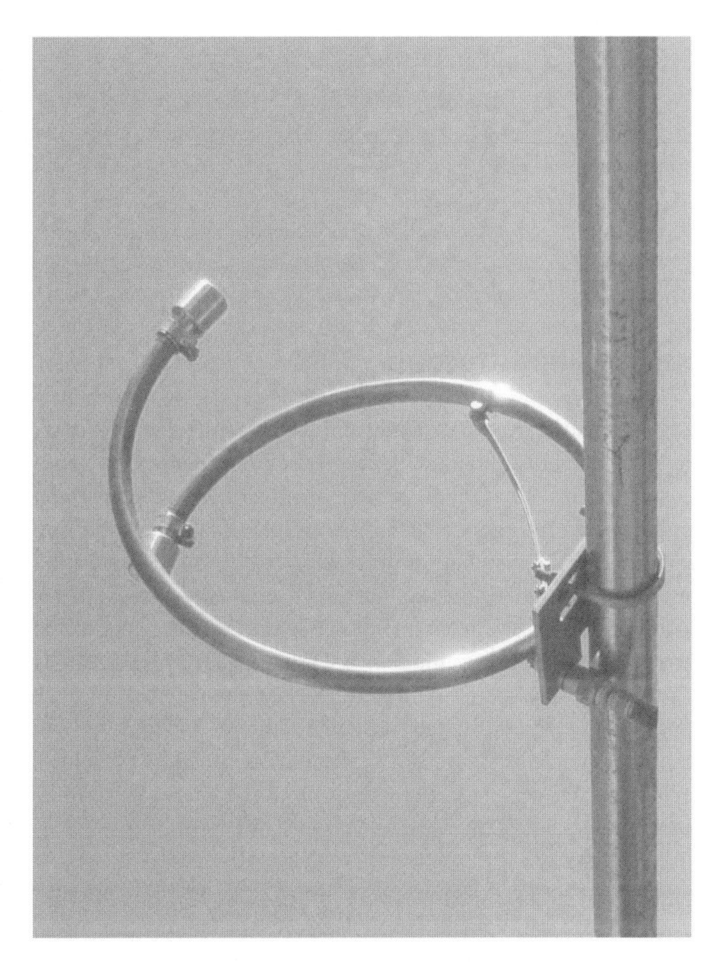

Figure 5.15 Antennas of circular polarization.

This is therefore an antenna of horizontal polarization. We can effectively double the number of channels by adding another antenna to make use of another polarization.

Antennas of circular polarization also exist, as the example shown in Figure 5.15. Here, the polarization plane rotates in a circular pattern such that it completes one rotation per wavelength. For example, the polarization would have rotated by 360° in 1 m if the wavelength were 1 m. Energy is radiated in all directions, including the horizontal (0° and 180°), and vertical (90° and 270°) planes. The propagation direction can be either clockwise (right hand circular, or RHC) or anticlockwise (left hand circular, or LHC). Generally, circular polarization is more suitable for non-LOS situations when passing through physical obstacles since the reflected signal bounced back to the transmitting antenna upon striking an obstacle will be different from that of the propagating signal.

Another way to achieve scalability with antenna placement is by "sectorization," where antennas are added on demand by increasing the number of available sectors, as shown in Figure 5.16. In this example, initial deployment is set up with one antenna providing omnidirectional coverage. As demand grows, three more antennas are added so that each only serves a 90° coverage, thereby providing three more channels for the same area. Further growth can be supported by further segmentation, by doubling from four to eight sectors (as shown in Figure 5.16).

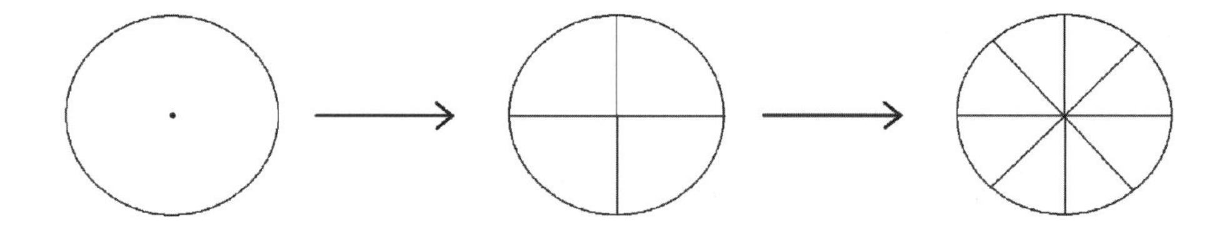

Figure 5.16 Sectorization from one cell into four and eight sectors.

5.3 Integration with Existing IT Infrastructure

Many telemedicine networks are built on existing networks. For example, by monitoring an asthma sufferer at home (Gibson 2002) the system might utilize the home network and its Internet connection via the ISP (Internet service provider). What we have here is an addition of the asthma self-monitoring system to an existing network in Figure 5.17. In this simple illustration, shutting down the home network temporarily during the installation process probably will not cause too much inconvenience. Network integration in more complex systems, such as an enterprise network of a hospital, would be a far more complicated issue since it is impossible to expect the network to be shut down for the integration process. As with any maintenance work, one key pre-requisite is to ensure any work carried out does not affect the operation of other parts of the network.

To facilitate any network work, a schematic of a building's structure depicting the location and details of all wiring, access nodes, and associated devices is an essential document. This allows physical integration for connecting all new devices to the right place. Both data network connectivity and power must be connected to the new devices as the vast majority of monitoring devices cannot draw power directly through the data cable.

Logical integration involves configuration of the new portion of an overall network. Different network architectures of the existing network will have different integration requirements. Most modern networks, including all IEEE 802.11 based WLANs, are Internet protocol (IP) networks

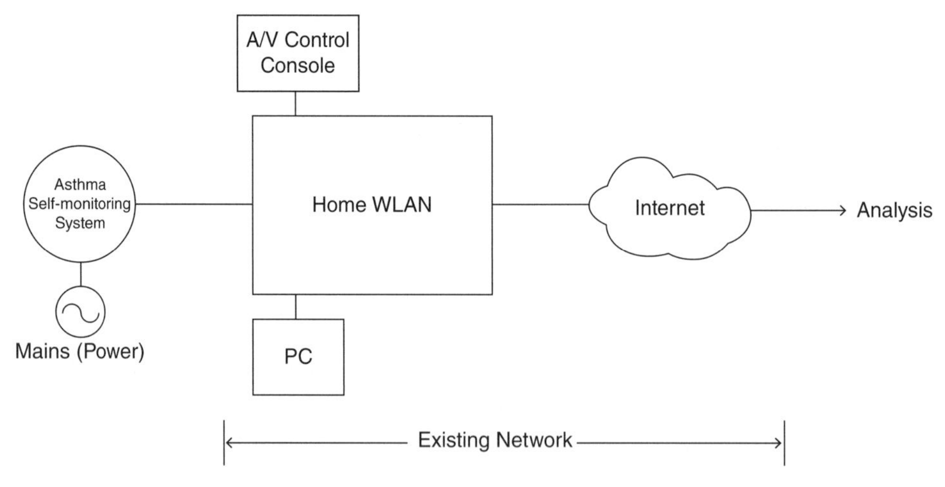

Figure 5.17 Asthma self-monitoring system.

that make connections easy with a comprehensive set of standards. Some older networks, however, may have legacy networking protocols that require additional work to be carried out during the integration process.

5.3.1 Middleware

One important and useful piece of tool in network integration is a "middleware." It is essentially a software bridge that links any system to a network by providing necessary services. The word "middle-" defines a piece of software that sits in the middle between the applications and the operating system (OS) of a computer. Middleware provides features such as communication, data access, and resource control for connection of devices with different platforms (Boubiche et al. 2002). Middleware with a wide range of features is designed for different healthcare applications (Spahni et al. 1999). The main purpose of the middleware is to facilitate the integration of computer systems, medical instrumentation, monitoring systems, databases, etc.

Middleware is frequently configured to access databases as it facilitates communication between applications that are often associated with the term enterprise application integration (EAI), describing the consolidation between different applications used throughout various entities within the healthcare system. EAI is designed to address problems such as integration and connectivity, where data integration is usually accomplished by incorporating application programming interfaces (APIs) for communication with legacy systems in order to ensure compatibility. An API provides an interface for a given application to obtain services from a computer's OS or libraries.

5.3.2 Database

A database stores a vast amount of information in the form of fields, records, and files in such a way that the collection of information is organized for easy retrieval as individual items or groups of items that satisfy certain user-defined selection criteria. A field is a single piece of information, a record is one complete set of fields, whereas a file is a collection of records. For example, a field may store information about the medication prescribed to a patient during that morning's consultation, and a record all information related to the consultation session, such as medication prescribed, nature of visit, symptoms, body temperature, etc.; and a file is the complete medical history of the patient. Database size can vary from a small clinic containing several hundred patients' medical histories, each stored in a file, all the way to a national health database containing millions of files of patients, as well as separate storage for information about suppliers, manufacturers, pharmacies, etc., totaling billions of files stored together in a logical structure. Information sharing among different databases of varying sizes and formats can be a nightmare. Take, for example, any attempt to link up a group of general practitioners in Canada nationwide. Entries in English and French, with different character encoding schemes, can exist that may have two sets of conventions. Such differences further complicate an already complex process of integrating with some legacy databases from different vendors.

Within a hospital, many applications may have been built for different purposes, each with its own database for a variety of information types. Integrating these applications for ease of information exchange would ensure information about each patient was shared while care was provided, and to assist better management. One example of integrating applications by connecting them to a shared database is shown in Figure 5.18. What matters most is consistency, so that a single piece of

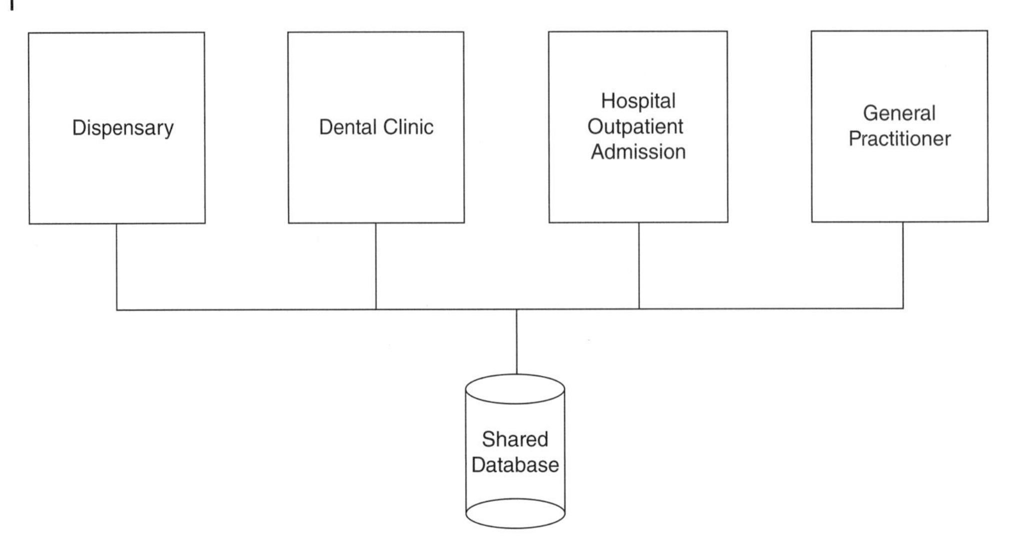

Figure 5.18 Database sharing in a hospital.

information about a given patient from different healthcare providers can be accessed and updated simply by some kind of transaction management system.

5.3.3 Involving Different People

As with any system, a healthcare information system cannot be completed without people such as end users, maintenance support technicians, designers, and engineers. They all perceive a system differently The basic rule of thumb is to retain the original user interface of all parts of a system as much as possible. Alteration to the way users interact with the system should not be made unless it is absolutely necessary. This is particularly important with healthcare systems, since there is no margin for error. Users should be able to continue using the system after integration in the same way as before. Designers should therefore incorporate any new functions or features without changing existing interaction functionality. Further, all interfaces should remain the same when integrating existing systems together. Successful system integration entails collaboration between installation engineers and both users and designers. Users need to be told of any temporary interruption that may be expected during the process in order to reschedule any tasks and to make alternative arrangements to cope with unexpected emergencies. Designers should ensure that anything which can possibly be done before shutting down an existing system is done in advance so that any new module can be installed as quickly as possible to ensure minimal interruption.

Testing is a vital part of system integration, as the process of testing enables any problems to be identified and rectified. Although we look at the details of system reliability and prognostics later in Section 10.4, we'll conclude this section by taking a quick look at system testing for the sake of completion.

Integration testing is an extension of "unit testing," where any modules to be integrated into an existing system are tested on their own before integration. Unit testing, prior to performing integration testing when carried out after putting everything together, ensures that any fault within a module can be detected and corrected prior to being put into an overall system that may otherwise cause serious damage to the system. Sometimes a module works well by itself but develops problems when integrated as part of a larger system. Exhaustive tests under all working conditions must be carefully carried out before and after system integration to ensure continuing reliability.

Compatibility is almost always an issue when adding new modules to an existing system. This is particularly niggling when integrating things from different manufacturers into existing infrastructure. Standards conformation helps ensure compatibility and interoperability among devices made by different manufacturers. The abbreviation CII is often used in system integration with three different but related meanings:

- *Common integrated infrastructure*: an integration model to integrate new and legacy applications within an enterprise for common cross-enterprise integration infrastructure (Helm et al. 2018). This is also applicable to healthcare enterprise systems.
- *Compatibility, interoperability, and integration*: a set of rules to ensure all these parameters are met, as described above.
- *Configuration identification index*: a manifestation to correlate documentation to the proper set of configuration for individual systems.

The final stage is the user acceptance test, which verifies the system's performance and usability. This step ensures that the new system after integration matches all user requirements. User training may also be necessary to provide information on what has been added to the old system.

5.4 Evaluating an IT Service and Solution Provider

Business opportunities are vast for IT companies seeking to enter the medical and healthcare industries since technology can be applied for caring for everyone from head to toe. Different modes of partnership between healthcare providers and technology firms exist. Also, owing to the fact that there are so many possibilities, there are also a number of important issues to address. Readers should gain a broad understanding of what is involved so as to prepare themselves to optimize resources.

5.4.1 Outsourcing

Many IT-related services are outsourced to third parties for both time and cost savings. It is very often true to say that money is best spent on letting the right people do the right job. People who are good at doing a specific task will accomplish the work in the most efficient manner. This is easier said than done. Finding the right people may not be that straightforward after all. The IT industry is so huge that there are many people providing essentially the same type of services, and of course some are good and some are bad. In cases of software development, outsourcing can even be offshore as the only thing needed is an Internet connection. Services such as diagnostic teleradiology, medical image processing, and electronic billing can be easily done by a company in a developing country where operating costs are much lower.

There are many disadvantages and risks involved in outsourcing, the most obvious one being the risk of disclosing confidential information to the service provider. Also, competitors such as other clinics and institutions may happen to hire the same provider, and it may not devote necessary attention to meeting certain needs. here may be hidden costs during the outsourcing process, among other possible problems such as delay and misunderstanding.

Before commissioning someone for a service, we need to go through a checklist of performance measurement and determine what potential risks exist. Although different areas may have a very different set of items on the checklist, there are some general guidelines to follow. For example, when choosing a service provider for providing a wireless network we need to look at parameters like bit error rate (BER), practical maximum data rate, and impact on network performance when more users are connected. These are just a few examples of what should be checked against. Others include any after-sales support such as mean time for repair, guaranteed response time, any loan units available as substitutes while removed for repair, etc.

5.4.2 Preparing for the Future

Keeping up to date with what is happening in IT-related industries is extremely important since technology has a significant impact on optimizing operational efficiency and revenue cycles. With environmental friendliness in mind, modern products are designed to be more power efficient, and the use of toxic substances is strictly limited. Emerging technologies are often associated with sustainability (Seele and Chesney 2019). This causes a number of other problems, such as more restrictive design requirements and how a system is laid out in certain sites due to management or compliance issues. Some regulatory concerns may lead to additional costs or delay, which a service provider should be fully aware of. The service provider should also advise on what contingency plans may be necessary in the event of any change in government regulations or unforeseen circumstances.

The ability of a service provider to keep up to date with the market is extremely important. Imagine what would happen if a service provider ended up having to deliver obsolete technology just to fulfil a long-term plan that had not considered technological innovation in the meantime. Advances in IT technology happen all the time, making many IT products obsolete in a very short time. Consider the following case study that looks at a portable glucose meter designed to be linked wirelessly to a home PC some two years earlier. The development project was outsourced as the "Vena" platform became available (Fong and Kim 2019), where compatibility of data exchange was supposed to be assured with appropriate procedures in place. Although there was nothing wrong with the development platform itself, the service provider did not realize that the IEEE 11073 standard had not been finalized at that time. As a consequence, the glucose meter failed to comply with the IEEE 11073-10417-2009 (2009) standard. The application profile standards had basically been neglected during the product development stage. This made a device update necessary in order to communicate with IEEE 11073 compliant PCs. Such an update, of course, incurred service interruption and additional labor costs.

Digital health development should take into consideration any standards and regulatory compliance that may come into effect in the foreseeable future. In particular, legislation related to environmental impact is becoming increasingly important in many countries. This could have an impact on initiatives such as the use of recyclable materials in consumer healthcare and medical devices concerning product design. Designers of both consumer healthcare and certified medical devices should check the applicable legislation of the countries of their intended markets.

5.4.3 Reliability and Liability

Reliability can be judged in a number of ways, either quantitatively by some kind of metric or subjective, such as by word of mouth. Generally, a reputable company who has been in the industry for a long time can be relied on. In spite of this, we cannot assume a good brand name will always provide reliable service, especially when supporting life-saving applications. Recall how many times our PC suddenly crashes and we have to press Ctrl-Alt-Del to get it restarted. Not everything can afford the time for a system that suddenly stops responding for no reason. If we use something like this for resuscitation the chance is that the patient will not be revived by the time the unscheduled reboot process is completed. As reliability is such an important topic in medical technology, we take a closer look at this topic in the next section, as well as introducing prognostics for healthcare in Section 10.4, where we shall look at some statistical analysis and modeling to determine the life expectancy of a medical device. For the remainder of this subsection, we focus on how to determine whether we can rely on a service provider in case we choose to outsource certain tasks to a subcontractor.

When we sign a contract to commission someone for a specific task to be completed on our behalf, we expect them to be reliable. This is exactly why we want to ensure that we assign the contract a person we can depend on. Among a long list of service providers that we may find from different sources, such as the Internet, we need to shortlist a small number of potential suppliers through a screening process. Some guidelines should be drawn up according to a set of criteria. If we refer to the example in Section 5.4.2, we may want to ask things like: how long does it take to get the first prototype for trial, what standards (if any) will be adhered to, how can firmware updates be performed, whether the cost is within budget, etc. These are just a few of many items that need to serve as guidelines for us to choose the right supplier for us.

The first logical step would be to radically shorten a long list of available suppliers so that we can hopefully end up with a dozen or so who appear qualified. An evaluation checklist should then yield an even shorter list, after going through processes such as: product preference, service quality, operational coverage, and financial stability. In the business world, many companies would initiate an RFP (request for a proposal) when a number of potential suppliers are selected from the process. A qualification checklist should be prepared prior to sending out RFPs to the last few remaining candidates. Any specific requirements –in our case study that the glucose meter must be wearable with a battery life of at least 72 hours, for example – must be investigated in more detail. Sometimes reference checks and enquiries to other institutions may be useful, although we should bear in mind that the information acquired may be biased. A simple ranking process can position the potential suppliers in order of capability and suitability for our requirements.

All this tedious work ensures we obtain a reliable service. There is more than just finding the right supplier to meet our requirements, and within budget. We also need to make sure our supplier is reliable but also allows us to offer our clients the correct service as well – there's no point in getting something that works but doesn't meet our clients' or users' requirements.

Liability can be a very important issue, especially in the healthcare sector where a patient may sue for (potentially huge) compensation should something go wrong. Defining who is responsible for what in these situations is extremely important. This should normally be stated clearly on the service contract and checked by a legal representative. Legal liability, in the context of providing healthcare services, can be a very serious matter. Let us take a look at an example of a simple individual health insurance policy. This entails a long list of terms and conditions attached with maximum liability and what is excluded under what circumstances. Sometimes a liability waiver is required before providing certain healthcare service. This is to limit any risk of being held responsible in the event of an emergency. A simple waiver form like that shown in Figure 5.19 would ensure written authorization is given when attending to an emergency.

It is worth noting that liability issues exist between the service provider and any supporting entities, such as outsourced contractors, equipment manufacturers, and solution architect – as well as with patients. Ultimately, we do not want to end up in a situation where we are held responsible for any issue caused by our service providers. And this is why quality assurance is so important, and why we discuss it in the next section.

MEDICAL/LIABILITY WAIVER FORM

This form must be signed by the patient before receiving medical services.
This form will be securely stored for up to six months.

Last Name: ____________________ First: ____________________ DOB: ____________

Parents/Guardian(if below 18): ____

Address: ____

____

Phone: ______________________________________ Email: ____________________

Medical Information and Release
In the event of an emergency, who should be contacted?

Name: ________________________________ Relationship: ____________________

Phone (Home/Office): __________________________ Mobile: ____________________

GP: ____ Phone: ______________

Which hospital would you prefer you and/or your child to be taken to in case of an emergency?

____

Medication currently taking and/or known allergies:

____

____

Other relevant medical information:

____

In the event of an emergency, if neither parent nor emergency person(s) can be contacted or if there is no time to make such contact, the following signature authorizes such emergency medical and surgical treatmentto be provided, including transportation to the nearest facility,as may be deemed necessary.

Signature ____ Date ______________

Full Name ____

Figure 5.19 A sample medical liability waiver form.

5.5 Quality Assurance

Quality assurance is a vital part of providing trustworthy healthcare services, and is certainly vital when providing wireless telemedicine technology, and so in this section we consider the issues that potentially affect the QoS of wireless communication.

The term "link outage" refers to the situation where a wireless channel is cut out. Communications and information theory is often accompanied by a statistical model that describes the probability of the successful reception of some kind of information, as it goes through the model of Figure 2.1, by a receiver. The information is sent out from a transmitter across a wireless channel. Main factors that can cause problems to the propagating signal (the signal that carries the information across from the transmitter to the receiver, through the channel) include (Fong et al. 2003b):

- *Attenuation*: weakening of signal strength over distance traveled.
- *Depolarization*: reduction of separation between two signal paths of different polarizations resulting from phase retardance.
- *Interference*: disruption of signal caused by other sources.
- *Noise*: unwanted additive energy that is inserted into the signal.
- *Scattering*: radiation toward different directions after hitting an object.

There are many other signal degradation factors, too. To illustrate the complexity of signal degradation, we take a look at interference. Interference that can affect the reliability of a communication system may include (Stavroulakis 2003):

- *CCIR*: also known as "crosstalk," effects of encountering signals from an adjacent channel of similar frequency.
- *Electromagnetic interference*: also known as "radio frequency interference" (RFI). Interruption due to signals from other sources. This is a technique intentionally used for "radio jamming," allowing one to disrupt a wireless link by emitting another signal of similar frequency. Solar radiation may also cause RFI in rare cases.
- *Intersymbol interference*: unwanted interaction between adjacent symbols (of data), usually caused by multipath. The net effect is quite similar to noise that is caused by the same signal being sent at slightly different times.

As we can see, there are several different types of interference. To measure the quality of a given wireless link, we have different parameters, these include:

- *Bit error rate (BER)*: measures the number of error bits that occur within a block of bit stream. For example, BER = 10^{-6} means statistically we can expect one corrupted bit per one million bits transmitted. This figure is normally considered as acceptable for general consumer electronics applications. However, telemedicine requires better quality than this (Shimizu 1999). To improve the BER performance of a given wireless link, reduction in data rate or allocation of adequate "link margin" should be considered. "Link margin" refers to the extra power necessary to combat signal loss due to different degradation factors. For example, a certain link margin is necessary to allow for attenuation due to rainfall. Since BER is a measure of data bit error, it is a performance measure against the E_b/N_0 (the energy per bit to noise power spectral density ratio) value for a given channel. E_b/N_0 can be viewed as the digital equivalent of the signal-to-noise ratio (SNR) in analog communication systems or, more appropriately, the SNR per bit. E_b/N_0 increases as BER improves (i.e. the BER value decreases, e.g. from 10^{-6} improves to 10^{-9}). BER is usually measured by a BERT (bit error rate tester), often in the form of a software package.
- *Signal-to-interference ratio (SIR)*: measures the ratio of signal power to that of the interference power in the channel to check the received signal quality at a receiver. It is sometimes called the "carrier-to-interference ratio." The SIR is similar to the SNR of a propagating signal before it is processed by the receiver. In this respect, the main difference between the interference "I" and noise "N" is that the former originates from an interfering transmitter source which is controllable through network resource management, whereas the latter comes from a combination of many manmade and natural sources. The SIR should normally be at least 18–20 dB to ensure quality reception. The SIR is usually improved by appropriate filtering algorithm for the receiver. Statistical modeling is usually used to measure the signal power in order to measure the SIR

(Van Der Bergh et al. 2015). This is essentially a process of developing an algorithm to analyze the signal power at the receiver before demodulation in relation to that of all interfering signals.
- *Carrier-to-noise-and-interference ratio (C/[N + I] or CNIR)*: is a measure of the amalgamate effect of both noise and interference in the context of CIR and SNR.
- *Co-channel interference (CCIR)*: nearby channels operating at the same frequency that cause interference between each other. Increasing the SNR will not only not improve the impact on interference it will also make the situation worse. Reduction of CCIR can be done by increasing distance between co-channels (Bhargav et al. 2017). In the USA, the Federal Communications Commission (FCC) regulates the "out of band" noise for radio transmitters in order to suppress sidebands that cause interference. To combat this problem, input filters with sharp cutoff are usually deployed at the receiver.
- *Link outage*: a statistical measurement of how much time within a year a wireless link is cut off. This is enumerated in minutes and second. For example, a system with 99.99% availability would have a maximum link outage or down time of 52 minutes per year. This is calculated by a simple equation as follows, given that there are 31 536 000 seconds in one year and t is the maximum permissible link outage time. In this particular example with 99.99% availability, Eq. (5.10) gives us 3153.6 seconds, so we simply divide this by 60 to convert our annual permissible outage to just over 52 minutes.

$$t = 31536000 \times (1 - \text{availability\%}) \tag{5.10}$$

We have discussed a number of major measurement parameters to quantify the quality of a wireless system. Ultimately, quality measurement ensures a wireless telemedicine system is capable of supporting its services. In most cases, practical systems will perform somewhat worse than what is theoretically computed, owing to many uncontrollable factors.

5.6 IoT and Cloud Integration

The Internet of things (IoT) provides a platform for connecting intelligent devices to people. A wide range of devices and systems can collect a wide range of information about a patient's state of health, daily activities, as well as the ambient environment within a telemedicine ecosystem. IoT connects people and devices together via the cloud concept and a software defined networking framework (Ansari and Sun 2018). Using appropriate data capturing methods such as wireless sensors and monitors, a wide range of telemedicine applications can be supported through tracking of relevant health metrics, generate alerts upon anomaly detection, and to provide informative feedback for health and disease management in assistive care and rehabilitation services. It should also be noted that IoT not only links devices and people together but the over IoT platform also entails IoT-driven software. Monitoring software not only analyses the health of patients but also the system health of devices such as carrying out self-calibration or to compensate for motion during measurement, as discussed earlier in Section 4.5.2.

5.6.1 IoT in Telemedicine

IoT helps capture a wide range of health-related information such as physiological, kinesiological, and contextual data (Leijdekkers and Gay 2015). The interconnected nature of IoT in telemedicine is shown in Figure 5.20. Central to the ecosystem is the patient, on whom health management and

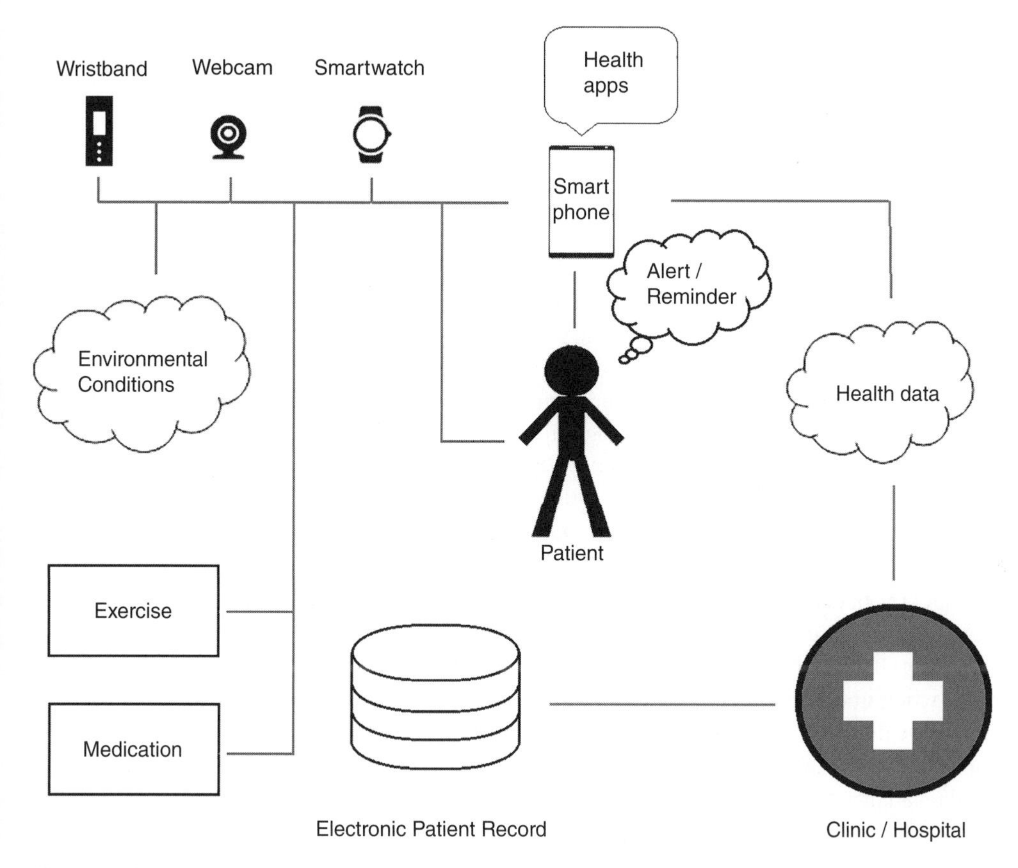

Figure 5.20 A telemedicine IoT ecosystem.

monitoring is based. Interaction with the health service provider, e.g. clinic or hospital, is provided via the patient's smartphone. A wide range of information about the patient is available to the caregiver so that remote consultation can be supported. There are both fixed and mobile devices and systems on the patient's end, all connected via IoT. There is fixed apparatus such as exercise equipment in a public gymnasium or the medication dispenser featured in Section 4.5.4. Environmental sensing can be important for chronic disease management, indoor air quality (IAQ) monitoring for COPD (chronic obstructive pulmonary disease) patients, as an example (D'amato and Molino 2018).

To better understand the significance of IoT, we must look at how the patient sees the benefits. Integrating telemedicine into a smart home environment allows IAQ to be maintained at a suitable level for the patient at all times (Fong and Fong 2011). The use of a smartphone as a control console not only provides the patient with readily accessible health information but also allows them to monitor health and ambient information so that the environment can be controlled, as well as to offer an interface with a healthcare support facility. Patients are therefore supported to take a more active role in their own disease management. The role of IoT is to connect the patient to various other parts of the telemedicine ecosystem via appropriate sensors and devices. Rapid technical advances in the design and implementation of devices and systems provide numerous opportunities for IoT in telemedicine (Fong and Kim 2019). In addition to tracking a wide range of electrical activities including heart (electrocardiography), retina (electro-oculography), brain (electroencephalography) and skeletal muscles (electromyography), IoT also supports location tracking via GPS (global positioning system), which can be extremely important in locating dementia patients and remote rescue operations (Koumakis et al. 2019).

5.6.2 Patient Location Tracking

Tracking the location of a patient is important in accounting for patients with cognitive impairments as well as to manage emergency evacuation in hospitals and nursing homes (Adam et al. 2017). The process of continuous tracking in an IoT network involves how to determine the subject's location based on analyzing the relative distance of fixed sensors with a known reference position. Implementation constraints include limitation in power resources that affects both operational durability and range of communication link (Vashist et al. 2015). This relies on a data structure that classifies the subject tracking into either one of the three approaches: agent-based, cluster-based, or tree-based (Gorce et al. 2007). Location discovery is analyzed through two separate datasets acquired from received signal strength measurement and arrival time between a reference sensor and adjacent sensors (Pahlavan et al. 2002).

A self-cognizant prediction methodology is one that can adopt itself autonomously when analyzing the tracking data (Fong et al. 2012). While the calculation of subject tracking is straightforward when at least three sensors can simultaneously communicate with the subject, as in the case of GPS tracking (Bajaj et al. 2002), real-time prediction (RTP) becomes necessary to improve tracking accuracy when utilizing available data from only one or two sensors (Sun and Ansari 2016). The cell that is nearest to the subject being tracked is assigned as a monitor cell that will forward the tracking data for analysis. In the self-cognizant approach, it is possible to use an RTP algorithm for the data analysis of tracking results in optimizing both tracking accuracy and energy efficiency. RTP is used for tracking a single moving target based on sensing nodes relative to a reference position of the device, as shown in Figure 5.21, which for simplicity is defined as the sensor worn by the subject to be tracked. The sensor nodes will detect the target and determine the location along with its adjacent sensors. Tracking commences by computing the round trip time (RTT) of the signal sent out from each sensing node to reach the subject and back. The angle of arrival (AoA) of the return signal is also computed. Each of the neighboring sensors needs to be simultaneously updated. Position computation is reasonably straightforward by calculating the distance between the subject and the sensors when the subject is successfully tracked by three sensors. However, when fewer than three sensors can pick up the signal, estimation can only be carried out using the RTP prediction algorithm.

In the most primitive case where only one subject is tracked, the process commences when the subject is detected within the sensing environment, which corresponds to the hospital's perimeter. The sensing node will start by initiating communication with the subject's wearable sensor to determine the initial location within the sensing network. For this purpose, the sensing network is defined as a collection of grids where each grid consists of a grid monitor, as shown in Figure 5.22. The grid monitor G1, which is the one where the subject is located within this grid when tracking commences, communicates with not only the sensing nodes within its only grid $\{0, 0\}$ but also with sensing node S1 in grid $\{0, +1\}$ as well as node S2 in grid $\{-1, 0\}$, since both S1 and S2 are located within reach of G1. Grid monitor G1 will wake the corresponding node to active when an adjacent sensor node is in the sleep state. Sensor nodes then communicate with its neighbors to track the current location of the subject which will be stored and forwarded through the grid monitor. The RTP algorithm is deployed in the grid monitor so that the tracking results can be analyzed.

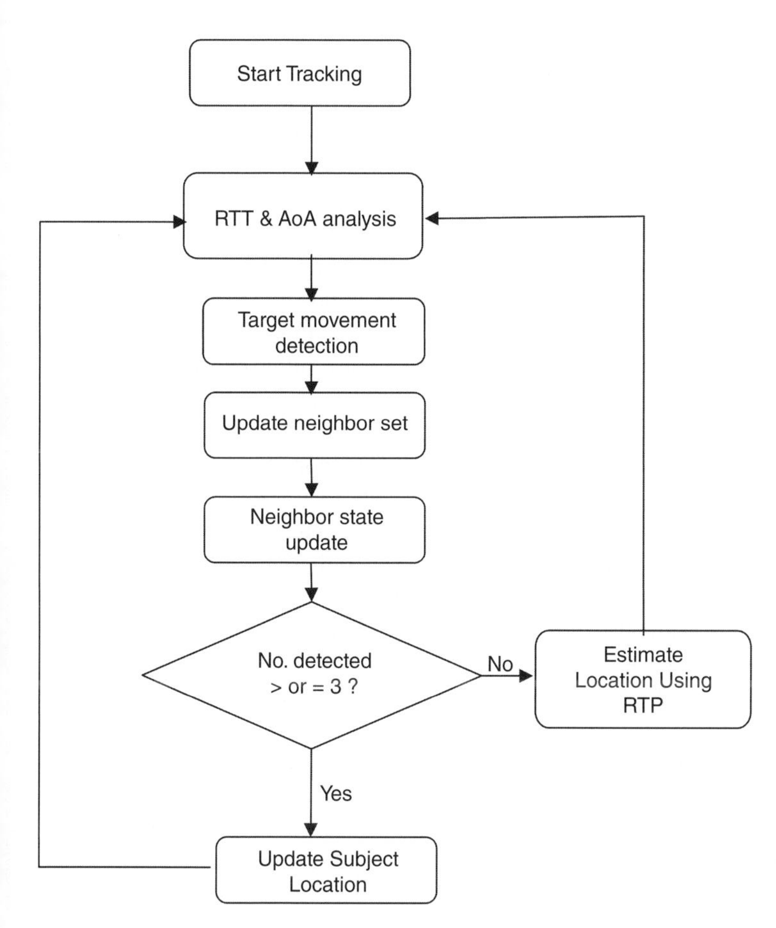

Figure 5.21 Real-time location tracking.

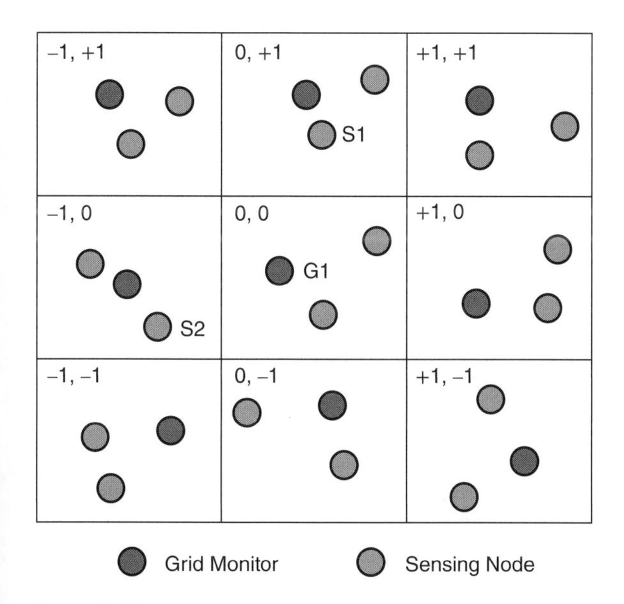

Figure 5.22 Tracking sensors' network grid.

5.6.3 Cloud for Patients and Practitioners

A "cloud" in technology terms refers to a global network of remote servers. Although these servers are interconnected, each of them is capable of functioning on its own. A public cloud allows information to be shared by different entities, such as in the case of sharing general health advice, public health information, etc. In contrast, a private cloud restricts information access within an internal network, for example staff rosters and medical resource management is only available to certain designated personnel within a hospital. A cloud can also be a hybrid such that it is configured to serve both as a public and as a private service provider, information such as electronic patient record and medication schedule are best served by a hybrid cloud so that privacy is kept within the hospital while certain information can be accessed externally by authorized entities such as individual patients and the NHS in the United Kingdom.

Cloud services form a vital part in a hospital environment in that the cloud facilitates data acquisition and analysis across different departments and supports service application for both patients and caregivers. We begin by looking at the cloud from a patient's perspective. The patient wears a selection of sensors that monitor daily activity as well as health parameters. A sensor in each shoe measures gait, walking distance, and speed. The data collected from these sensors is stored in the cloud. Figure 5.23 gives a visual representation of IoT in telemedicine applications, which shows how a healthcare provider can collect data using smart personal smart technology.

Cloud-based healthcare service delivery utilizes either the Platform-as-a-Service (PaaS) or the Software-as-a-Service (SaaS) cloud resource provisioning scheme (Hwang et al. 2016). PaaS uses a platforms that provides cloud services through APIs, programming languages, middleware, and frameworks. This is effectively a set of functions and procedures that facilitate data storage functionality, device management, as well as access control. An SaaS is centered on "data mashup" using cloud computing facilities, where cloud mashup is an intelligent service enabler of on-demand healthcare services across interconnected consumer healthcare and medical devices (Malik and Om 2018).

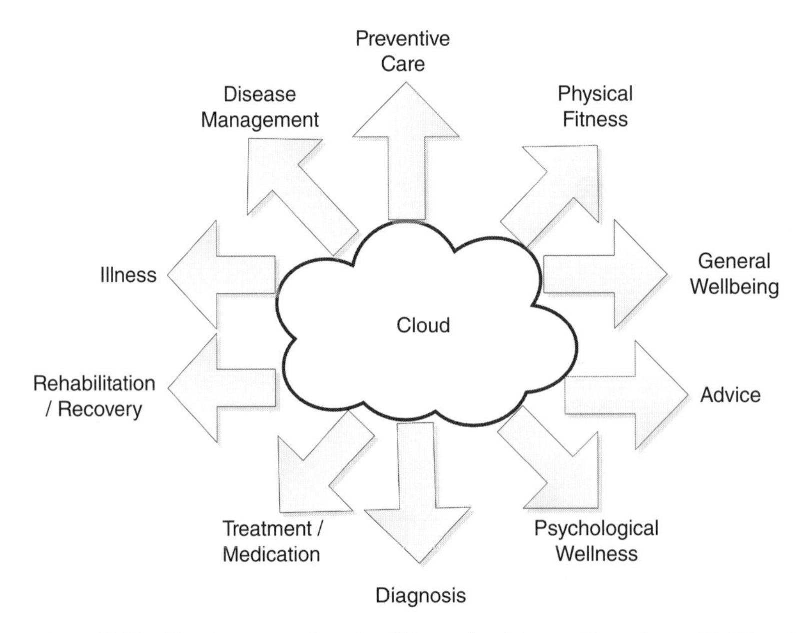

Figure 5.23 Model representing the different healthcare dimensions of IoT in telemedicine.

In summary, IoT provides ease of access to medical resources for both patients and health practitioners from anywhere with Internet access. The major drawbacks of using the cloud instead of local storage include higher latency, which leads to slower access, and concerns over the privacy and safety of patient records. Cloud-based telemedicine services are scalable and flexible. This potentially yields cost savings for service providers on resource sharing and allows new services to be more easily adopted. Interconnectivity through cloud healthcare services supports a wide range of functions from patient diagnosis to public health management through big data analytics for risk identification and disease tracking (Sharma et al. 2018). A cloud platform also provides vital support for rural care.

References

Adam, S., Osborne, S., and Welch, J. (2017). *Critical Care Nursing: Science and Practice*. Oxford University Press.

Ansari, N. and Sun, X. (2018). Mobile edge computing empowers Internet of things. *IEICE Transactions on Communications* 101 (3): 604–619.

Bajaj, R., Ranaweera, S.L., and Agrawal, D.P. (2002). GPS: location-tracking technology. *Computer* 35 (4): 92–94.

Bhargav, N., da Silva, C.R.N., Chun, Y.J. et al. (2017). Co-channel interference and background noise in kappa–mu fading channels. *IEEE Communications Letters* 21 (5): 1215–1218.

Boubiche, D.E., Pathan, A.S.K., Lloret, J. et al. (2002). Advanced industrial wireless sensor networks and intelligent IoT. *IEEE Communications Magazine* 56 (2): 14.

D'amato, V.C. and Molino, A. (2018). Environmental control of asthma, COPD, and asthma-COPD overlap. In: *Asthma, COPD, and Overlap: A Case-Based Overview of Similarities and Differences* (eds. J.A. Bernstein, L.-P. Boulet and M.E. Wechsler). CRC Press.

Fong, A.C.M. and Fong, B. (2011). Indoor air quality control for asthma patients using smart home technology. In: *2011 IEEE 15th International Symposium on Consumer Electronics (ISCE)*, 18–19. IEEE.

Fong, B. and Kim, H. (2019). Design and implementation of devices, circuit and systems. *IEEE Communications Magazine* 57 (2): 66.

Fong, B., Rapajic, P.B., Fong, A.C.M., and Hong, G.Y. (2003a). Polarization of received signals for wideband wireless communications in a heavy rainfall region. *IEEE Communications Letters* 7 (1): 13–14.

Fong, B., Rapajic, P.B., Hong, G.Y., and Fong, A.C.M. (2003b). Factors causing uncertainties in outdoor wireless wearable communications. *IEEE Pervasive Computing* 2 (2): 16–19.

Fong, B., Ansari, N., Fong, A.C.M., and Hong, G.Y. (2004). On the scalability of fixed broadband wireless access network deployment. *IEEE Communications Magazine* 42 (9): S12–S18.

Fong, B., Ansari, N., and Fong, A.C.M. (2012). Prognostics and health management for wireless telemedicine networks. *IEEE Wireless Communications* 19 (5): 83–89.

Gibson, P.G. (2002). Outpatient monitoring of asthma. *Current Opinion in Allergy and Clinical Immunology* 2 (3): 161–166.

Gorce, J.M., Jaffres-Runser, K., and De La Roche, G. (2007). Deterministic approach for fast simulations of indoor radio wave propagation. *IEEE Transactions on Antennas and Propagation* 55 (3): 938–948.

Helm, R., Schuler, A., and Mayr, H. (2018). Cross-enterprise communication and data exchange in radiology in Austria: technology and use cases. *eHealth* 248: 64–71.

Hummel, S. (2007). *Cisco Wireless Network Site Survey*. Charleston, NC: BookSurge Publishing.

Hwang, K., Bai, X., Shi, Y. et al. (2016). Cloud performance modeling with benchmark evaluation of elastic scaling strategies. *IEEE Transactions on Parallel and Distributed Systems* 27 (1): 130–143.

IEEE 11073-10417-2009 (2009). *Health informatics – Personal health device communication part 10417: Device specialization –Glucose meter*. IEEE Standards. http://ieeexplore.ieee.org/servlet/opac?punumber=4913383 (accessed 20 January 2020).

ITU Recommendation X.200 (1994). Information technology – Open Systems Interconnection – Basic Reference Model: The basic model. Art. E 5139. https://www.itu.int/rec/T-REC-X.200-199407-I (accessed 20 January 2020).

Koumakis, L., Chatzaki, C., Kazantzaki, E. et al. (2019). Dementia care frameworks and assistive technologies for their implementation: a review. *IEEE Reviews in Biomedical Engineering* 12: 4–18.

Leijdekkers, P. and Gay, V.C. (2015). Improving user engagement by aggregating and analysing health and fitness data on a mobile device. In: *International Conference on Smart Homes and Health Telematics (ICOST)*. Springer https://opus.lib.uts.edu.au/bitstream/10453/41684/5/ICOST%20paper%209%20Leijdekkers%20Gay.pdf (accessed 20 January 2020).

Malik, A. and Om, H. (2018). Cloud computing and internet of things integration: architecture, applications, issues, and challenges. In: *Sustainable Cloud and Energy Services* (ed. W. Rivera), 1–24. Springer.

NIST (2006). *General Purpose Link Budget Calculator*. National Institute of Standards and Technology http://www.itl.nist.gov/div892/wctg/manet/prd_linkbudgetcalc.html (accessed 20 January 2020).

Pahlavan, K., Li, X., and Makela, J.P. (2002). Indoor geolocation science and technology. *IEEE Communications Magazine* 40 (2): 112–118.

Seele, P. and Chesney, M. (2019). Toxic sustainable companies: a critique on the shortcomings of current corporate sustainability ratings and a definition of “financial toxicity”. *Journal of Sustainable Finance & Investment* 7: 139–146.

Sharma, S., Chen, K., and Sheth, A. (2018). Toward practical privacy-preserving analytics for IoT and cloud-based healthcare systems. *IEEE Internet Computing* 22 (2): 42–51.

Sheikh, K., Gesbert, D., Gore, D., and Paulraj, A. (1999). Smart antennas for broadband wireless access networks. *IEEE Communications Magazine* 37 (11): 100–105.

Shimizu, K. (1999). Telemedicine by mobile communication. *IEEE Engineering in Medicine and Biology Magazine* 18 (4): 32–44.

Shmueli, G., Minka, T.P., Kadane, J.B. et al. (2005). A useful distribution for fitting discrete data: revival of the Conway–Maxwell–Poisson distribution. *Journal of the Royal Statistical Society: Series C (Applied Statistics)* 54 (1): 127–142.

Spahni, S., Scherrer, J.R., Sauquet, D., and Sottile, P.A. (1999). Towards specialised middleware for healthcare information systems. *International Journal of Medical Informatics* 53 (2–3): 193–201.

Stavroulakis, P. (2003). *Interference, Analysis and Reduction for Wireless Systems*. Boston: Artech House.

Sun, X. and Ansari, N. (2016). EdgeIoT: mobile edge computing for the Internet of things. *IEEE Communications Magazine* 54 (12): 22–29.

Van Der Bergh, B., Chiumento, A., and Pollin, S. (2015). LTE in the sky: trading off propagation benefits with interference costs for aerial nodes. *IEEE Communications Magazine* 54 (5): 44–50.

Vashist, S.K., Luppa, P.B., Yeo, L.Y. et al. (2015). Emerging technologies for next-generation point-of-care testing. *Trends in Biotechnology* 33 (11): 692–705.

Zou, Y. and Chakrabarty, K. (2005). A distributed coverage- and connectivity-centric technique for selecting active nodes in wireless sensor networks. *IEEE Transactions on Computers* 54 (8): 978–991.

6

Safeguarding Medical Data and Privacy

Information has always been a precious asset to society. Thousands of years ago, ancient people started sharing information about where to find food and shelter. As society becomes more complex, some information is shared as knowledge, while some is kept with strict confidence for security reasons. For example, books exist for the purpose of sharing existing knowledge and to consider new ideas built upon known facts; bank vaults are designed for locking up private effects so that whatever is stored inside is only accessible to authorized persons. In the past, most medical information was stored in physical formats, such as cards and logbooks. As a result of rapid expansion in the amount of information being collected and created, there are more incentives to use computer-based data storage media for the safe-keeping of medical information. The topic "information security and privacy" is simply the course of protecting information availability, data integrity, and confidentiality, so that it will only be accessible to authorized personnel, to ensure data cannot be tampered with or leaked.

We talk about the importance of data security and privacy from time to time in the previous chapters. Its significance needs no further discussion as it should be well understood by now. There are two main issues: keeping information related to an individual in strict confidence (for example a patient's medical history) and the collecting anonymous data for statistical analysis (for example by conducting a healthcare survey). For both cases, it is vitally important to ensure that any data collected cannot be used to identify a person or where the data comes from.

Many countries already have legislation governing the privacy of individually identifiable information and the confidentiality of electronic patient records (EPRs) so that information access is strictly limited to authorized personnel with the consent of the patient, for example the Health Insurance Portability and Accountability Act (HIPAA) in the USA. Conversely, regulations governing patient privacy in the UK are moving toward the use of such information without a patient's consent (Spencer et al. 2016). Such initiatives cause even greater concern about the security and privacy of health information. In this chapter, we look at security and privacy in two areas, namely safeguarding patients' medical history and the use of biometric features for identification. The former is important for the interest of the general public, whereas the latter concerns technology widely used in personal identification such that an individual can be uniquely identified.

6.1 Information Security Overview

Information security involves compromise between security and usability. To illustrate this, we look at an example where a little girl called Melody wants to safeguard a candy from being taken by others. Melody's idea is to make it as safe as possible. She has drawn up a simple plan in Figure 6.1.

Telemedicine Technologies: Information Technologies in Medicine and Digital Health, Second Edition.
Bernard Fong, A.C.M. Fong, and C.K. Li.
© 2020 John Wiley & Sons Ltd. Published 2020 by John Wiley & Sons Ltd.

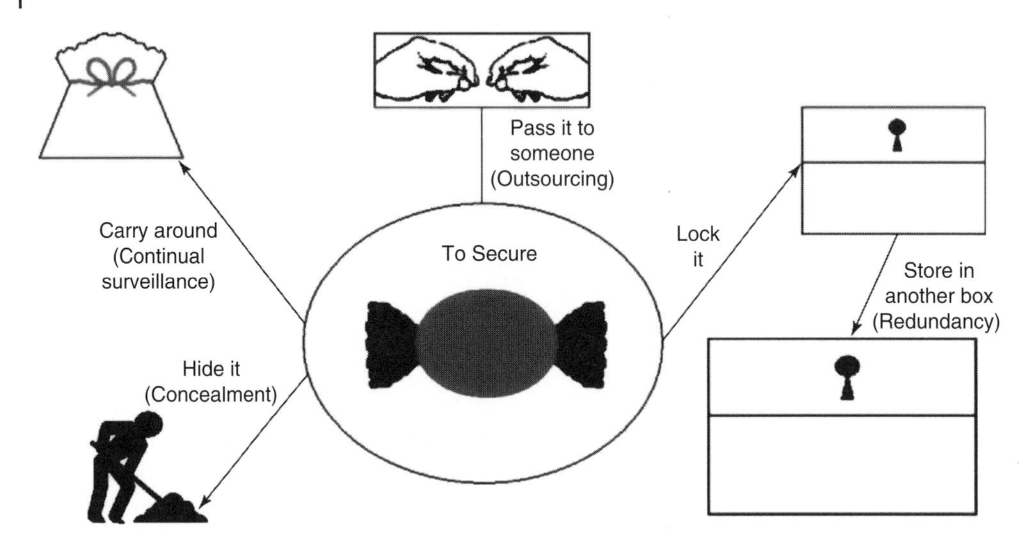

Figure 6.1 A simple safe-keeping plan.

Melody realizes that if she places the candy in a treasure box, and this box is then concealed inside a bigger box, there is a smaller chance the candy will be found by others. Alternatively, she can dig a hole to hide the candy in the ground. She realizes that the deeper she digs, the smaller is the chance of its being taken by others. So, Melody has hidden the candy in a safe place. However, when she wants to eat the candy she finds that it becomes more troublesome and time-consuming to retrieve the candy. From this scenario, Melody learns that the more security is in place, the more difficult it is to access the candy and it will also take longer to retrieve it. Information (or data) security works in exactly the same way.

Now, Melody decides to put her candy in a little pouch and she passes the pouch to her brother, Vale. Vale then slips the pouch into his bag and passes it on to their mother. While their mother drives her children to the mall, Melody decides to eat her candy. She asks Vale for the pouch. As Vale gave his bag to mum, he asks his mum for the bag. As mum is concentrating on the road she asks Vale to open the armrest storage compartment for the bag. So he does and puts his hand into the bag, after a few seconds he holds up Melody's pouch and passes it to her. Melody suddenly shouts, "Where is my candy?" What has happened? Well, to help Melody, we investigate every possibility. A security alert is first initiated when the zipper is found to be open, which leads to a number of possibilities: (i) did Vale open it or did Melody forget to zip her pouch? (ii) Was the bag packed and handled by Vale properly? Or (iii) has mum done anything to it? Here, we can see in this simple example that *security is everyone's responsibility*. Every party, including Melody, Vale, and mum is involved. No one in the system can deny a part when ensuring security. In this example, although only Melody deals directly with security, everyone can be held responsible for the loss.

According to the famous quote "A chain is only as strong as its weakest link," security is weak if there is one single point that exhibits any form of weakness no matter how strong a security system is. Therefore, security relies on everyone to safeguard everywhere.

6.1.1 What are the Risks?

IT (information technology) applications, including those supporting medical and healthcare services, often need to meet the conflicting requirements of users. Problems such as secured access to data and applications may arise when users of different roles attempt to share something, such as when the police require information about a patient's medical records during a criminal investigation. How to share the information, whether direct access should be allowed, and what to share are simple questions that need to be asked. These kinds of situations, when different users have different requirements due to different perspectives, may cause security issues that make a system more vulnerable to attacks. Security risk is related to both the likelihood of a security breach of any form and its impact. There are many types of risks, some more serious than others. Among these dangers are viruses that can erase everything in the system, breaking into your system and tampering with the stored data, someone impersonating you or using your computer to assault others, Unfortunately, it is an unrealistic expectation to guarantee with absolute certainty that this will not happen despite all the best provisions. We can only do whatever is necessary to minimize the risk and any consequential impact. Although we cannot totally eliminate risks, there are ways to control or manage them with appropriate policies, procedures, and practices involving managerial, legal, technical, and administrative aspects (Rindfleisch 1997). Before we go further, here is a brief description on some common terms:

Hacking: activities that explore weaknesses in software and computer systems. Some may be motivated by benign curiosity, while others may be looking to steal or alter data illegally.
Malware: also known as "malicious code," is a small piece of software written for attacking computers. These are things like viruses, worms, and Trojans.
Phisher: spear phishers are specifically designed to attack individuals.
Spam: detestable advertising materials that flood the Internet and waste network resources such as bandwidth and mailbox storage with junk, usually sent with malware.

Having talked about these manmade problems, we cannot overlook risks caused by natural events such as storms, flooding, fire, and earthquake. Since the early age of computers people have been very conscious about data backup. Backup is the process of making exact copies of the data in another storage medium so that the data can be retrieved in the event of a loss or failure of the original copy. In the past, bulky backup tapes, as those shown in Figure 6.2, were used. There were a few major problems with these tapes in addition to the slow data retrieval speed. As shown in Figure 6.2, many tapes took up a great deal of space. Another problem about storage was that magnetic tapes are prone to humidity and mold; therefore, these tapes were usually stored in a controlled environment where the temperature and humidity remained more or less unchanged. Before networking became popular, off-site redundant storage was a logistical nightmare as the frequent update of each backup copy made storage in different sites impractical. Imagine what you need to do if you have to distribute tapes to a few locations on a daily basis. Storage in more than one location is extremely important for the prevention of fire and flooding. In the event that a fire breaks out, another copy can still be retrieved from another geographical location.

Network attached storage (NAS) has become a preferred option for backup storage in recent years. As shown in Figure 6.3, an NAS device typically contains two or more internal hard disk drives (HDDs), a control circuitry that coordinates the read and write operation of the internal HDDs. NAS can be configured to work as RAID (redundant array of independent disks) such that multiple HDDs are installed in an NAS, with the main advantage being the simultaneous backup of identical data on all internal HDDs so that in the event of a physical hard disk failure the information can still be retrieved from the other HDDs inside the NAS.

One thing to take into consideration is that in a hospital environment, where the NAS is expected to operate continuously 24/7, the constant read and write operation can cause wear and tear of the internal components of the HDD's mechanism. An annual failure rate as high as 3% can be

Figure 6.2 Backup tapes that were used for many decades until around the early 2000s.

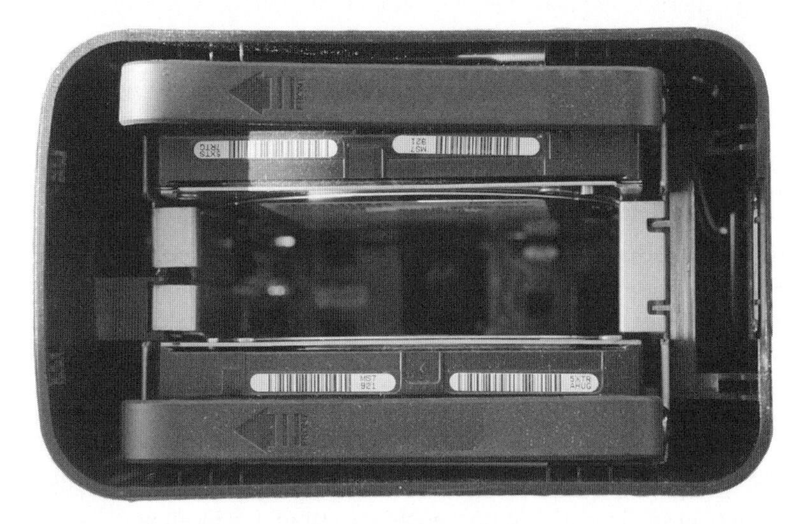

Figure 6.3 Top view of the inside of an NAS that contains a control circuitry at the bottom and two identical HDDs mounted vertically that store exactly the same data as backup.

expected, so the risk of two HDDs simultaneously failing cannot be completely ruled out when considering the safe storage of critical data (Queiroz et al. 2016).

In a networked system, as shown in Figure 6.4, frequent data backup in various location is very easy. Data is simply sent to "mirror site" backup facilities via the network with appropriate synchronization. As its name suggests, a mirror is an exact copy of the data in computer terms. So, a mirror site is simply an exact duplicate of another site. All mirror sites can be synchronized to be automatically updated once the original data changes.

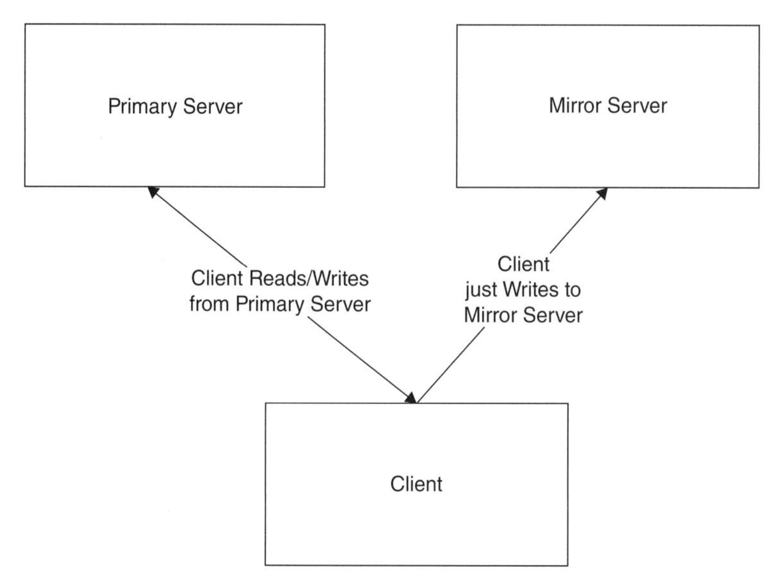

Figure 6.4 Backup with a mirror site.

Data vulnerability can be related to main areas within the entire communication system. The weakest link can be anywhere in the system. Repeated cases of careless hospital staff members losing their USB (universal serial bus) thumb drives containing patient information have been reported anecdotally many times in our experience. Such irresponsible acts can make even the strongest security system useless.

Security also depends on network configuration. In a peer-to-peer (p2p) network, there is normally trust between servers so that a user who has access to one server will automatically been granted access to another. An intruder can therefore move freely throughout the network once access to one server is gained.

6.1.2 Computer Viruses

The threat of cyberterrorism has been expanding rapidly over the past couple of decades. Someone opening an email attachment risks spreading a virus throughout the entire hospital network and beyond. Some viruses can unleash themselves without the host file being opened. Just like viral infection in the human body, a computer virus can sneak into the system and spread to other computers by copying itself. Since computer viruses are willfully created by people, they can mutate and be destructive, malicious, or bothersome in nature.

Similar to biological viruses, there are other software codes that people create for malicious purposes. These include:

Worm: a self-replicating program that spreads itself across the network. Its main difference from a virus is that it is self-contained, whereas a virus normally attaches itself to another program or file, such as a script or an image. Also, most viruses are written to attack computers, whereas worms are designed to attack networks.

Spyware: software written to observe the interaction between the computer and its user that sends such information to a third party via the network. This can be risky as it can also steal data, including confidential information stored.

Trojan: malware that disguises itself in perceivably harmless application software that includes code to allow access to a computer and its stored data. This can be serious and includes the theft of information and unauthorized people seizing control of the computer.

6.1.3 Security Devices

Many devices are available for making a network more secure. These may include purpose-built devices or software installed on a computer. In a communication network, there are many places where we can install something to safeguard security, and this something can perform different types of security, such as identifying an individual user; granting access to part or all of a system; logging activities during a session; filtering incoming and outgoing data based on types, origin, or destination; and enforcing the use of certain keywords, etc. To understand more about the features of security devices, we shall look at some commonly used security devices:

Firewall: a rules-based device that filters certain types of data from entering a network. A firewall can be implemented either as a hardware box plugged into the network, or as software installed on a computer; it can also be a combination of both.

Front end processor: a host computer that manages the lines and routing of data in the network. It can also authenticate a user attempting to login from a remote location.

Proxy server: a type of firewall operation that specifically filters out anything that either enters or leaves a network. It essentially hides the actual network addresses so that any attack will be made more difficult without knowledge of information about the network.

So, all these devices have one thing in common: either allowing or denying access of users to a network or data to pass through that network. This may sound simple enough by applying a number of rules, but in the real world, the implementation of security plans is far more complex than just setting up some security devices as damage to a network can be done from a large number of locations, as we discuss below.

6.1.4 Security Management

Having looked at some basics of information security, we should have a broad understanding that security is about the management of:

- Integrity: data remains in its intended form, without being tampered with in any way.
- Privacy: patient information is not released to unauthorized entities.
- Confidentiality: safeguard personal and corporate information; assets such as patient records, development plans, and work schedules are all kept with strict confidence.
- Availability: system is kept available at all times, without being affected by sabotage or breakdown.

To facilitate all of this, security needs to be addressed in the following areas:

- Computer system security: hardware and software, access to computers and the data stored in them. Protection against virus infection is also an important issue to address.
- Physical security: areas where equipment is installed should be monitored. This applies to both restricted and open access areas. For example, doctors should not leave their computers unattended so that someone can sneak in with a USB drive to copy data from them. Protection of portable equipment is an increasingly important topic. Laptop computers are stolen by organized criminals for the data stored on them (Sileo 2005).

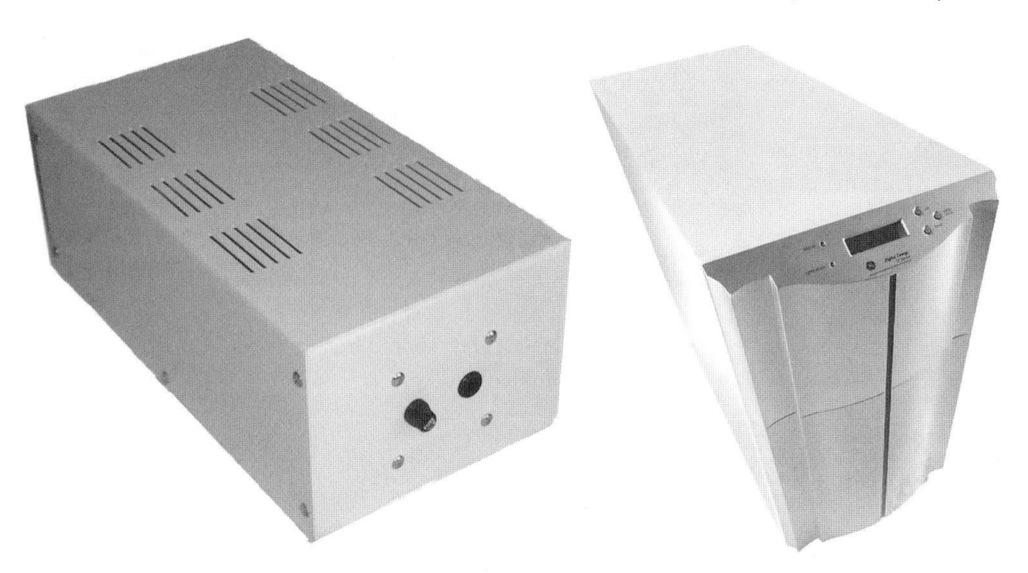

Figure 6.5 Uninterrupted power supply with an external battery that supplies power for a short period of time in the event of a mains power outage.

- Operational security: operating conditions and usage logging. Conditions include ensuring clean power supply, such as using an uninterruptable power supply (UPS). A UPS has a mechanism to ensure that power is not interrupted but still available for a short time even in the event of a power failure, so that users are given sufficient time to save all data before performing a proper shutdown. A typical UPS has an external battery and power filtering mechanism to ensure any power surge is removed. Figure 6.5 shows an example with two boxes that consists of a battery on the left and a power outlet on the right. A small display on the top of the power outlet shows information such as remaining battery life, power currently drawn, and estimated time of available power. A UPS for computer systems usually has accompanying control software so that features such as automatic file saving followed by shutdown can be configured. Also, a fairly accurate estimate of remaining battery life can be displayed on the screen when backup power is activated and users will be alerted of a mains power cutoff.
- Communications security: protection of network and communication equipment including computers, routers, and personal digital assistants (PDAs); network access ports as well as possible attack points should be monitored. These vulnerable attack points include installing an access point in a location where it can be physically accessed. Tampering is as easy as resetting the access point to its factory default settings. Attack point also opens up when using legacy a wired equivalent privacy (WEP) protocol that simply does not provide adequate security.

Other types of wireless attacks include fake Wi-Fi access points, often using the name of an establishment in the service set identifier (SSID). An example is like someone sets the mobile hotspot on a mobile phone with an SSID name "Hospital Free Wi-Fi" that would be very easy to get patients to connect. Patients can still access the Internet without realizing that all data traffic can be monitored by the person operating this mobile phone hotspot.

Consumer electronics devices are increasingly becoming a high risk point of attack when IoT (Internet of things) capability is incorporated into devices ranging from cameras to fridges and televisions. These IoT devices are particularly vulnerable to supply chain attacks when the devices can be reprogrammed as a tool to attack Wi-Fi networks (Lysne, 2018).

Privacy and confidentiality leads us to the issue of information classification. Data can be classified into different categories according to risk, data value, or any other specific criteria. Information may have different value or use, and it is subject to different levels of risk. We should therefore implement different protection procedures for different types of information. For example, the consequence of leaking patient records will be far greater than that of having information about drug usage stolen.

The use of proper preventive measures reduces the risk of security attacks. Information security management involves a combination of prevention, detection, and reaction processes. Contingency planning is vital and should provide details of how to respond to a threat and how to combat a problem when it arises. Proper security management should ensure risks from all sources be minimized – even though they cannot be eliminated.

6.2 Cryptography

Cryptography is the process of converting meaningful data into a scrambled code for transmission across any communication channel that can be "deciphered" or converted back into the original data. The primary function of cryptography is to hide the original information so that it appears to be meaningless while in transit. It is such a vast topic that Schneier (2007) presents a set of three volumes totaling 1664 pages exclusively on cryptography. Cryptography involves applying algorithms to convert the data before transmission, and when the "encrypted" data reaches the receivers, it needs to be "decrypted" back to its original data for interpretation. The process of encryption followed by transmission and subsequent decryption is illustrated in Figure 6.6, where a "key" is generated (a number code) and distributed to both the transmitter for encrypting the original message (expressed in plain text) and the receiver for decrypting the cyphertext to extract the original message. Although cryptography may not be able to achieve absolutely 100% security, it does serve as an essential part of a secure communication system, owing to its effectiveness and capabilities. It is very widely used in virtually all aspects of communications, including but not limited to patient records, medical images, supply order processing, e-prescription, transactions processing, etc.Referring back to Figure 6.6, Melody wants to send a secret to Vale, without letting their mum know. They first agree on a key pair $k = (c,d)$ so that both Melody and Vale keep a copy of the same key. When Melody sends her secret message m to Vale, she uses the key k to generate the ciphertext $c = \mathrm{E}(m,k)$ and sends c to Vale. She knows that the message m by itself can be read by her mum, and if her mum picks up c it will make no sense to her since she does not have the key to "decode"

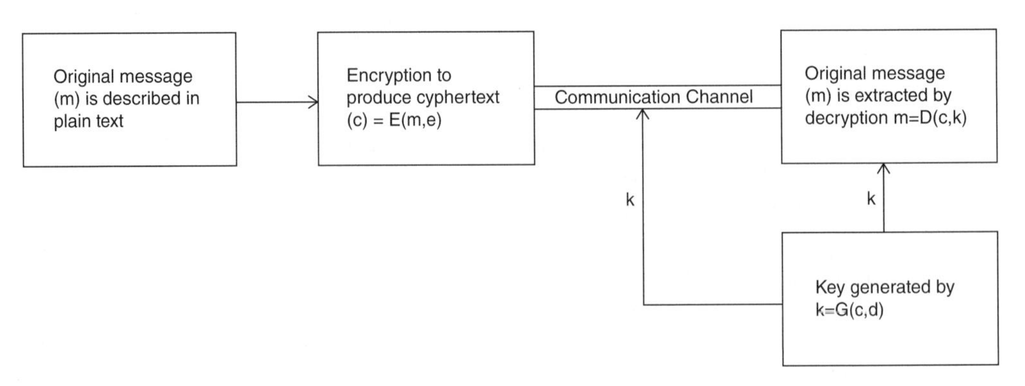

Figure 6.6 Cryptography.

the message from c. When Vale picks up c, he uses the inverse method that Melody used to encrypt the message with the key, so that $m = D(c,d)$ to extract the original message m from the received encrypted message c. Melody trusts this method because she knows her mum does not have the key or any knowledge about the method she used to generate c.

So, here is the basic principle; but why does Melody need a key? In hindsight it may appear as if Melody can just choose any encryption method and Vale applies its corresponding decryption method. Here, the main purpose of the key is that even if their mum finds out the method they use, they can still use it without redesigning the method very simply by using a new key. It is therefore only a matter of changing the key every now and then.

Loosely speaking, there are two approaches to encrypting data, either symmetrical or asymmetrical. The former uses the same key for encryption and decryption, whereas the latter uses one key for encryption and another different key for decryption. In this section, we lconsider both approaches. To illustrate how these algorithms work, we shall bring in our two little children, Melody and Vale, to explain to us the underlying mechanisms.

6.2.1 Certificate

A "digital certificate" is an electronic document for the identification of a user or a server. Like any form of personal identities (IDs), a digital certificate serves as a proof of a personal ID. It is issued by a certificate authority (CA), which is an entity that validates IDs and issues certificates. The CA is just like any government agency that issues personal IDs, where certain checks are performed to authenticate a person's ID before issuing a valid ID for that person. Methods used to validate an ID can be different depending on the individual CA's applicable policies. This is similar to different policies imposed by, for example, the UK's Driver and Vehicle Licensing Agency (DVLA) for driver licensing and the Identity and Passport Service for ID cards and passports.

"Client authentication" is the process of identifying a client by a server so that the identification of a user on the client can be checked, whereas "server authentication" is the opposite process, where the ID of a server is verified by a client so that a user can be assured that the server is indeed one that the user wants to access, such as in the case of ensuring that a bank's website is legitimate but not one that is forged by criminals attempting to steal login information.

Certificate-based authentication is widely used on the Internet. It is where the client signs a "digital signature" (see Section 6.2.4) and then attaches the certificate to the signed data to be sent across the network. The server validates the signature and the certificate upon receipt. The entire process of certificate-based authentication is shown in Figure 6.7. Often used in software distribution and EPRs, a digital signature is a code that proves the authenticity of a message. Digital signatures use "asymmetric cryptography" for messages sent across an unsecure network in ensuring the true ID of a sender.

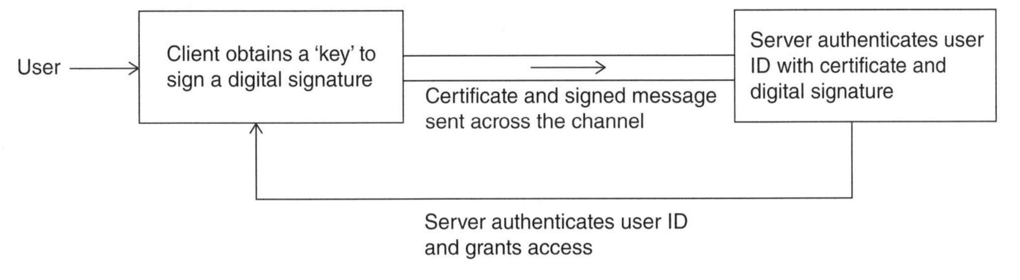

Figure 6.7 Certificate-based authentication.

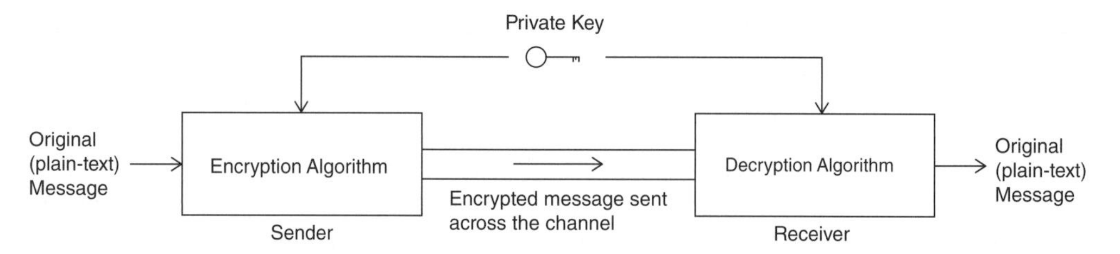

Figure 6.8 Private key encryption.

6.2.2 Symmetric Cryptography

Symmetric key encryption is also known as "private key encryption" or "secret key encryption" because the key used is never made available to parties other than the sender or receiver. As shown in Figure 6.8, its operation is fairly simple. Suppose Vale sends a message to Melody using private key encryption, the process is as follows:

1. Melody creates a key and sends a copy of this key to Vale.
2. Vale uses this key to encrypt his message.
3. The encrypted message is sent to Melody via the network.
4. Melody gets the encrypted message.
5. She decrypts it with the key.

This mechanism is fast and simple to implement as Melody only needs to generate one key and send it to Vale for encrypting the message. The first obvious problem with this mechanism is that Melody has no way of checking the integrity and authentication here. So, from the received message alone she cannot find out if the message has been corrupted or if it was indeed sent by Vale. Also, Melody has to advise Vale in advance of which key to use. They have to have identical keys on both sides for this to work. To overcome these fundamental problems, most modern key systems use "asymmetric cryptography."

6.2.3 Asymmetric Cryptography

Also known as "public key encryption" or "shared key encryption" because a key is generated and placed in the public domain so that potentially anyone can get this key. The term "public key" refers to encryption that uses a key that is published so that it is basically available to everyone. Since the public key (for encryption) is used to generate a private key for decryption, everyone has a pair of different keys. Anyone can publish a public key so that whoever wants to send a secret message to the person who publishes the public key can do so by using this key.

To look at how public key encryption works, let's consider the mechanism behind when Vale sends a message to Melody:

1. Melody generates a public key and places it on the table (everyone can access this key because the table is unsecure).
2. Vale grabs the public key from the table.
3. Vale uses this key to encrypt the message.
4. The encrypted message is sent to Melody via the network.
5. Melody gets the encrypted message.
6. She decrypts it with her private key (the private key has never been made available to anyone, including Vale).

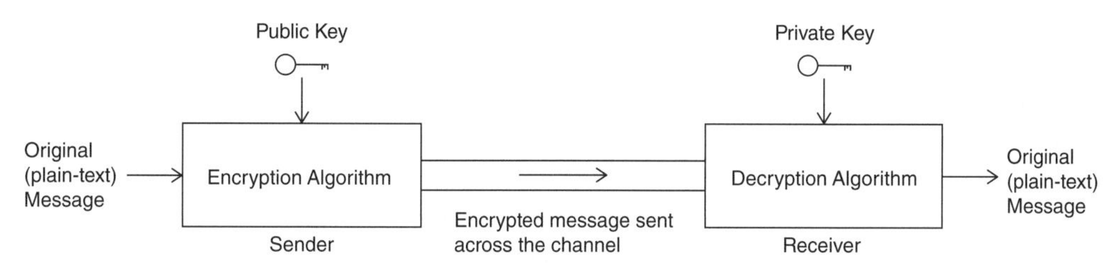

Figure 6.9 Public key encryption.

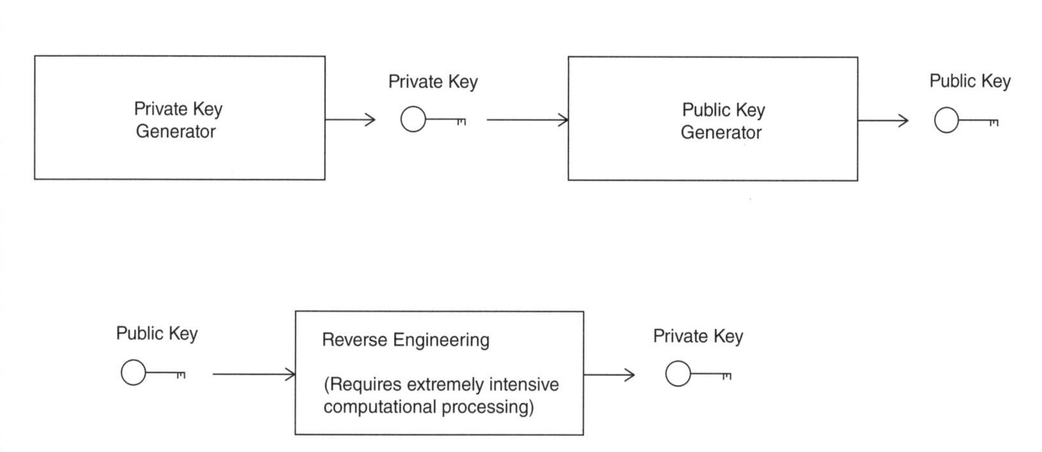

Figure 6.10 Key encryption process.

This process is summarized in Figure 6.9, where we can see the private key is securely stored in the receiver. Melody has two keys: a private key that she keeps and a public key that she places in an unsecure place so that anyone can access it. The public key is only used in encrypting the original message. The encrypted message is sent out, without any key attached, to the receiver and the receiver then uses the private key for decryption. The process of key generation is summarized in Figure 6.10.

The main advantage of this key system is that it avoids the risk of having a shared secret key being stolen when it is passed over a communication channel. The public key was originally designed to eliminate the need for exchanging a key over a communications network. The public key concept was originally developed by British mathematician Clifford Cocks in 1973. The original work does not have a reference as it was classified as a British government secret. The public key system was later formally defined by Ron Rivest, Adi Shamir, and Leonard Adleman in 1978 (Rivest et al. 1978), commonly known as the RSA algorithm, named after the three developers' surnames.

In a public key system, everyone who is connected to the network can access the public key, and so this key can be used to encrypt a message for sending the encrypted message to the person who publishes the key. Only the receiver can decrypt the message with a private key not available to the public. The system's main feature is that the secret key does not need to be sent via the network, thereby eliminating the risk of being stolen while in transit. This system eliminates the need for agreeing on which key to use, as it will always be the one that has been made publicly available. The message will remain secure as long as the receiver keeps the private key secret. Public key cryptography uses certificates to uniquely identify a person, and so authenticity can be guaranteed. However, the system is prone to plain text attacks such that such encryption can be decoded by hackers because the private key (for decryption) can be generated by using the public key that everyone can obtain. While such threats can be minimized by the proper design and implementation of the cryptographic process, it is in fact possible to generate the private key from the public key using some algorithm with the necessary computational power. So, public key cryptography by itself is not 100% foolproof. The principles behind breaking the public key system are not within the scope of this text; however, an overview of attacks on the RSA system is given in Wong (2005).

6.2.4 Digital Signature

Public key encryption uses digital signatures for data integrity assurance. The digital signature, in the form of data codes, is attached to a message. It can be used to check whether the message has been tampered with during transmission. The sender uses a unique signature so that the message is encrypted with a "message digest algorithm." The set of codes that forms a digital signature is computed based on both the message and the sender's private key.

Like a handwritten signature, a digital signature relies on the slim statistical probability that two identical signatures created by different entities will never exist. When a public key system is used to generate a digital signature, the sender encrypts a "digital fingerprint" based on the message along with the private key. The signature can be verified with the pubic key by anyone. To see how this works, let's see how Vale sends a signed message to Melody. The process illustrated in Figure 6.11 is as follows:

1. Vale generates a message digest by using a message digest algorithm applied on the message.
2. Vale encrypts the message digest with his own private key to generate a digital signature.
3. Vale sends the message with the encrypted message digest attached.
4. Melody authenticates the signature by applying the same message digest algorithm.
5. Melody decrypts the message digest using Vale's public key to compare with (4).
6. Digital signature verified if the results of (4) and (5) are identical.
7. Authentication fails if (4) and (5) do not match. This tells Melody that the message is either sent by someone impersonating Vale or the message has been tampered with.

An obvious advantage of creating a digital signature by encrypting only the message digest but not the entire message itself is computational speed, as the message digest is much shorter than the

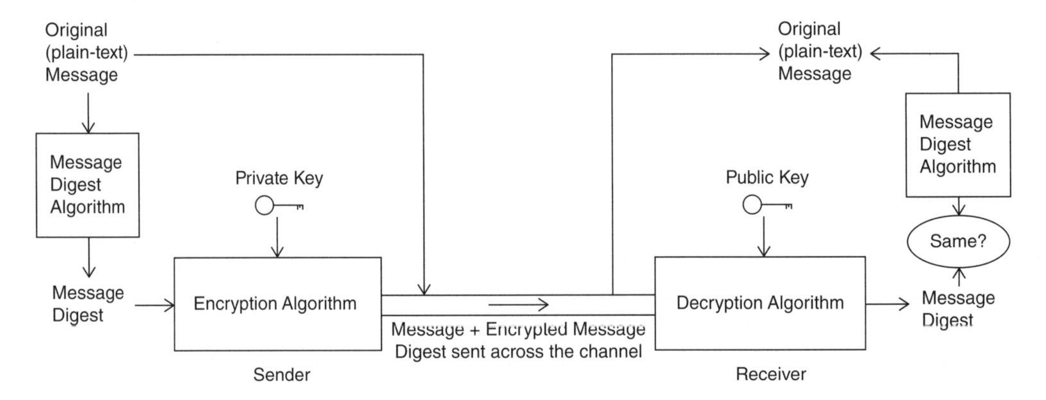

Figure 6.11 Digital signature.

message. The major problem is the possibility of "*collision*," which occurs when the sender signs another message with the same message digest. Message digest algorithms should be designed to avoid collision.

6.3 Safeguarding Patient Medical History

Recognized by the UK Parliament, EPR systems can benefit both patients and practitioners by improving clinical communication efficiency, reducing errors, and assisting in diagnosis and treatment (Barron et al. 2007). The report describes NHS Care Record Service (NCRS), which created two EPR systems, the national Summary Care Record (SCR) and local Detailed Care Record (DCR). As their names suggest, the SCR contains general information for the entire UK, whereas the DCR contains all-inclusive clinical information in a local context. Almost a year after the launch of the NCRS Greenhalgh et al. (2008) presented a study on patients' attitudes to the SCR, which found that the general public was unclear about the policies on shared records, while most surveyed viewed the system as a positive development. The SCR is now available for UK residents via the HealthSpace website. Anyone living in England of age 16 or over can register for a basic account, which lets users book a hospital appointment. Registration for an advanced account would require participation in NCRS by a user's local NHS service. Only the advanced service provides users access to their SCR.

The NHS initially launched the HealthSpace website to provide a range of services where users can scroll down for the following services in addition to account registration:

- Booking a hospital appointment even without an account.
- Health and lifestyle information on various parameters, such as blood pressure, cholesterol levels, and medications.
- Keeping appointments and location of clinics, pharmacies, and NHS offices.
- Access to SCR by logging in with an advanced account.

This initial offering was designed to help users access information and but was closed in March 2013 after the NHS's first implementation of the EPR organizer. Its closure has been linked to difficulties in using HealthSpace. Health records in then NHS were last updated on 24 May 2018. This included many different online service providers, as listed in the screenshot in Figure 6.12. By accessing relevant information from respective service providers, the process of health information retrieval should now be more efficient than the previous HealthSpace offering.

6.3.1 National Electronic Patient Record

All these services, from HealthSpace to current offerings such as myGP and Medloop, share the goal of providing residents in the UK with a tool to access their own health information online. Important information for emergency treatment, such as drug allergy and ongoing therapy, can be retrieved by healthcare service providers when needed. Owing to the useful features of EPR, the Social Insurance Institution of Finland also operates a system similar to HealthSpace. Finland's main feature is the inclusion of a telemedicine system for medical image archiving. Entries by all medical professionals across Finland will be able to append information in a patient's archive. It also supports the automated delivery of prescriptions, which assists with the prescribing and dispensing processes.

National EPR systems do have their drawbacks. Problems with participation have been reported in the Netherlands (Weitzman et al. 2009). Comments posted on researchblogging.org suggests that

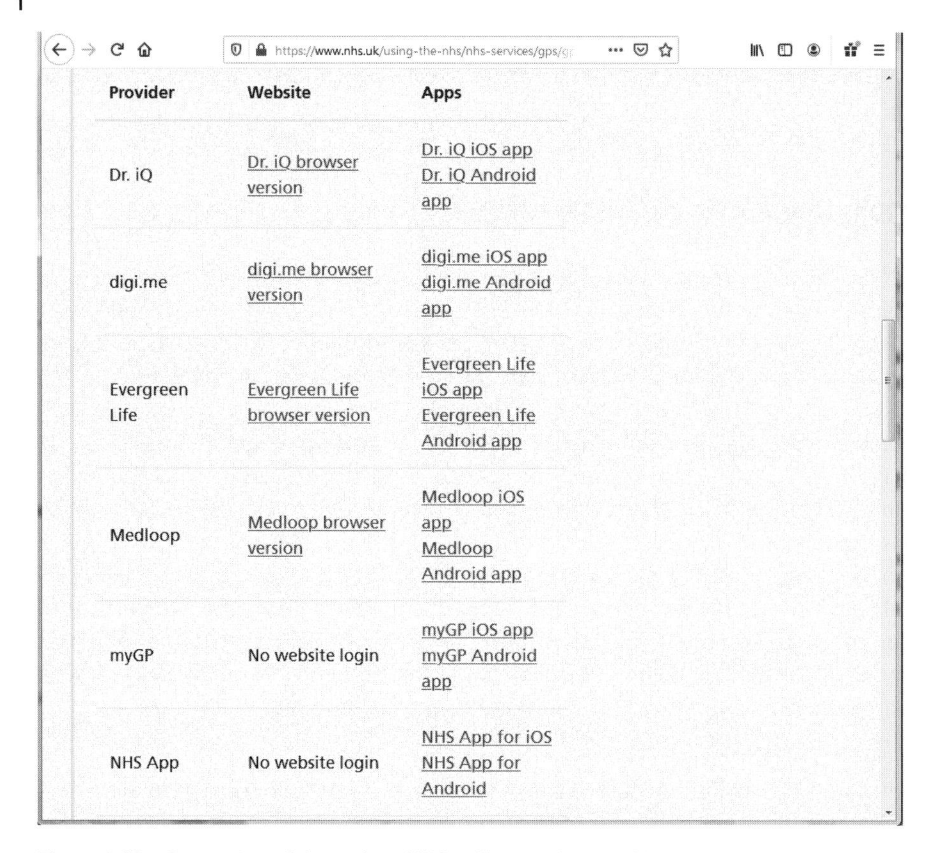

Provider	Website	Apps
Dr. iQ	Dr. iQ browser version	Dr. iQ iOS app Dr. iQ Android app
digi.me	digi.me browser version	digi.me iOS app digi.me Android app
Evergreen Life	Evergreen Life browser version	Evergreen Life iOS app Evergreen Life Android app
Medloop	Medloop browser version	Medloop iOS app Medloop Android app
myGP	No website login	myGP iOS app myGP Android app
NHS App	No website login	NHS App for iOS NHS App for Android

Figure 6.12 Screenshot of the various NHS online service providers.

31% of Dutch doctors are reluctant to subscribe to the national EPR, with a further 25% considering it. What is wrong with the Dutch system? The first fundamental problem with their system is a virtual EPR, meaning that medical data will remain physically where it originates but not on a national server. This therefore requires all participating doctors to have their services accessible online at all times. Linking up servers across all clinics would mean there will be a cost involved for the software and hardware for network connection, which the Dutch authority will not fully subsidize. In addition to this problem, the system was initially implemented as an amalgamation of two separate units: an electronic medication record and a deputy GP record. The former stores information about each patient's visit, including prescription details, and the latter provides after-hours access of patient data. The arrangement is prone to unauthorized access to an individual patient's data since anyone who can access a clinic's computer can access all patient records. In theory, patient privacy is respected by means of contacting a supervisory agency in the event of a suspected infringement of privacy. However, such reporting methods require patients to initiate an inquiry after a patient suspects their data has been accessed without consent. This would be virtually unheard of in practice since no mechanism exists to alert a patient of any unauthorized data access anyway.

6.3.2 Personal Controlled Health Record (PCHR)

A PCHR is a form that allows a patient to control their health data's access rights and content. User control is accomplished by subscription and appropriate access control mechanisms (e.g. password access). The system enables patients to own and manage a complete and secure electronic copy of their medical records. Patients can choose to connect their record to entities such as clinics and pharmacies at will. This can improve the management and analysis of their medical data.

A similar approach is the personal health record (PHR), accessible through the Internet for a compilation of EPR containing a wide range of medical information and history as well as other personal information such as age, address, etc. The two well-known systems are Google Health and Microsoft HealthVault.

6.3.3 Patients' Concerns

The sheer size of a national EPR database may cause concern to many about whether their data is securely stored. Looking at Finland's example again, its population of 5.2 million is expected to occupy as many as 500 petabytes (each petabyte is equivalent to 1024 TB). We are roughly talking about half a million huge (as of end-2009) 1 TB hard disks to store the Finnish EPR. Roughly speaking, this is about 100 GB of storage for each patient. Security is always the biggest concern. In the Finnish case, access is restricted to users in possession of a certificate issued by the National Authority for Medicolegal Affairs. The digital signature is used for user identification. All access is logged. Even if the access right issues are sorted out, there is no guarantee that EPR systems are infallible: entry error is certainly possible. Also, deficient software reliability can undermine a well-designed EPR system. ERPs that depend on certain operating systems (OSs) may let down the entire system because of software bugs (Cohen 2005).

There are several potential issues with the Dutch example (Spaink 2005). The general public does not seem to endorse the idea of the civil service number (CSN) system being used uniformly among all government services, including healthcare, law enforcement, education, and taxation. It has been reported that the system, without accompanying EPR software support for Dutch citizens, is used more for promoting biometric identity cards than for letting citizens view their EPRs. This turns out that benefits brought to patients by EPR are not widely accepted. However, EPR is supposed to bring benefits, such as the reduction of the number of medical errors and in cost, as well as a reduction of bureaucracy in the national health system.

Throughout the world problems with entry error and privacy issues are major factors that discourage public acceptance. Authorities need more comprehensive plans and public education well before EPR rollout.

6.4 Anonymous Data Collection and Processing

The Finland EPR system means roughly an individual take up of 100 GB of data storage for one medical record. Considering that a typical laptop manufactured in 2020 has at least 4 TB of storage with conventional magnetic hard disk drive (HDD) or 256 GB if a laptop is equipped with a solid state drive (SSD), the need for masses of extra storage space that clinicians can access (sometimes wirelessly) with their laptops is unavoidable. Throughout the book we have looked at many aspects of health information, these include vital signs, images, records of each doctor visit, and medications taken– essentially everything related to health that begins from a person's birth throughout their entire life. Such information tells a lot of details about an individual.

The vast range of information makes medical records extremely useful in many areas, for example marketing, government planning, and pathology analysis. Companies utilize plentiful resources to find out the state of individuals for the purpose of segmentation marketing. Although marketing is an important topic in promoting healthcare technology-related services, it is not within the scope of this text and readers are encouraged to read Hung et al. (2009) for details.

Let us concentrate on matters directly related to national healthcare services. Serving an entire country's healthcare needs is not an easy task. It involves a very complex governance structure. For example, in the UK alone the following authorities exist: the NHS (including separate entities for NHS Wales, Scotland, and Northern Ireland), 10 Strategic Health Authorities serving different areas, the Health Education Authority, and 10 Special Health Authorities responsible for different health-related issues. The healthcare structure is simplified in Figure 6.13. In addition to many authorities in England alone, there are also hundreds of trusts throughout the UK. These cover areas such as ambulance, mental health, and numerous primary care trusts (PCTs). With so many entities employing over one million people throughout the healthcare system, who can access what information and how data can be shared remains complicated. In case any unauthorized leaking of information, mechanisms must be in place for finding out what has happened and identifying any employees accountable for the leak.

6.4.1 Information Sharing Between Different Authorities and Agencies

Before the IT era, when paper was the main medium for circulation, information sharing between authorities and agencies took place in a very limited manner on a case-by-case basis. Back in the 1980s, medical information sharing was not supported in real time (Thacker et al. 1983). Data that flowed between different entities such as clinics, hospitals, laboratories, and insurance companies in the provision of healthcare services was commonly shared at an aggregated level. Today, data sharing is guided and restricted by both excessive legal regulations and derisorily written guidelines. Many hurdles must be cleared in order to establish a proper process of sharing information between agencies. Most of these hurdles are not related to technical issues. Political and social hurdles are difficult but important to address. In many countries, bureaucracies as well as differences in state and federal laws may greatly hinder the prospects of properly establishing such processes.

The increased threat of bioterrorism underscores the need to develop disease surveillance and information sharing mechanisms for supporting real-time data analysis and protection, and for circulating information about outbreaks, both for naturally occurring and for manmade viruses (Clinton 1999). As EPRs transform from paper-based log cards to electronic database management systems, issues such as interoperability, flexibility, accessibility, and scalability become increasingly important to consider. The West Nile Virus-Botulism is an example of a national infectious disease information infrastructure designed for capturing, accessing, analyzing, and visualizing disease-related information from various sources to support real-time reporting and alerting functions. The system is so vast that it contains data from humans, animals, and insects that can potentially carry diseases, as well as botulism data. This involves many agencies associated with public health and safety, animal and pest control, and the National Institute of Allergy and Infectious Diseases that is responsible for botulinus intoxication in the US. Tracking *Clostridium botulinum*, the bacterium that causes botulism, is jointly accomplished by all participating agencies. Other related information such as climate pattern and bird migration is also recorded for analyzing and tracking disease occurrence and spread. In the US, state and local regulations govern information sharing between agencies that may require prior approval from the governing hierarchy of the agencies involved, prohibiting informal information sharing agreements between those agencies. These regulations may differ in terms of confidentiality requirements and how long data can be kept for. Owing to privacy issues, certain regulations may prohibit the unique identification of individual persons or actual locations, making disease tracking more difficult.

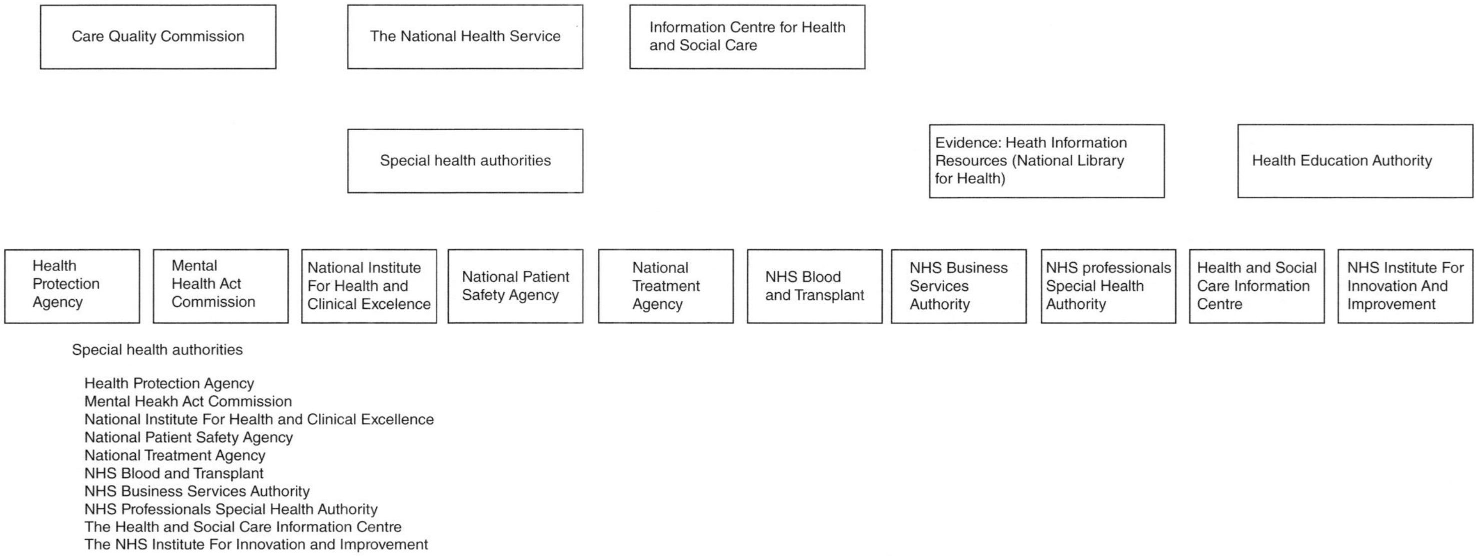

Figure 6.13 Healthcare service infrastructure.

6.4.2 Disease Control

Disease outbreaks can spread across states, countries, and continents. An outbreak can be "overt" and "covert" in nature, meaning it can be easily observed as natural cause or covert as surreptitious on purpose. Overt outbreaks may be initially noticed and managed by a public health agency, whereas covert outbreaks may not be identified and hence initially managed by a public health agency then passed on to law enforcement agencies to track down the source (Butler et al. 2002). Overt diseases can often be tracked initially through a geographic spread by animal or human movement. A covert disease outbreak triggered by a large-scale bioterrorism event whose occurrence may appear random, and so one requiring information sharing involving national and international entities, may involve the establishment of data sharing mechanisms of a far more complex nature than in our above case study. The analysis of spread pattern, for both natural virus and bioterrorism, entails collection of data about the time and place of occurrence for each case.

In epidemiology, disease spread usually exhibits certain spatial patterns over time. The spatial and temporal dynamics of a virus during an epidemic are usually examined to predict the rate of virus spread. A long-term study by Viboud et al. (2006) shows that virus outbreaks exhibit hierarchical spatial spread evidenced by higher pairwise synchrony between areas of higher population densities through quantifying long-range dissemination of infectious diseases. A statistical model that describes the spatial spread pattern can be computed by a number of methods, such as that proposed by Pasma (2008). The process is simplified in Figure 6.14. This chart shows the sequence of manipulating spatial analysis statistics by performing some general clustering tests based on the principle of whether the clustering location of a disease is known or not (Besag and Newell 1991). A "cluster" is defined as closely grouped cases of disease with a well-defined spread pattern over space and/or time (Yan and Clayton 2006). Clustering tests of unknown locations are divided into global and local tests. The former is useful in determining any relationship between reported cases and clustering exists if cases occur in such close proximity that they spread out of control, such as in the case of an avian influenza outbreak (Lycett and Segato 2019); local clustering tests show the rates of cases and identify affected areas as clusters by comparing rates of diseases in different clusters. Clustering tests are also performed for all areas in order to draw a clearer picture about the spread pattern. Temporal analysis helps the detection of clusters of disease over time and spatial–temporal

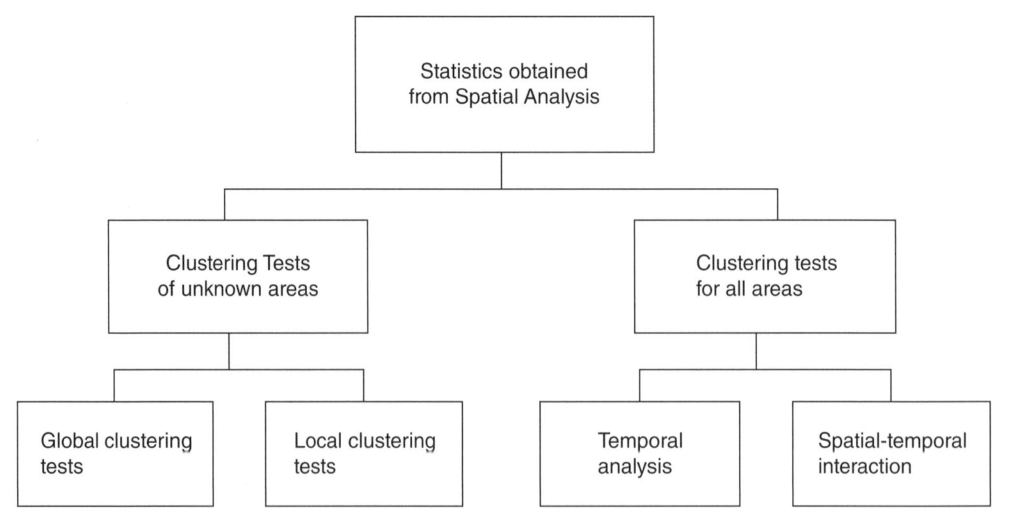

Figure 6.14 Process for infectious disease spread pattern analysis.

interaction analysis uses both space and time information on cases close to each other that occur around the same time.

Cluster analysis and detection involves detection of clustering of disease in historical disease data, focused cluster analysis, and spatial cluster detection. The main purpose of deducing a clustered spatial structure is to group what may otherwise appear to be randomized cases of diseases together to reveal a pattern with data description for visualization. Apart from spatial and temporal information that can be used to build a statistical model for predicting the pattern of disease spread, other useful information for containing and controlling disease includes demographic factors of infected persons and in cases of evaluating the severity of an epidemic medical history about pre-existing conditions and prescribed medications are also extremely useful.

The accurate prediction of disease spread as well as balancing data capturing and respecting patient privacy are equally important in disease control. Let us take a look at one single case of swine flu that put over 300 healthy people into enforced detention for one week (Yuan 2009), where the local authority decided to quarantine all 340 hotel guests and employees as a result of one guest having been diagnosed with the H1N1 virus that saw 10 000 cases in just over four months afterwards with a mortality rate that is statistically similar to that of most other influenza strains. While we are not in the position to debate whether the compulsory detention was justified, let us take a look at some facts gained over the five months after the incident so that we can get a good understanding of what prediction of virus spread pattern can do for us and what needs to be compromised between public health and privacy:

- Comparing the H1N1 swine flu with H5N1 avian flu in the same vicinity, H5N1 is statistically much more deadly, whereas H1N1 exhibits a faster rate of spread.
- H5N1/H1N1 recombinant virus poses a serious hazard yet there is no scientific proof of such occurrence.
- In the 1918 flu pandemic, the "Spanish flu" that spread to virtually everywhere on earth, H1N1 was blamed as a deadly virus that killed no fewer than 50 million people, mainly healthy young adults (Mitka 2005).
- Antiviral drugs such as Tamiflu™ are adequately effective with only isolated cases of drug resistance (Community Central 2009).
- Fairly high prevalence with more than 1 in 1000 infected within four months of the first reported outbreak, carriers can travel in and out undetected by planes, ships, and trains.
- Regular seasonal flu virus, H3N2, poses greater threats than H1N1 (Higgins et al. 2009).
- A second strain of H1N1 may have mutated around the same time of the incident (Ushirogawa et al. 2016).

Further to various attributes of the virus itself, there are also humanity issues:

- Hundreds of tourists have their vacations ruined.
- Sudden loss of business opportunities without proper planning can cause a wide range of problems (Cheng et al. 2009), which affects both the hotel itself and guests' business trips.
- People are involuntarily detained in a confined room for one week, for example Weaver (2009) reports that a prostitute was owed a great deal of money as she was forced to share a room with a paying customer as opposed to being given her own room.
- Logistics support from potentially hundreds of people, from basic necessities to entertainment.
- Businesses surrounding the hotels suffer as the area was cordoned off.

Was this a publicity show, was there a genuine need, or was there a political agenda? To answer these questions, we can find some clues from the points listed above. Digging deeper into the story,

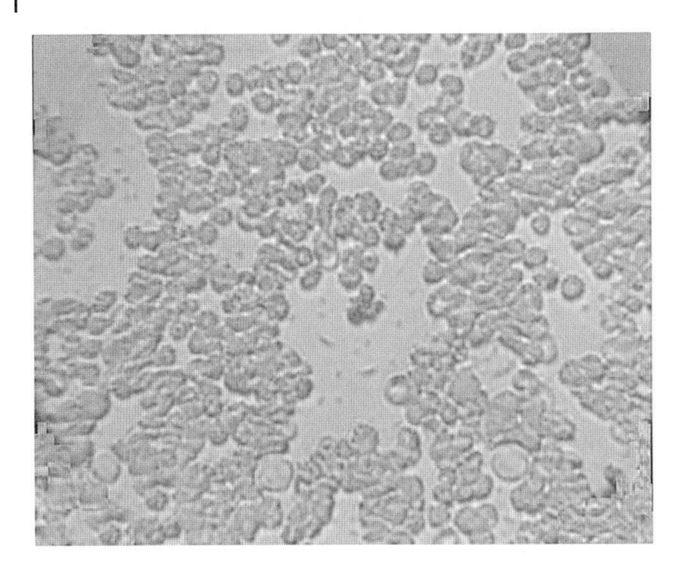

Figure 6.15 Blood sample of a swine flu patient.

we notice that if the authorities had known earlier that the virus would still spread throughout the entire world irrespective of how this case was handled, they would probably have sent the only patient into hospital while letting the remaining 300 people go about their business as usual. In the end, this is back to the knowledge of the disease spread pattern.

More knowledge about the disease would also provide authorities with the necessary information about prevention and diagnosis. The swine flu virus may increase acidity of blood, as Figure 6.15 shows the blood sample of a patient with H1N1 magnified by 300 times. This sample is likely to suggest chronic fatigue syndrome for the patient. It may also appear similar to a mycoplasma infection, which also looks like pneumonia or severe acute respiratory syndrome (SARS) (He et al. 2003). Such symptom may suggest urgent respiratory treatment is required.

Now we understand the importance of finding an accurate prediction of how a virus may behave when it spreads. This is often much easier said than done. Tracking the space and time changes of the spreading process may depend mainly on anonymous information about each reported case. Gaining a more comprehensive picture of the disease may, however, involve analyzing information of individual patients, including where and when the patient has been, who they have been in contact with, and even the transportation mode of a journey. As Yuan (2009) reports, even taxi drivers were caught up in the incident in the above case study. Acquisition of personal information may affect policy planning, as we discuss in the next subsection.

6.4.3 Policy Planning

Authorities often collect information for planning. This involves many different agencies and departments for matters related to education, prevention, healthcare system infrastructure, and emergency services. Traditionally, statistics related to specific types of accidents have been used by authorities so that they can design education campaigns for certain groups of people that statistically have higher risks, for example statistics about alcohol-related traffic accidents have been analyzed to find information about times and locations of likely occurrence and campaigns are set out to target those at higher risk of drink driving. This seemingly simple course of combating drink driving involves many entities. In the UK, the Department of Transport organizes campaigns and promotional materials for prevention, conduct surveys of attitudes to road safety, and provides related information; local drug and alcohol departments may also assist with rolling out campaigns. Of course, the police are out to catch the drink drivers. The Royal Society for the Prevention of Accidents (RoSPA) takes care of driving safety and training, and the Campaign Against Drinking & Driving (CADD) was established as a charity to support crash victims and their families. Each of these entities has its own specific functions and they collect data in different ways and for different purposes. They all share one common responsibility: keeping data with strict confidence and ensuring that data will not be lost or stolen, regardless of how the data is treated.

Statistics are gathered and some shared by various entities. Obviously, data collected cannot be sent "as is," because it may contain sensitive information. For example, the result collected from a breath test is associated with a specific driver, whose detailed information including address, driving license number, and vehicle registration details are all gathered. Information shared for statistical analysis may include time and place of the test, age group of the driver, and the breath test result. Nothing that can uniquely identify the driver being tested should be transferred from one entity to another without the driver's consent.

Health statistics for policy planning sometimes involves agencies of a certain region such as a county or a province, or at a national level that handles statistics from all areas within the country where each local government may have its own agencies with different functional duties for planning. At the same time, they also interact with national agencies in a broader context. For example, each state has its own agency with different functions in the US. In our case study, we look at the structure of two adjacent states in the western US, California, and Nevada. California has its Center for Health Statistics (CHS), whose primary function is to manage the collection and distribution of health-related statistical data. It consists of several offices and sections, as shown in Figure 6.16. These offices are each responsible for a number of functions, such as maintenance of a system for registration for all births, deaths, and marriages, and include issuance of appropriate certificates. Vital records also include over one million miscellaneous incidents each year. The health information gathered is used for research on the general health status of the state's residents. California's CHS is also responsible for disease control, local health services, and regulation of the public drinking water systems where the vast geographical coverage of the state denotes the need for different field operation branches for the northern and southern parts of the state, in conjunction with the Technical Operations Sections, Monitoring and Evaluation Unit, and the Infrastructure Financing and Infrastructure Funding Administration Sections. Within the California state, health data including information about information of individual residents is also used together with agencies such as the Department of Health Care Services and the California Health and Human Services Agency.

In the state of Nevada, health information is administered by the Bureau of Health Statistics, Planning & Emergency Response (HSPER), under Nevada's Department of Health and Human Services. It too is responsible for birth, death, and marriage registration; vital and health statistics analysis; and public health monitoring. Unlike California, where the Office of Statewide Health Planning and Development, under the California Health and Human Services Agency, is responsible for the health planning, health planning in Nevada is also handled by the HSPER. Public health planning is extremely important in ensuring the optimal utilization of available resources and residents are well cared for.

Health planning, irrespective of organizational structure, involves the collection, storage, processing, validation, analysis, and distribution of health and related data. Such data allows authorities to plan for future healthcare services such as prediction of future needs and preventive education. For example, using information about population growth and statistics on usage of

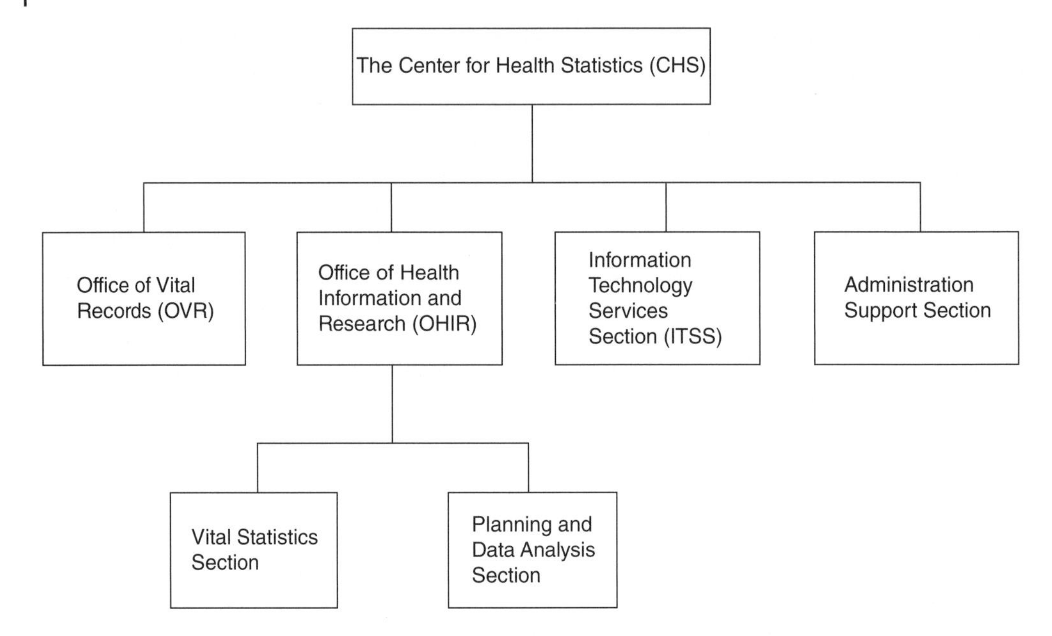

Figure 6.16 Healthcare service organizational structure.

various units of surrounding hospitals would enable planning for any new hospital with an appropriate size of each unit in it.

Many countries conduct censuses for prediction and planning a for wide varieties of needs. These are usually carried out by a government agency dedicated to collecting and analyzing data once every few years (United Nations' recommendation is once every 10 years). The process of acquiring information about population dates back to the eleventh century when the Domesday Book of 1086 appeared in England with information about individual households. An census is now conducted by the Office for National Statistics (ONS) in the UK for planning, including healthcare policies. In the US, this is handled by the Census Bureau, whose functions are similar to the UK's ONS.

Another piece of important information that the census gives is about an aging population. This growing problem in most developed countries means demand for healthcare will rise over the next two to three decades. A snapshot of the UK's National Statistics Online in Figure 6.16 shows a "population pyramid," which conveys some basic information that is helpful for long-term healthcare policy planning.

1. A clear spike around the age of 60 results from the postwar baby boom that shows a large group of people are around the retirement age; over the next decade these people will very likely need more healthcare assistance.
2. There was another baby boom around the mid-1960s, now around the early to mid-1940s. We can deduce that an influx of retirees will utilize the healthcare system in about 20 years' time.
3. It also gives information about migration, a net increase in migration (i.e. the number of immigrants settling down exceeds that of the number of people emigrating from the UK to settle permanently overseas). Immigrants may settle in certain areas within these countries. Therefore, an increase in demand may suddenly affect individual hospitals.
4. This pyramid shape reflects a similarity in number between both genders among the UK population. The balance shifts toward the female side from around the age of 70, which is in line with the fact that life expectancy of females is higher than that of males. Older women and men

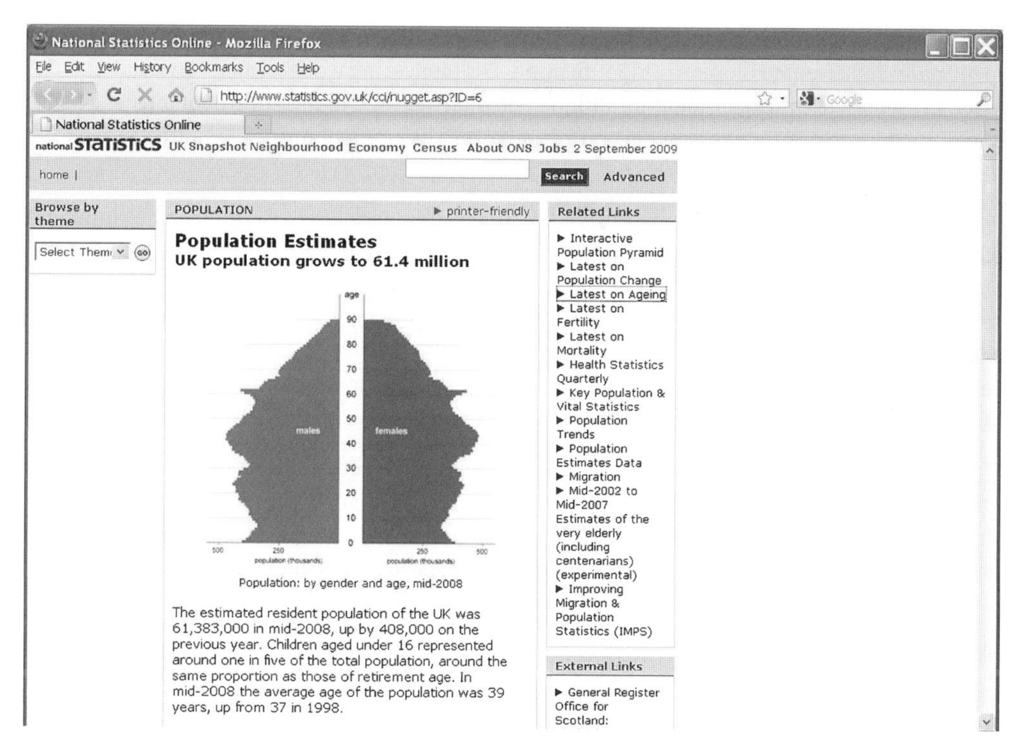

Figure 6.17 Screen shot of a population pyramid.

may have different chronic diseases that require different treatment. Such trends can be used to estimate the needs.

5. By comparing population pyramids of different years, an upward trend confirms population aging is becoming a greater problem. Such study allows authorities ample time to develop policies to meet the anticipated growth in healthcare demand.

It is worth noting that the process of analyzing population statistics is a very lengthy one that may take more than one year. The information in Figure 6.17 pulled out from National Statistics Online was released at the end of August 2009, which shows data for mid-2008. The lack of timely information and the complexity involved in statistical analysis together explain why best effort healthcare policies adopted throughout the world may not provide the best solution to the nation's needs. No matter whether the data collected is important or not, the single most important point to observe is privacy issues when collecting data.

6.5 Biometric Security and Identification

"Biometrics" is the term given to measuring and analyzing the physical characteristics of the human body; it can also be used for security and identification purposes. Voice recognition software has been around for decades, although it is only relatively recently that speech recognition and filtering algorithms have improved enough to be of service in the field of telemedicine as modern computers have the necessary processing power. Although biometric security is used in many areas outside the medical domain, it is a topic that warrants noteworthy discussions, owing to its popularity in different areas of healthcare and telemedicine-related applications.

French anthropologist Alphonse Bertillon (Dirkmaat et al. 2018) was perhaps the first person who formally documented biometric identification. His pioneering work created "anthropometry," a systematic biometric measurement for unique personal identification. Anthropometry is accomplished by taking measurements of certain body parts. The original anthropometrical system of identification consisted of three parts:

- Physical measurement: precision measurement under some stipulated conditions. Measurement includes characteristic dimensions of parts of a human body (e.g. size of the ear).
- Morphological description: the shape and contour of the body, which entails a characteristic description of mental and moral attributes, something that relates to the movement of a certain parts, such as limbs.
- Peculiar marks description: observation of any signs on anywhere of the body that may be left from an accident, disease, or disfigurement. These include scars, moles, and tattoos.

The rather complicated process proposed by Bertillon may even result in two sets of different results being obtained from one individual when the process is repeated. To simplify Bertillon's identification process, it was later said that the ear was sufficiently unique such that he said, "It is, in fact, almost impossible to meet with two ears which are identical in all their parts" (McClaughry 1896). It described a series of four visual features of an ear as follows:

- Helix: three portions of the border of the ear, and the extent of openness.
- Contour: degree of adherence to the cheeks and the ear lobe size.
- Profile: inclination from horizontal and the amount of reversion in front of the antitragus.
- Fold: dimension and pattern of antihelix (anteroinferior to the helix).

The details documented in were so extensive that McClaughry (1896) gave over 15 pages to the anthropometric description of every single feature of the ear, including the earlobe, the tragus, the antitragus, the concha, and the superior fold. Such in-depth description of the ear, although never applied to personal identification, did, however, serve as an important milestone for the systematic identification by bodily measurements.

One obvious application of biometric identification is restricting access to different parts of a hospital; where staff members can enter an area by being uniquely identified without using an identity card, which can be lost or stolen. Apart from behavioral uniqueness such as the signature and voice of an individual person, there are certain physiological attributes of a human body that are unique in nature with practically zero chance of finding two persons, even twins, who possess identical properties. These include finger and palm prints, iris patterns, facial patterns in terms of optical and thermal outlines, and deoxyribonucleic acid (DNA). However, DNA is not commonly used, because the current technology still requires some form of tissue to be analyzed.

6.5.1 Fingerprint Recognition

Fingerprint identification is perhaps the most-well-known method of biometrically identifying a person. A fingerprint impression consists of patterns that exhibit physical differences between the ridges and valleys on the finger's surface (Lee and Gaensslen 2001), where "ridges" and "valleys" refer to the upper and lower segments of the skin, respectively. Where each ridge ends, it forms a "minutia point," where the size and shape of can also differ. Also, a "ridge bifurcation" refers to where a ridge splits into two, branching out. The positions of these unique features as shown in Figure 6.18 are used to identify a person. Ridges and valleys are show in black and white, respectively. There are five fundamental fingerprint patterns, namely: whorl, arch, tented arch, left loop,

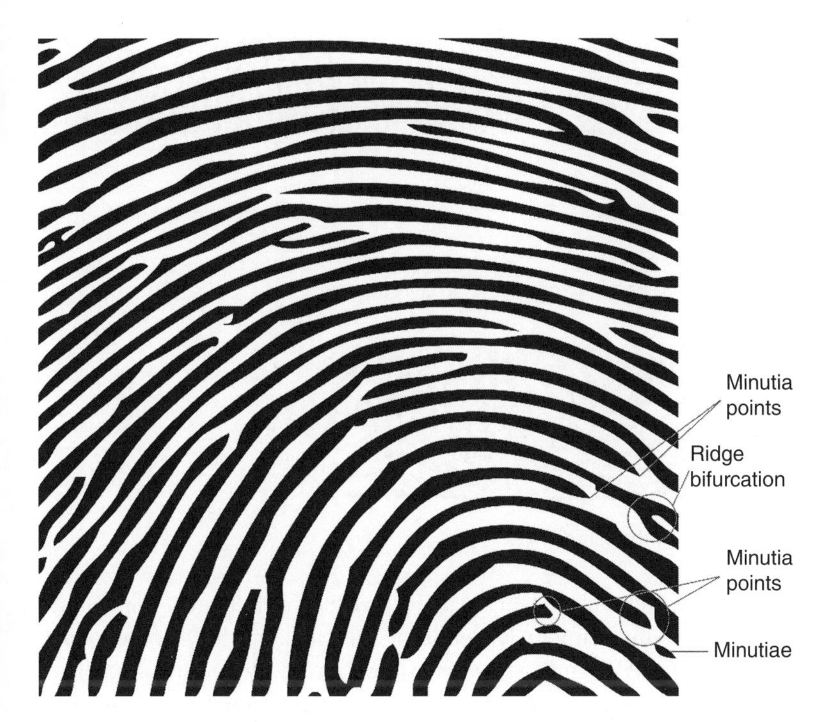

Figure 6.18 Fingerprint impression.

and right loop. Generally, loops cover about two-thirds of the fingerprint and whorls cover about a quarter of the finger. The remaining 10% is covered by arches. In a loop, one or more of the ridges enters on either side of the impression. A loop consists of a "core" and a "delta." These are circular and triangular patterns that when grouped together form a loop. In a whorl, some of the ridges turn several times. A portion with at least two deltas is considered to exhibit a whorl pattern. Classification can generally be accomplished by counting the number of deltas. The lack of a delta is an arch, one delta is a loop, whereas at least two deltas make a whorl. In an arch, the ridges run from one side to the other across the pattern.

For more than a century, fingerprints have been widely accepted as an infallible means of unique identification. Fingerprint analysis has put countless criminals behind bars throughout the world. This is a proven technology based on the comprehension that no two fingerprints have ever been found to be identical among billions of people who have ever existed. Fingerprint impressions were first formally studied by Czech physiologist Jan Evangelista Purkyně (1823). His work followed the much earlier work of Grew (*De Extemo Tactus Organo*, 1685) according to the classical study by Penrose (1968), who had illustrated nine fingerprint patterns that were later used for identification purposes.

Powerful image processing algorithms and the lower cost of high resolution scanners make such application so popular that fingerprint identification has been seen in many consumer electronic devices. Many medium to high end laptop computers now include a narrow strip of optical scanner that scans a portion of the user's fingerprint. When used in these kinds of consumer electronics devices, it relies on comparing the portion of scanned fingerprint with the impression that has previously been stored. Since repetitive scanning of the exact same portion of the finger during different occasions is practically impossible, authentication is only done by using a small portion of the finger, as illustrated in Figure 6.19. Compromise is made between the physical size of the

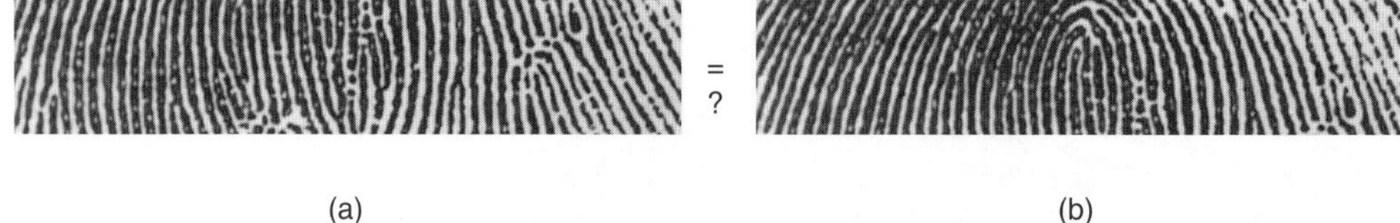

Figure 6.19 Scanned image of a portion of a finger under different alignment.

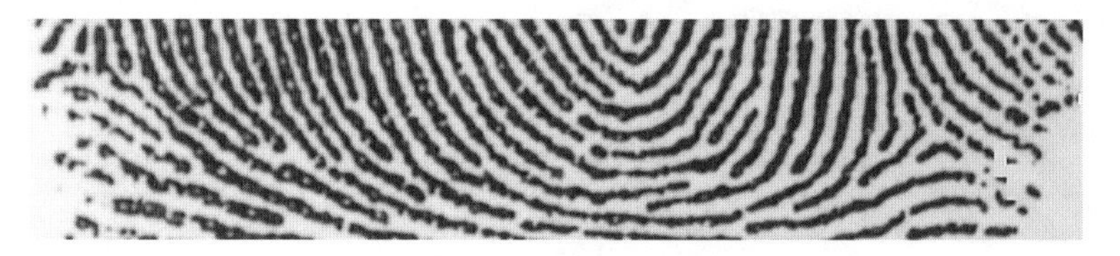

Figure 6.20 Another scanned portion of the same finger.

optical scanner (where the user puts only a portion of the finger on) due to space saving and manufacturing cost reduction versus the ability to uniquely identify an individual person. When we look at Figure 6.19, the impression in (a) was originally stored as the reference. When the same finger was scanned again at a later time in (b); with reference to the stored impression, the finger was placed a little further to the right, and a somewhat below. So, only a fraction of the image in (b) is identical to that of (a). As the stored reference does not contain the entire finger's impression, the authentication algorithm needs to extract the certain portion of the scanned fingerprint image in order to make a comparison to the reference. In this particular example, there is a sufficiently large portion that "overlaps," so that authentication is successfully performed when comparing the two images. There may also be circumstances where alignment is so far off that the scanned portion would appear too different, such as that scanned in Figure 6.20, obtained from a narrow portion of a fingerprint scanner which shows a lower portion of the same finger. Almost the entire scanned image is out of range relative to the stored reference. This kind of authentication problem can potentially be fixed by using a sufficiently large scanner such as the one shown in Figure 6.21. This particular fingerprint scanner is generally far more efficient than those found in consumer grade smartphones. Note, incidentally, that the presence of dirt and surface scratches on this sample can also severely impact authentication performance.

6.5.2 Palmprint Recognition

In addition to palmprint patterns that exhibit certain similarities with fingerprints, veins inside the palm also have unique patterns that can distinguish an individual person. Radiation of near-infrared rays can construct an image with the differing absorption rate by deoxidized hemoglobin in the palm vein. So, the veins will appear as black lines, owing to less reflection of the rays. The main advantage of scanning the vein pattern is that the user does not have to touch the scanner during the image capturing process, making the scanning process faster.

The earliest recorded case of printing human hands and feet impressions was during the pyramid construction era in Egypt some 4000 years ago. Long before this, a small portion of palmprint has been reportedly found in Egypt that dates back 10 000 years with an impression on hardened mud. In the modern world, there are situations where palmprint scanning is more convenient than

Figure 6.21 An industrial grade fingerprint scanner with dirt and surface scratches.

using fingerprints, particularly in telemedicine applications where the entire hand is involved in operating a system as intervention by the user is minimal. Palmprint is more appropriate in situations such as robotic surgery involving haptic sensing, where the surgeon can be identified while performing an operation. Also, in telecare systems where older patients move around, logging of exit and entry by placing the entire hand on the scanner would be faster and easily than to align one finger on a fingerprint scanner.

Identification of a palmprint is usually accomplished by local feature extraction through a voting scheme that combines a set of fuzzy k-nearest neighbor (k-NN) classifiers (Hennings-Yeomans et al. 2007). Image preprocessing is performed with global histogram equalization for the scanned image of size $M \times N$ with G gray levels and cumulative histogram $H(g)$, whose transfer function is:

$$T(g) = \frac{(G-1) \times H(g)}{M \times N} \tag{6.1}$$

The local histogram equalization is then applied by cropping the image starting in the upper-left corner having a predetermined window size, followed by applying the histogram equalization function to the cropped image. The same process is then repeated by moving the crop all over the image and for each one applying the equalization. This mathematical description of the scanning process may sound fairly complicated, but the process itself is actually quite simple: first, the palm impression is scanned to generate a monochrome image of the palm and certain features throughout the palm are identified and extracted, then a certain portion of size $M \times N$ number of pixels is extracted and represented by varying levels of gray shades. A repetitive equalization process is performed progressively throughout the image, starting from the top left corner of the impression.

Similar to fingerprint authentication, the user to be authenticated has a digital image of the palm captured from a scanner stored as a reference. The scanner resolution must be high enough so that palm lines are detectable in subsequent processing stages and image analysis. Also, the background of the palm image should be as plain as possible without any pattern that may be misread as palm lines for the convenience of edge detection. Prior to the authentication step, the background information should be removed by the image processing algorithm.

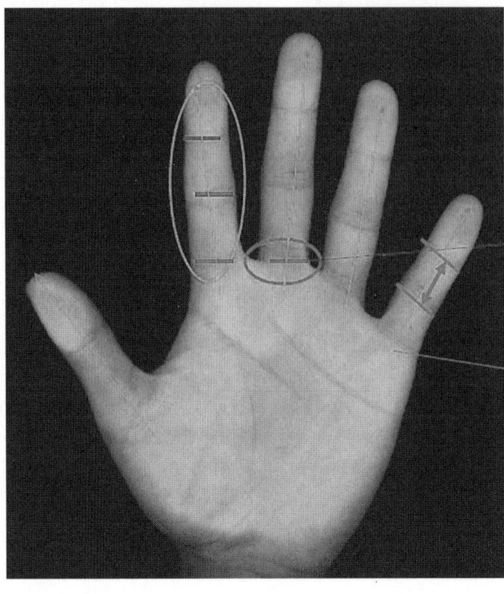

Figure 6.22 Palmprint scanning.

The process first involves palm alignment. To the human eye identifying where the palm is may sound easy. Machinery whose optical sensors attempt to identify the palm must first identify where the palm is placed. This involves removal of any background information that does not represent any part of the palm. Having removed the background, the position of the palm will be identified. Once the position of the palm is identified, subimages of the palm center and fingers can be transformed from the original palm image using the coordinate transformation technique: palm partition. Since finger length and length of parts of a finger are of great importance in finger analysis, a method of finding creases separating the finger into three parts is shown in Figure 6.22.

When data is sent across a telemedicine network, there will be tradeoff between power usage versus palmprint recognition. This means we must look to evaluate the percentage of power necessary to achieve an acceptable rate of scan so that the scanner's power consumption can be minimized while maintaining adequate performance. While hand geometry is also adopted for user authorization without the necessity of special equipment (Klonowski et al. 2018), it is less secure compared to analyzing palmprint.

6.5.3 Iris and Retina Recognition

Eye scan depends on the unique pattern of blood vessels. Retina scanning is accomplished by shining a low intensity light beam into the eye and the reflected light generates a pattern of the "capillary" (the minute connections between arteries and veins) in the retina. The blood vessels on the retina absorb more light than the surrounding tissues. This allows a monochrome image of different gray shades representing the darker blood vessels and the brighter background to be formed. A scanner is normally placed around one to two centimeters (or 0.5 in.) away from the eye. Light that reflects back from the retina will form an image, as shown in Figure 6.23. The image, captured by the retina scanner shown in Figure 6.24 exhibits a unique pattern of lines.

As blood vessels run within an eye in a 3D space, a scanned image that effectively produces a 2D representation may lose certain details that compose the overall impression of the eye. Such loss in

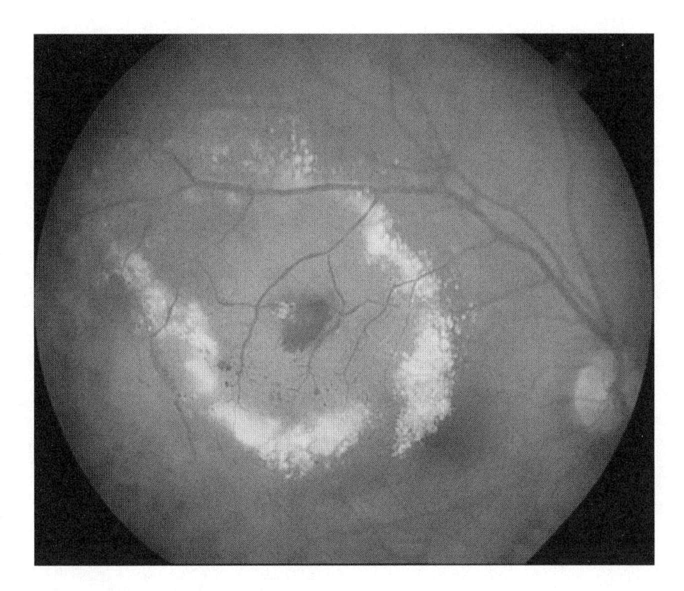

Figure 6.23 Image of the retina.

Figure 6.24 Retina scanner with a built-in low-energy infrared light source.

image quality is unlikely to be as serious as any changes due to diseases such as cataracts, diabetes, and glaucoma. Owing to the high resolution details necessary for identification, retina scanning has only became popular over the past couple of decades when affordable 3D scanning became possible despite its being known about as early as the 1930s (Simon and Goldstein 1935). Although retina scanning is traditionally used by government agencies for authentication and more recently expanded to civilian use, one of its important medical uses is early disease detection as potentially fatal illnesses such as chicken pox, malaria, and sickle cell anemia as these conditions affect the eye even during the very early stages of prevalence. Similarly, chronic diseases like atherosclerosis also affect the eye at an early stage.

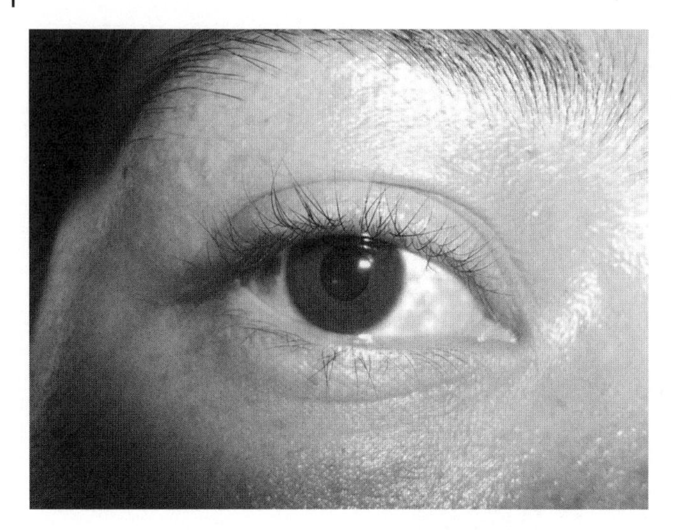

Figure 6.25 Contact lens does not affect the performance of retina scanning.

Another ocular scanning technique is iris scanning. Instead of using veins, iris scanning uses its intricate structure, which is widely accepted as a unique feature of the eye. One major convenience is that its effectiveness is not affected by glasses or contact lenses. The subject shown in Figure 6.25 wears a contact lens and the iris features of her eye can still be clearly observed. With reference to a scanned photographic image of the subject's eye, it operates by first extracting the boundaries of the iris and the pupil. This identifies the portion of the image that forms part of the iris. Image recognition is performed with part of an eye because the iris is normally partially covered by eyelids. Therefore, it would be impractical to make use of an image of the entire eye. Less restrictive than retinal scan where the subject needs to align an eye very close to the scanner, iris scanning can normally be performed around 1 m (or 3 ft) away, so the subject does not have to sit right in front of the scanner.

Iris scanning is becoming increasingly popular as the related optical technologies become more mature. The UK Border Agency started running an iris recognition immigration system (IRIS) at major international airports in England a decade ago, so that registered persons can enter the UK by looking at installed cameras while walking through the immigration arrival hall.

Eye scanning, although widely considered as a more accurate method than using fingerprints, does have a major problem as it is considered unsuitable for frequent scanning. Prolonged exposure to light emitted by the scanner can be harmful to the eye, hence frequent use is not recommended. Although both techniques entail scanning of typically around 10–20 seconds, iris scanning would be less harmful, owing to the distance separation between the eye and the scanner.

6.5.4 Facial Recognition

This is not to be confused with the face detection function on many modern digital cameras, where algorithms are designed to recognize certain areas within a scene as a human face. Facial recognition uses certain features of an individual person for unique identification using techniques in computer vision and image processing. Unlike the three methods described above that principally involve comparison with a reference still image stored on record, facial recognition also works with video (a collection of constantly changing images at a certain frame rate per second). Recent developments in facial recognition algorithms allow 3D representations of the shape of a face, such that contours of eyes, nose, lips, and chin can be expressed in digital format. While most of the methods described above use some kind of optical sensors that require the subject to remain stationary for a short period of time, expression recognition can be accomplished from different angles with 3D (Bonsor and Johnson 2019). In addition to face shape, skin texture with details such as patterns, and unique visual features can also be used together for more relevant features to describe a subject's face.

This technique is already used by the US Department of State, where photographs for entry visa applications are stored for the facial recognition of each applicant. Among some half a million closed-circuit televisions (CCTVs) installed around London to help combat criminal activity and meet CCTV legal requirements (Webster 2019), the City of London Police introduced their Public Space Surveillance Camera System, where the control room monitors 100 public space surveillance cameras across the City of London, with the capability of moving 360 degrees. Using facial recognition, individual persons that enter the image captured by a CCTV can be identified. It is reported that the average person living in London will be recorded on camera 300 times on a single day and there have been debates on privacy versus crime reduction and tackling antisocial behavior (Lapsley and Segato 2019).

Following the first textbook exclusively on facial recognition by Hallinan et al. (1999), several volumes on facial recognition were published in the early 2000s, covering a wide range of topics related to recognition from facial image and the effects of demographic factors. There are also suggestions that the subject's gender may also contribute to the effectiveness of facial recognition. In recent years, facial recognition has been featured in many smartphones where a user can be authenticated through the front camera (Thavalengal et al. 2016).

6.5.5 Voice Recognition

Voice recognition is the identification of the vocal modalities by analyzing the sounds a person makes when speaking. Note, incidentally, that voice recognition is something different from speech recognition, which only deals with translating what someone speaks to a computer into either a command or a string of text that corresponds to the meaning of what the persons says, which has nothing to do with the scope of biometric security and identification. In voice recognition, acoustic analysis is based on software that extracts the unique biological factors associated with someone's voice as picked up from a microphone, which usually consists of an input with additive ambient noise. Similar to fingerprints, a person's voice is digitized as a "voiceprint," such as that shown in Figure 6.26, which represents a chart of the digital representation of the voice from the microphone as a function of time. During the process of performing identification through a voice generated from the vibration of vocal folds, as the person speaks through the microphone the digitized input audio sound is compared against the pre-recorded reference stored in a database. This is, again, similar to the case of fingerprint or palmprint identification. In voice identification, variations in vocal cavities and movement of mouth while speaking establishes the uniqueness of a person's voice (Rilliard et al. 2018).

As a person speaks via a microphone, the frequency components of the voice (bottom) are compared against a pre-recorded reference sound (top) from the same person. In addition to applying voice recognition in biometric security, it is possible to use the same principle for the diagnosis of respiratory system diseases (Priftis et al. 2018).

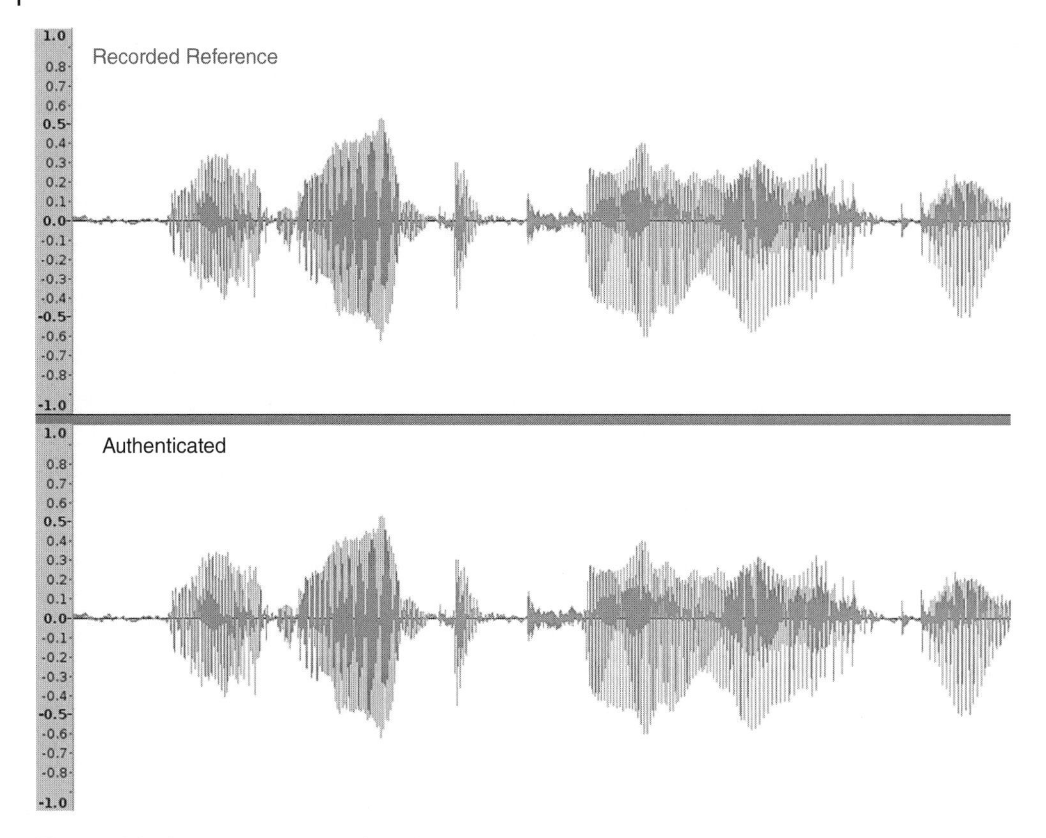

Figure 6.26 Sound spectrogram of a voiceprint.

6.6 Conclusion

This chapter has discussed a number of areas in information technology where patient information can be safeguarded. We conclude by taking a look at Figure 6.27, where a generalized block diagram shows how various biometric authentication techniques can be linked up to a telemedicine network for the identification for user access as well as telecare applications.

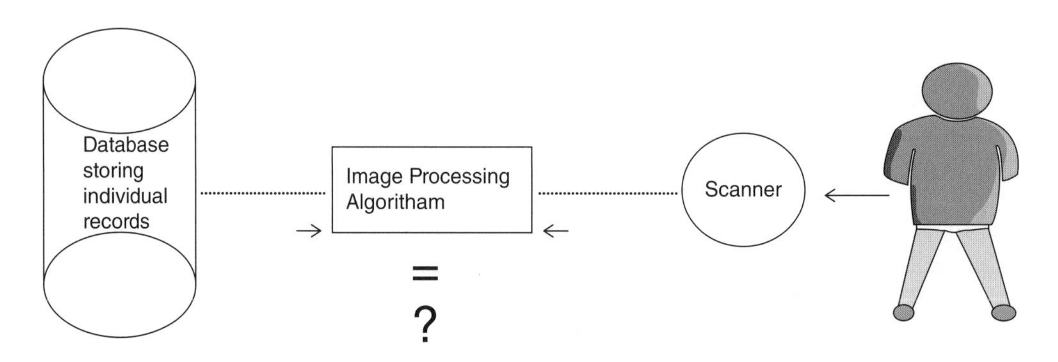

Figure 6.27 Framework for user identification over a telemedicine network.

References

Barron, H.K. et al. (2007). *The Electronic Patient Record: House of Commons Health Committee- Sixth Report of Session 2006–07: Vol. 1*. London: The Stationery Office.

Besag, J. and Newell, J. (1991). The detection of clusters in rare diseases. *Journal of the Royal Statistical Society: Series A (Statistics in Society)* 154 (1): 143–155.

Bonsor, K. and Johnson, R. (2019). How facial recognition systems work. *HowStuffWorks*. http://electronics.howstuffworks.com/gadgets/high-tech-gadgets/facial-recognition.htm (accessed 20 January 2020).

Butler, J.C., Cohen, M.L., Friedman, C.R. et al. (2002). Collaboration between public health and law enforcement: new paradigms and partnerships for bioterrorism planning and response. *Emerging Infectious Diseases* 8 (10): 1152.

Cheng, L., Leung, T., and Wong, Y. (2009). *Financial Planning & Wealth Management: An International Perspective*. McGraw-Hill.

Clinton, W.J. (1999). *Remarks by the President on Keeping America Secure for the 21st Century*. Washington, DC: National Academy of Sciences.

Cohen, A. (2005). Software is too buggy and unreliable, *PC Magazine* (3 August). http://eusesconsortium.org/docs/PCMagazine-aug03-2005.pdf (accessed 20 January 2020).

Dirkmaat, D., Garvin, H., and Cabo, L.L. (2018). Forensic anthropology. In: *The International Encyclopedia of Biological Anthropology* (ed. W. Trevathan), 1–17. Wiley.

Greenhalgh, T., Wood, G.W., Bratan, T. et al. (2008). Patients' attitudes to the summary care record and HealthSpace: qualitative study. *British Medical Journal* 336: 1290–1295.

Hallinan, P.W., Gordon, G., Yuille, A.L. et al. (1999). *Two- and Three-Dimensional Patterns of the Face*. A.K. Peters Ltd.

He, M.L., Zheng, B., Peng, Y. et al. (2003). Inhibition of SARS-associated coronavirus infection and replication by RNA interference. *JAMA* 290 (20): 2665–2666.

Hennings-Yeomans, P.H., Kumar, B.V., and Savvides, M. (2007). Palmprint classification using multiple advanced correlation filters and palm-specific segmentation. *IEEE Transactions on Information Forensics and Security* 2 (3): 613–622.

Higgins, R.R., Eshaghi, A., Burton, L. et al. (2009). Differential patterns of amantadine-resistance in influenza A (H3N2) and (H1N1) isolates in Toronto, Canada. *Journal of Clinical Virology* 44 (1): 91–93.

Hung, H., Wong, Y.H., and Cho, V. (2009). *Ubiquitous Commerce for Creating the Personalized Marketplace: Concepts for Next Generation Adoption*. IGI Global https://www.igi-global.com/book/ubiquitous-commerce-creating-personalized-marketplace/1011 (accessed 20 January 2020).

Klonowski, M., Plata, M., and Syga, P. (2018). User authorization based on hand geometry without special equipment. *Pattern Recognition* 73: 189–201.

Lapsley, I. and Segato, F. (2019). Citizens, technology and the NPM movement. *Public Money & Management* 39 (8): 553–559.

Lee, H.C. and Gaensslen, R.E. (eds.) (2001). Methods of latent fingerprint development. In: *Advances in Fingerprint Technology*, vol. 2, 105–176. Boca Raton, FL: CRC Press.

Lycett, S.J., Duchatel, F., and Digard, P. (2019). A brief history of bird flu. *Philosophical Transactions of the Royal Society B* 374 (1775): 20180257.

Lysne, O. (2018). *The Huawei and Snowden Questions: Can Electronic Equipment from Untrusted Vendors be Verified? Can an Untrusted Vendor Build Trust Into Electronic Equipment?* Springer.

McClaughry, R.W. (1896). *Signaletic Instructions: Including the Theory and Practice of Anthropometrical Identification*. Kessinger Publishing.

Mitka, M. (2005). 1918 killer flu virus reconstructed, may help prevent future outbreaks. *JAMA* 294 (19): 2416–2419.

Pasma, T. (2008). Spatial epidemiology of an H3N2 swine influenza outbreak. *Canadian Veterinary Journal* 49 (2): 167.

Penrose, L.S. (1968). Medical significance of finger-prints and related phenomena. *British Medical Journal* 2 (5601): 321.

Priftis, K.N., Hadjileontiadis, L.J., and Everard, M.L. (2018). *Breath Sounds: From Basic Science to Clinical Practice*. Springer.

Purkyně, J.E. (1823). *Commentatio de examine physiologico organi visus et systematis cutanei*. Prussia: University of Breslau https://ci.nii.ac.jp/naid/10031047394/ (accessed 20 January 2020).

Queiroz, L.P., Rodrigues, F.C.M., Gomes, J.P.P. et al. (2016). A fault detection method for hard disk drives based on mixture of Gaussians and nonparametric statistics. *IEEE Transactions on Industrial Informatics* 13 (2): 542–550.

Rilliard, A., d'Alessandro, C., and Evrard, M. (2018). Paradigmatic variation of vowels in expressive speech: Acoustic description and dimensional analysis. *Journal of the Acoustical Society of America* 143 (1): 109–122.

Rindfleisch, T.C. (1997). Privacy, information technology, and health care. *Communications of the ACM* 40 (8): 92–100.

Rivest, R.L., Shamir, A., and Adleman, L. (1978). A method for obtaining digital signatures and public-key cryptosystems. *Communications of the ACM* 21 (2): 120–126.

Schneier, B. (2007). *Schneier's Cryptography Classics Library: Applied Cryptography, Secrets and Lies, and Practical Cryptography*. Wiley.

Sileo, J.D. (2005). *Stolen Lives: Identity Theft Prevention Made Simple*. Da Vinci Publications.

Simon, C. and Goldstein, I. (1935). A new scientific method of identification. *New York State Journal of Medicine* 85 (7): 342–343.

Spaink, K. (2005). Hacking health: Electronic patient records in the Netherlands. 22nd Chaos Communication Congress, Berlin, Germany (27–30 December 2005).

Spencer, K., Sanders, C., Whitley, E.A. et al. (2016). Patient perspectives on sharing anonymized personal health data using a digital system for dynamic consent and research feedback: a qualitative study. *Journal of Medical Internet Research* 18 (4): e66.

Stertz, S., Duprex, W.P., and Harris, M. (2018). A novel mutation in the neuraminidase gene of the 2009 pandemic H1N1 influenza A virus confers multidrug resistance. *Journal of General Virology* 99 (3): 275–276.

Thacker, S.B., Choi, K., and Brachman, P.S. (1983). The surveillance of infectious diseases. *JAMA* 249 (9): 81–85.

Thavalengal, S., Nedelcu, T., Bigioi, P., and Corcoran, P. (2016). Iris liveness detection for next generation smartphones. *IEEE Transactions on Consumer Electronics* 62 (2): 95–102.

Ushirogawa, H., Naito, T., Tokunaga, H. et al. (2016). Re-emergence of H3N2 strains carrying potential neutralizing mutations at the N-linked glycosylation site at the hemagglutinin head, post the 2009 H1N1 pandemic. *BMC Infectious Diseases* 16 (1): 380.

Viboud, C., Bjørnstad, O.N., Smith, D.L. et al. (2006). Synchrony, waves, and spatial hierarchies in the spread of influenza. *Science* 312 (5772): 447–451.

Weaver, M. (2009). Quarantine brings love in the time of swine flu at Hong Kong hotel, *The Guardian* (8 May). https://www.theguardian.com/world/2009/may/08/swine-flu-hotel-hong-kong (accessed 20 January 2020).

Webster, W. (2019). Surveillance cameras will soon be unrecognisable: time for an urgent public conversation. https://www.storre.stir.ac.uk/retrieve:/ab777844-8edc-4e2d-98dc-a38ce2ff948f/Webster-Conversation-2019.pdf (accessed 20 January 2020).

Weitzman, E.R., Kaci, L., and Mandl, K.D. (2009). Acceptability of a personally controlled health record in a community-based setting: implications for policy and design. *Journal of Medical Internet Research* 11 (2): e14.

Wong, W.H. (2005). Timing attacks on RSA: revealing your secrets through the fourth dimension. *ACM Crossroads* 11 (3): 5.

Yan, P. and Clayton, M.K. (2006). A cluster model for space–time disease counts. *Statistics in Medicine* 25 (5): 867–881.

Yuan, E. (2009). 1 swine flu case leads to 340 quarantines in Hong Kong, *CNN* (4 May). http://edition.cnn.com/2009/WORLD/asiapcf/05/04/hk.flu.hotel/index.html (accessed 20 January 2020).

7

Information Technology in Alternative Medicine

The wireless healthcare market was valued at $73.97 billion in 2019 and is expected to reach $316.0 billion by 2025, at a compound annual growth rate of 27.38% over the forecast period 2020–2025 (Mordor Intelligence 2020). Such growth in demand is fueled by the combined effect of telemedicine technology advances and increase in general health awareness. Consumer healthcare technology and alternative medicine will certainly become increasingly important in this regard. Alternative medicine is defined by (Bratman 1997) as healing practice that "does not fall within the realm of conventional medicine." Examples such as acupuncture, biofeedback, herbal medicine, hypnosis, and yoga are collectively referred to as complementary and alternative medicine (CAM). It is not difficult to realize that they have one thing in common: they do not entail drug prescription. Most of what we currently use for wireless home healthcare and maintaining fitness matches the description of alternative medicine. Many of these provide us with information about our health state and possibly with suggestions of how we can improve our own health. However, almost all of these consumer health monitoring devices provide us with a direct link to medication. In the consumer electronics market, there are many healthcare-related products, literately covering from head to toe. Some are said to improve a user's health and metabolism, while others claim to keep a user in optimal shape.

In the official website of the US National Center for CAM, it classifies traditional Chinese medicine (TCM) as part of CAM. This leads to the concept of blending TCM practices such as acupressure and herbal medicine with consumer healthcare technology. With over 5000 years of history, the regulation of various aspects of the human body can help with both prevention and treatment (Yuan and Lin 2000). Given the diverse range of benefits CAM/TCM provides, technology that complements CAM practices would certainly be of significant value to the general public. As outlined above, TCM is only a subset of what CAM has to offer. The enormous business opportunities related to CAM warrant thorough study into how various aspects of information technology and telemedicine can make CAM more economically viable. This chapter aims to explore how some popular CAM remedies can be provided by consumer healthcare products and where technology comes into improving practices that have been around for millennia. Given its vast coverage, our intention is not to go into the details of CAM. Instead, we seek to give a brief introduction to several mainstream topics inside the vast CAM market. According to a new report by Grand View Research, the CAM market is expected to generate revenues of $210 billion by 2026 (Grand View Research 2019). Business opportunities in CAM applications in the West are certainly way behind countries across the Far East, where CAM is widely practiced often in conjunction with mainstream medicine.

Telemedicine Technologies: Information Technologies in Medicine and Digital Health, Second Edition.
Bernard Fong, A.C.M. Fong, and C.K. Li.
© 2020 John Wiley & Sons Ltd. Published 2020 by John Wiley & Sons Ltd.

7.1 Technology for Natural Healing and Preventive Care

In unrecorded prehistoric medicine, it is generally believed that plants have long been used as healing agents on a trial and error basis. In his translation of the famous *Herodotus* Rawlinson (1956) describes a public health system that was supported by the practice of medicine. In addition to shamanism, ancient Egyptian medicine also made use of clinical diagnostics and anatomy (Nunn 2002). Babylonian also introduced diagnosis, prognosis, physical examination, and medical prescriptions (Horstmanshoff et al. 2004), leading to the evolution of modern medicine. While these focus primarily on targeting specific symptoms, ancient Chinese medicine paid more attention to the general health state and well-being of the human body with empirical observations of behavior forming the basis of TCM (Veith 1972). As an alternative medicine covering regulation of the entire body from head to toe, TCM practices mainly rely on herbal medicine, acupuncture, and dietary treatment. TCM is said to rely on thorough observation of both the human body and nature (Unschuld 2003).

"Biofeedback" refers to a collection of methods that relieve stress-related symptoms, and phobias. Electronic monitors are used to gauge a patient's response by altering the output signals and physiological signs and/or environmental conditions. By increasing one's awareness of physiological activity in their muscles, one can be trained to control what are otherwise natural physical responses to tension and stress, such as heartbeat, blood pressure, and breathing. The use of biofeedback intervention for treating high blood pressure was established in clinical trials in the 1990s (Nakao et al. 1997).

7.1.1 Acupuncture and Acupressure

Acupuncture is perhaps the most popular TCM practice widely accepted throughout the world. Relying on small areas across the anatomy associated to specific organs or parts of the body, known as "acupoints," or *tsubo* from the Japanese, there are hundreds of these acupoints scattered across the body with varying healing properties and effectiveness (Lu and Needham 1980). Acupoints distributed throughout the body, as shown in Figure 7.1, are linked to different organs or parts of the body. It is not necessarily true that an acupoint is located close to the respective organ. For example, application of a needle to an acupoint on the foot is said to be effective for relieving digestion problems. Given that so many acupoints are interconnected, its complexity makes a 2D chart such as Figure 7.1 extremely confusing and difficult to read. It is given for the sole purpose of illustrating the chart's sophistication rather than giving any useful information about individual acupoints. Both acupressure and acupuncture rely on the flow of energy across the body through "meridians" (Maa et al. 2003). Each meridian is a circuit that links some point on the external body to an internal organ with some kind of related physiological functions.

The aim of this book is not to discuss acupuncture/acupressure practice or properties of individual acupoints but to explore how telemedicine and related technologies can be applied to assist such practices. For this purpose, let us briefly look at some general relationships between acupoints and the human body before going into technology applications. Although we do not intend to discuss any details of these practices, it is worth noting that both acupuncture and acupressure use the same points for healing. The former relies on the insertion of fine needles, while the latter is a noninvasive method that is stimulated with pressure exerted by a finger. In the context of our discussion on technologies that support such practices, we make no distinction between the two.

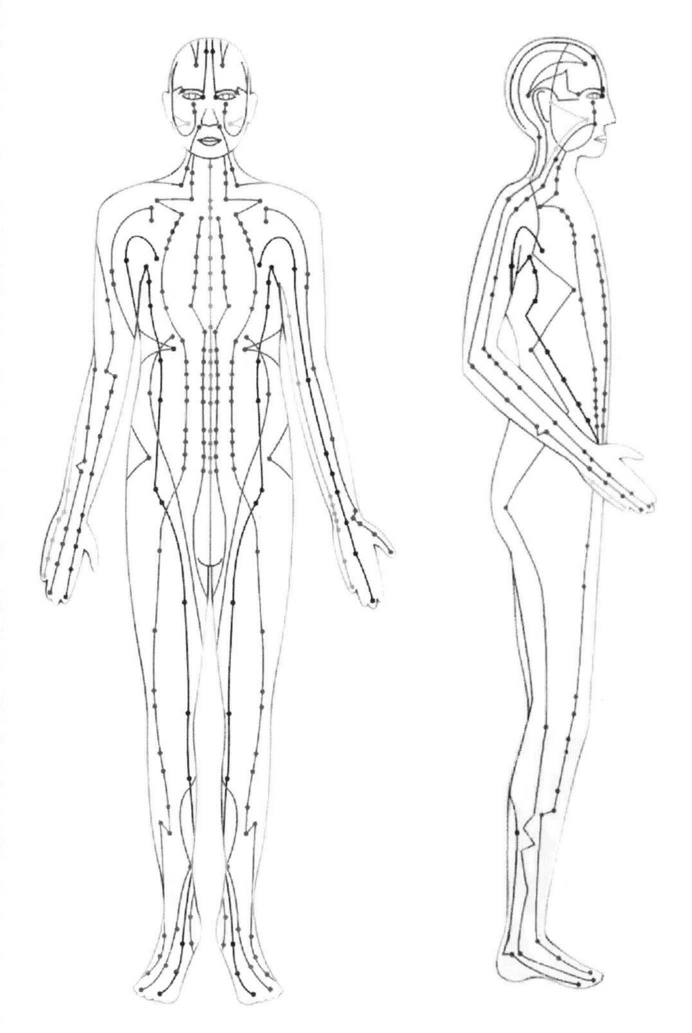

Figure 7.1 2D acupoint chart.

Any efforts that maintain a person's health should commence with the immune system. The concept of linking lifestyle and living longer healthily results in attention to self-care (Barrett 1993). Its importance has gained increasing attention over recent years, owing to the combining effect of aging population, change in lifestyle, and stress induced by irregular work schedules among many people (Marshall and Altpeter 2005). Much of CAM emphasizes balance, as in optimally balancing between various attributes for making the most of one's metabolism. The process of metabolism determines the rate at which food is digested hence calories are burnt. The basic idea is to maximize the efficiency of the body by strengthening the immune system. Stress from various daily activities causes shoulder and neck tension as well as anxieties that affect digestion. Adding to the overall picture of an "unhealthy lifestyle" are insufficient rest and exercise that tend to make a person feel tired after a day's work. Many appliances are commercially available to provide some form of relief that partially addresses the problems caused, as we discuss in Section 7.4. To understand how technologies help, we explore the effects on the immune system due to excessive overstressing of certain acupressure meridian pathways. This is often caused by prolonged or repetitive activities, and by commonly encountered occupational hazards such as

- Continuous computer monitor usage: results in emotional imbalances that affect the small intestine; relieved with an acupoint located at the center of the breastbone.
- Prolonged sitting on a chair: results in anemia, and digestive and stomach problems, relieved with an acupoint on the leg slightly below the knee.
- Excessive standing: backaches and fatigue, also causes bladder and kidney problems; various acupoints on the upper chest below the collarbone, along both sides of the spine at the lower back, and as far down as the insides of the ankles, can relieve the tension caused.
- Physical exertion: causes cramps and spasms that can eventually lead to liver failure, an acupoint on top of the feet will ease the problems caused.

The above are among countless examples where acupressure can improve a person's well-being. They are realized by applying firm and steady pressure on the appropriate acupoints. This can be done on a daily basis just like a regular workout (Gerogianni et al. 2019).

7.1.2 Body Contour and Acupoints

To facilitate the application of acupressure, any automated system needs to identify each apposite acupoint for a specific purpose. A reference chart only provides some indication on roughly where an acupoint is. Physically locating it is far more difficult for an inexperienced person, and even more difficult for machines. Individual body size and shape can be very different. For example, the location of a specific point on a 5′ tall thin person can be very different from that on a 6′ tall fat person. What a human eye perceives can be very different from what a machine perceives. Further, the body contour can vary significantly from person to person. Some appliances, such as a massage chair, automatically search for the approximate location of acupoints (pinpointing would be virtually impossible, owing to the precision involved) by first scanning across the user's back to obtain the positions of the neck and spine. Several reference points can be established by visually aligning the anterior superior iliac spine (ASIS) and posterior superior iliac spine (PSIS) (Figure 7.2). Indeed, this figure can be used to construct the user's body profile as shown in Figure 7.3.

How a machine sees things is governed by "computer vision" technology. The term "computer," incidentally, refers to any computation machinery here, from simple consumer electronics to sophisticated high precision medical image scanners. Computer vision is about identifying and

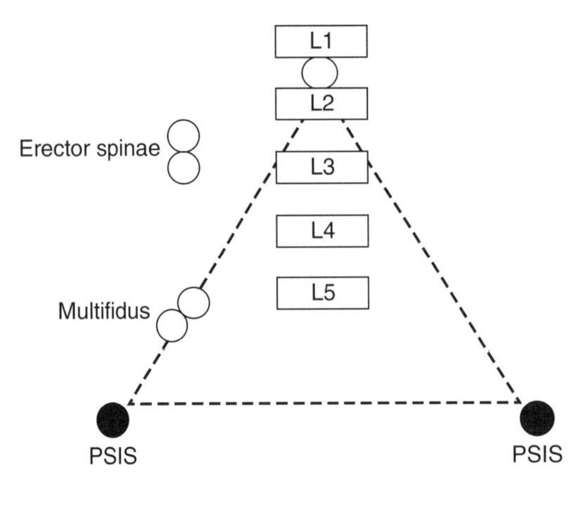

Figure 7.2 Reference points with reference to anterior superior iliac spine (ASIS) and posterior superior iliac spine (PSIS).

Figure 7.3 Body profile.

extracting information from digital images through learning and object recognition. To begin, how can a computer differentiate a human body from the background? The human body does have some kind of generic shapes, but they can differ significantly, as shown by the three examples in Figure 7.4. All three seem very obvious to us that they are images of a person standing. To us, we can easily tell that the left and center features the same person, and the image on the right is the sketched figure of that in Figure 7.1. However, a computer visualizes things very differently. A computer relies on algorithms that extract features, objectives, and any specific activities.

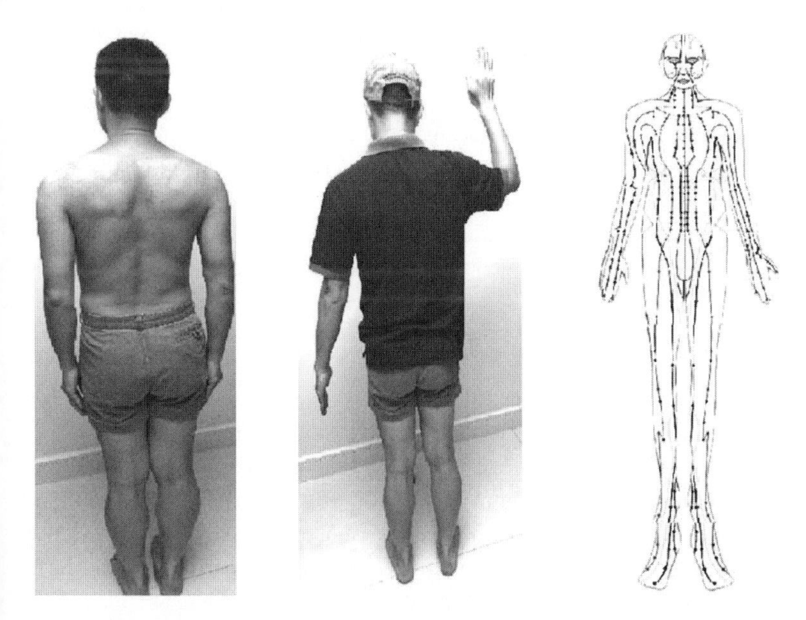

Figure 7.4 Three human pictures not easily recognized by computational algorithms as factors such as posture and clothing need to be analyzed.

The body shape, perceived as a 3D image to the machine vision algorithm, can be manipulated by pattern recognition and feature extraction mechanisms, such as those described by (Ezquerra and Mullick 1996). Given the number of options available, the transmission efficiency should be thoroughly considered particularly for users who move around making detection even more difficult. Feature extraction is usually necessary when the image is too large so that any information not related to the user's body is extracted and removed, leaving behind only relevant information for analysis.

The remaining data contains a description of the body's curvature, which conveys information such as edge direction and shape information (Hoshiai et al. 2009). The information is then mapped to a generic body contour profile. Apart from imaging methods that reassemble what an eye sees, other methods such as sensors that press against the user so that the distance of various points can be measured against the relative position to a flat plane. This can produce a set of information about the body shape.

So how do these technologies benefit the medical and healthcare industry? Various consumer appliances for general fitness and well-being discussed later in this chapter use these to apply certain therapies to specific areas of the user's body. Another major application found throughout the world is the removal of excess fatty deposits under the skin for body trimming in cosmetic surgery. Many are willing to spend hundreds or even thousands of dollars on removing a few pounds of bodyweight. Ultrasonic and laser liposuction have been used for shaping body contour by tightening the skin surface and removing fat. The former injects a saline fat loosening fluid into the area concerned and the ultrasonic wave provides energy for melting the body fat. Ultrasound is also used in "vaser liposuction" as a noninvasive surgery using local anesthesia to numb the area before flushing out the fat. Laser liposuction inserts a tiny tube into the area where the excess fatty deposits are located and eradicated by the laser beam. Owing to its highly focused beam, surrounding tissues are not affected. A laser is mainly used for areas not as accessible such as the back of an arm or the inner thighs. As with most types of surgeries, the patient's medical history should be retrieved prior to operation, since patients with ailments such as high blood pressure, diabetes, or cardiopathy may risk complications in addition to pulmonary fat embolism, perforation of viscous tissue, edema, or swelling.

The acupoint detection process is even more complicated than computing the body contour as acupoints are relatively small and can sometimes be located very close to each other. Apart from mapping graphically based on a reference chart, Liu et al. (2007) describe a method using the electric characteristics of an acupoint's various anatomic layers for detection. Having briefly discussed the technology related to identify body contour and to locate acupoints, we now turn to using acupressure for providing temporary relief for an emergency with reference to (Kausar et al. 2017).

7.1.3 Temporary On-scene Relief Treatment Support

Acupoints have different healing properties. Bock (2009) reports that some acupoints can be accessed as part of general warm up and stretching routines to help the body prepare for training. Acupoints also differ in perceived effectiveness, i.e. some can be readily felt with swift response upon application of force, while others may yield more long term but slower effects. More than one acupoint may be linked to an organ and these points for the same organs may not necessarily be located in close proximity. As such, any effort in providing temporary relief using acupressure requires thorough knowledge of properties for many acupoints.

As an example, we look at a case study on treating sea sickness, a type of motion sickness that is a normal response to movement to the body. Different individuals respond to perceived or actual movement differently (Riccio and Stoffregen 1991). In some cases, the inner ear may sense rolling motions that the eyes do not see. Sometimes symptoms may return momentarily even after the motion stops. Motion sickness can cause anxiety, dizziness, nausea, or vomiting. Although medication can control motion sickness, there are times when someone may not have been prepared for adverse motion, such as unusually rough seas in a normally calm area. Medications for motion sickness such as scopolamine and promethazine are not always suitable, owing to side effects such as blurred vision, drowsiness, and impaired judgment. Although biofeedback and cognitive behavioral therapy are reported to be effective for managing motion sickness (Dobie and May 1994), the former utilizes instruments to record skin temperature and changes in muscle tension and the latter relies on exposure to a stimulus and usually involves a specially designed chair; these are not readily available tools that one can easily access when the need arises. Acupuncture at the P6 or Neiguan point to relieve motion sickness is reported to be effective (Stern et al. 2001). In addition, Barsoum et al. (1990) report that the need for unplanned antiemetic injections could be reduced by acupressure.

Although (Miller and Muth 2004) did not find any concrete scientific evidence of linkage between acupressure and motion sickness, acupressure is still widely used by yachtsmen (Shupak and Gordon 2006). This is an area where telemedicine can be very helpful providing temporary relief treatment support. Offshore support can only be provided by wireless links, as this is the only practical method of obtaining anything instantaneously. Having learned the underlying concepts of telemedicine in the earlier parts of this book, readers should be able to sketch the support block diagram shown in Figure 7.5. Here, we can see that treatment information can be delivered to the yacht through wireless communications. In this particular example we use a satellite link to do the job. So, why should we spend time discussing this example if the system is so simple? To answer this let us conclude this subsection by going deeper into what happens when someone wants to provide offshore acupressure treatment.

In Section 7.1.1 we mention that acupoints serving the same purpose have varying healing properties: some provide swift response, while others may not produce an immediate response. Any attempt to memorize the characteristics of a set of acupoints would be impractical, as this would be analogous to learning an entire pocket dictionary by heart. In practice we need some kind of database that is accessible onboard. Essentially, a database containing information about various acupoints should be searchable and accessible on the Internet. The information received includes

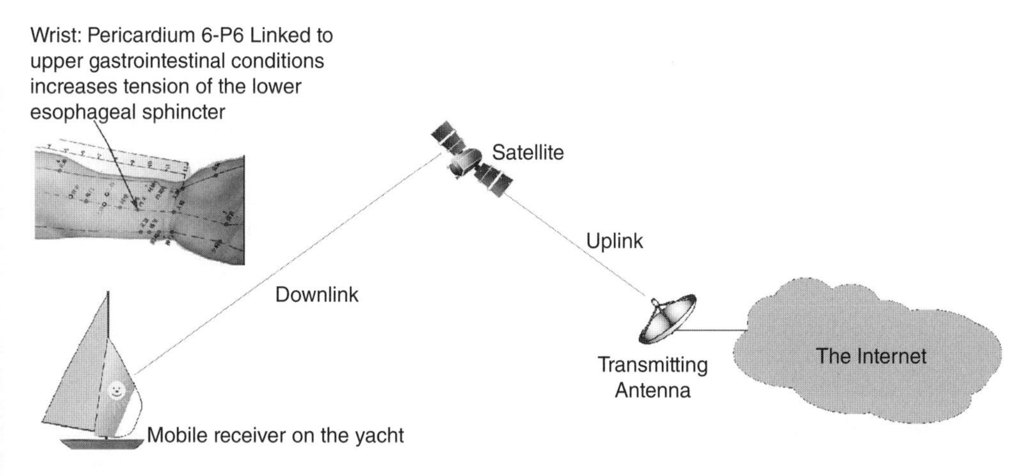

Figure 7.5 Telemedicine for offshore relief support.

location of the appropriate acupoint, where it is linked, and what it relieves, as shown in Figure 7.5. Such information is given to yachtsmen who may not have any practical knowledge on acupressure, but owing to its popularity among the boating community the first time they attempt practicing may well be when a genuine need arises. Information may therefore be delivered in the form of an interactive tutorial, which shows how and where pressure is exerted based on needs. However, the amount of data shown, after compression, is likely to be in the tens of megabytes (MB) for this kind of illustrations. This may not be practical with the inherent delay of satellite communication.

7.1.4 Herbal Medicine

Telemedicine on herbal medicine is arguably one of the most important applications that IT supports. Although it is not necessarily true that telemedicine is widely used in herbal medicine, its origin certainly forms the basis of remote support for modern medical science, as we describe in Section 1.1. The need to deliver medicine away from its source existed thousands of years before the creation of telecommunications.

A herb is a plant, or part of one, that possesses medicinal, aromatic, or savory qualities. Many popular medications used today were developed from ancient healing traditions that provided cures with specific plants (Weiss 2000). The healing components of a given herb are extracted and analyzed for pharmaceutical exploitation. One good example that directly resulted in respiratory treatment came from the use, written about in 2735 BCE, by the Chinese emperor Shen Nong of the bronchodilator "ephedrine," which his servants extracted from the plant *Ephedra sinica* and used as a decongestant. This evolves to pseudoephedrine, as its synthetic form, now applied in many allergy, sinus, and cold-relief medications and mass produced by the pharmaceutical industry. Around two decades ago when CAM became widely practiced, it was reported that the link between herbs and modern pharmacology was as much as 40% of the prescription medicines dispensed in the US contain at least one active ingredient derived from herbs (Wilson 2001). The vast majority of these drugs are either made from plant extracts or synthesized to fabricate a natural plant compound.

The first formal documentation of herbal medicine is probably the translation of the dispensatory entitled *A Physical Directory* by Nicholas Culpeper (*c.*1649). The use of herbs for treatment became popular throughout Europe and beyond since it was written. In the modern world, telemedicine allows information about herbal materials to be gathered from remote forests for analysis of substances that act upon the human body and study for contraindications and any possible side effects. The compositions of plants contain many substances, including vitamins and minerals. One important aspect of technology is to ensure that the amount of intake of any component does not exceed a toxic level that may result in damage rather than benefit to health. According to the Natural Resources Conservation Service of the US Department of Agriculture (NRCS 2009), hundreds of thousands of plant species exist. The identification and subsequent isolation of active ingredients would entail thorough studies of individual plants. The vast numbers of plant species mean only a small number have been studied for their healing properties. Further to active ingredients, synergistic interactions between different components within an individual plant also need to be studied in order to grasp a comprehensive understanding of their medical value. Botanical study will continue to play a significant role in pharmaceutical research in the foreseeable future. Likewise, related telemedicine technologies will be an important part in supporting such research work. This is due to the fact that herbal medicine does not have the same level of acceptance in all countries (Chitturi and Farrell 2000). Cross-border study facilitated by information exchange would certainly accelerate the lengthy process of botanical study.

7.2 Interactive Gaming for Healthcare

Over the past couple of decades, many children and adults alike have become addicted to videogames. A good workout on the videogame console is likely to keep a user there for longer than that on a treadmill. Many videogames make users concentrate on playing them for an extended period of time. This prolonged continual exposure to a computer or TV screen will very likely degrade eyesight and increase the risk of glaucoma, an insidious disease that affects particularly those who are short-sighted. The impact of glaucoma may result in the loss of peripheral vision over time. In addition, (Kasraee et al. 2009) also reported a confirmed case of idiopathic eccrine hidradenitis, a skin disorder that affects the game player's hands. The gripping of a game controller may well contribute to sweating, which causes swollen sweat glands in the palm. Further, the user's posture may also contribute toward acute tendonitis and back pain. The physical symptoms added to the known psychological effects of videogame addiction combine to make videogaming notoriously unhealthy. While eSport serves as a venue for exercising as an alternative venue for variously sport activities, it is worth pointing out that there are negative health effects like eye strain (Sheppard and Wolffsohn 2018). Most display units used in eSport is lit by an array of blue LEDs that are neutralized with a yellow phosphor coating (Fong et al. 2012).

7.2.1 Games and Physical Exercise: eSport

eSport, involving physical interaction between human and computer, usually engages a controller such as a joystick that has been widely used for decades, of which a user presses several buttons in rapid succession. For other different control methods, there are workout activities, strength training with a balance board, and so on (Robertson 2008). The concept of fitness gaming is not something new, several major Japanese videogame manufacturers have already launched a number of workout games over several years (Brandt 2004). Some workout games have features that record calories burned and the distance one would have covered to get the equivalent results during a gaming session. In a small room, one can play a range of simulated sporting activities by standing on a mat with an array of sensors. The player's movement will be relayed to the TV screen via a game console. For example, skiing can be simulated by detecting body lean as the movement and translates to the direction the player is leaning which will then be relayed and displayed on the TV screen, as illustrated in Figure 7.6. The mat is connected to the game console via a wireless link so that there is no risk of tripping over tangled wires while playing.

7.2.2 Monitoring and Optimizing Children's Health

Although some game consoles feature the automatic tracking of parameters related to a user's fitness, such as body mass index (BMI), measuring size of chest, biceps, waist, and thighs, etc. Technology can do far more than this. A comprehensive health monitoring system can be built on the paradigm shown in Figure 7.7 that consists of:

- Gaming Console: basic features include computing the amount of physical exercise completed during school on the day, such as inclusion of scheduled activities sessions. These games will also be categorized into individual and group activities so that children can play alone or together either at their own homes physically separated or gathered together. Unlike most off-the-shelf computer games (where user control is accomplished through joysticks involving movement of very limited parts of the body), these games will be controlled by a small body area network (BAN)

Figure 7.6 Virtual skiing videogame.

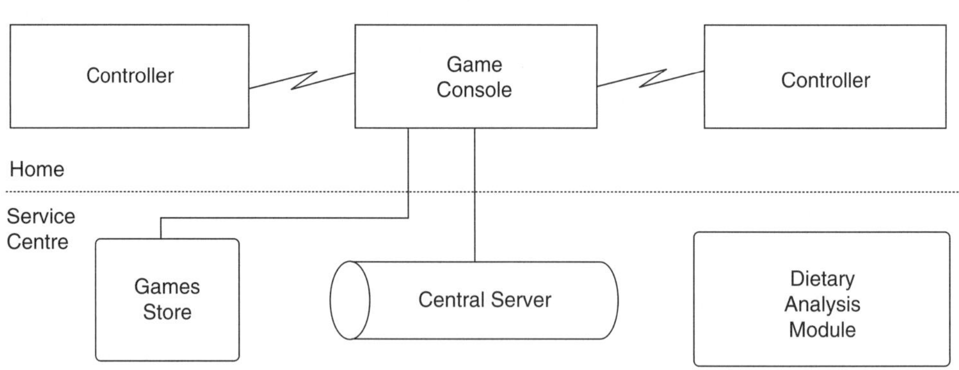

Figure 7.7 eSport gaming system for physical exercise.

installed in various positions on the user's limbs to track their movements, thereby providing their scoring and feedback on the amount of exercise completed. This module mainly consists of three major parts:

- The games (software), supported by appropriate controllers, have to be made suitable for different ages with varying intensities of physical demands.
- Also, graphic design must be tailored for the intended age of users. The games are played by body movement when users undertake physical activities.

   - Artificial intelligence (AI) is part of the underlying technology that determines the most suitable game for an individual user, selected based on user parameters such as age, BMI, time since the previous meal, or whether a physical exercise lesson at school has been scheduled on the day. Whiles games are generally developed for children according to age groups, an option should be provided for boys and girls since children of different genders may prefer games of different themes.
- Controllers: Game control for fitness and workout involves complex detection of movement with various parts of the body in order to serve the primary purpose of promoting physical exercise. A BAN with sensors will be set up for movement tracking. Haptic sensing for touch will also be used for the control console and menu navigation. A wireless receiver captures the data from the BAN. In addition to body sensor networks, video imaging techniques that analyze the user's movement based on object extraction and frame analysis can also be used on a commercial scale. This is not practical for consumer applications because the camera installed and complication in evaluation of amount of force exerted by the limbs during the exercise may be very costly.
- Dietary analysis module: recommended levels of essential nutrient intake. This can be accomplished on the basis of information enlisted by the Food and Nutrition Board of the National Academy of Sciences to be adequate to meet the nutritional needs of a child of a certain age to ensure consumption of adequate amounts of essential nutrients. References such as the Food Guide Pyramid (FGP) can be used as a tool for healthful food choices (Welsh et al. 1992). Key guidelines include not exceeding 30% of total energy intake from fat and getting less than 10% from saturated fats. The FGP for young children (two to six years old) identifies recommended portions of foods from grains (six servings), vegetables (three servings), fruit (two servings), milk (two servings), and meat (two servings), as well as recommending limiting the intake of fats and sweets. The nutrient needs of teens can be determined using the FGP for adults. These will be used as references when calculating the optimal dietary recommendation for a user. Users will be asked to enter what has been provided by school. Without any knowledge of quantities of each type of food, estimation needs to be computed from a list of pre-programmed combinations so as to derive an estimate of the intake from a school meal. This module will be a computer program that takes user input on the contents of a school meal, estimates the amount of each component and therefore the nutrition composition, and thereby produces a list of recommended food for snacks and dinner for optimal health.
- Central Server: in addition to storing all necessary information supporting the above modules, this module accommodates a database serving individual users through the Internet. It provides main administrative and support functions such as information update, games maintenance, and data protection.

7.2.3 Wireless Control Technology

Wireless communication plays a vital role in fitness gaming since users cannot be tangled with wires as they exercise. Controllers may include mats and a variety of handheld controllers. Control can be accomplished by conventional handheld units, such as those used in Figure 7.6, or an array of sensor networks. All these have two fundamental attributes in common: (i) they are all battery powered with an onboard power source, i.e. each unit or sensor is self-contained and (ii) each has its own wireless transmitter, usually implemented in an SoC (system-on-chip) configuration, i.e. control data is transmitted to the console through a wireless link by a single IC (integrated circuit) chip that contains components such as transceiver, data buffer, and filters.

Bluetooth is a good choice for wireless gaming control as discussed earlier in Section 2.2, its properties such as low power and eliminating the need for line of sight certainly make it suitable for game controllers. However, there are certain requirements and limitations with the Bluetooth standard. The videogames industry is so huge that eSports sales are expected to surpass $1.5bn by 2023 (Business Insider 2020). This enormous sales figure can well support the development of proprietary standards for data communication that can be customized exclusively for specific applications, with regulatory compliance being just about the only constraint. This leads a challenge to balancing between transmission power and data throughput. The transmission power should be minimized to prolong battery life while maintaining adequate data throughput for the data collected from movement. As these controllers are battery-operated, the system must be developed to eliminate the risk of sudden loss of control due to battery exhaustion. Rechargeable batteries for consumer electronics are primarily made using nickel cadmium (NiCd), nickel metal hydride (NiMH), or lithium ion (Li-ion) cells.

Before we conclude this section, we take a quick look at different types of batteries that can be used in these wireless controllers. Compared to the oldest NiCd type found in the 1970s, NiMH has about twice the energy density and it does not have any "memory effect," a term meaning a battery that only retains a proportion of its maximum capacity after repeatedly only partially discharging before recharging. In addition to weight and capacity advantages, NiMH batteries are also more environmentally friendly than their NiCd as they do not contain heavy metals and the chemicals used are less toxic. Li-ion, which have become increasingly popular in recent years and are found in most of the latest appliances, produce the same energy as NiMH batteries but weigh one-third less. The physical movement of the controller may be a charging method to recharge the battery while in use. However, each type requires a different charging pattern to be properly recharged such that some may require continuous electric current while others may be bursty with pulses at a certain time interval. The life of a rechargeable battery operating under normal conditions is generally less than 1000 charge–discharge cycles, such that the user will experience a decline in the running time of the battery over time. Mechanisms that provide battery health monitoring would be helpful for determining the health status of the battery (Berecibar et al. 2016). Emerging technology utilizing ruthenium oxide may soon become available, which has many advantages over the three major cells discussed above. This type of high capacity thin-film battery has been the main type of power source found in most wearable consumer and medical devices in recent years.

7.3 Consumer Electronics in Healthcare

As the emergence of next-generation wireless devices provide capabilities for supporting more healthcare services, market growth will certainly continue in the years to come. Consumer healthcare technology is not restricted to linking between manufacturers and consumers. Entities such as government agencies and insurance companies also have direct interests, as the former promote general health awareness to reduce avoidable hospitalization and other healthcare services and the latter will certainly benefit from fewer insurance claims by addressing fitness and well-being.

Telecare products business may be more challenging under which many consumer healthcare products will be wireless enabled, for example, a wide range of healthcare applications are available for mobile phone apps. The availability of existing network infrastructure and wireless connection especially in rural areas would certainly affect sales growth. Also, these services rely on cooperation from vendors such as Internet service providers (ISPs) who may not be willing to guarantee connections for data transmission and availability.

7.3.1 Assortment of Consumer Appliances

A very wide range of consumer electronics products are readily available from stores all around us. These range from small electric toothbrushes for $30 to a giant luxury massage chair with a $10 000 price tag. Literally everything that covers the entire body from head to toe, the selection is so vast that even appliances like hair dryers and shavers are claimed to be healthcare products. Loosely speaking, other devices readily available from a local appliance store – like blood pressure monitors, digital thermometers, massaging apparatuses, and grooming devices – are all healthcare related. Whether these devices are really related to healthcare is not a discussion within our scope. With so many healthcare products around we are unable to cover every single category in a single chapter. We have no intention to provide extensive coverage for each type of product; instead, our primary focus is on the underlying technologies. There are some common attributes for consumer electronics products. First, they are mass produced, meaning that minimizing the manufacturing cost of each unit while maintaining maximum reliability is vitally important. Parts selection therefore plays a vital role in this (Fong and Li 2011). Unlike most consumer electronics appliances, healthcare devices have more direct physical contact with the user. The inherent safety risk is therefore comparably higher. Let's look at a massage chair with the product specifications listed in Table 7.1. What does it have for us here? First, its current can be over 1 A; it must be mains powered. Another potential fire hazard is the of 30 minutes of continuous massaging. This is, in fact, very common among these products. The unit should be left to cool down after half an hour to avoid overheating. Reclining must be clear of all physical obstacles that can result in motor damage. Next comes the bulkiness of the unit: ergonomic design for handling should be considered to minimize the risk of back injury with such large products. Finally, the material used can be damaged by sharp objects, for example a user might sit on the chair with metal objects in their trousers that cut through the upholstery. Any damage to the device may lead to exposure of hazardous parts that could lead to physical injury or even electrocution. Also, many devices are made for operation in the bathroom, which may increase the risk of electric shock. So, running down this seemingly short list reminds us of the many possible risks that need to be carefully considered during the product design stage.

Table 7.1 Product specifications of a massage chair.

Operating Voltage:	**110 ~ 120 V/50–60 Hz**
Power Consumption:	120 W to 300 W/5 W standby
Massaging Rated:	30 min (continuous)
Recline Angle:	115~175°
Massage Stroke Length:	30 in./76 cm
Dimensions:	W33 in. × D45 in. × H49 in. (upright)/83 cm 114 cm × 124 cm
	W33 in. × D75 in. × H30 in. (recline)/83 cm × 190 cm × 76 cm
Net Weight:	190 lb/86 kg
Material:	Fire retardant PVC leather

Although many small devices are battery operated so that the electric shock is not an issue, there are many other risk factors that must be considered, as we discuss next.

7.3.2 Safety and Design Considerations

A baby monitor is probably the best product to use as an example since it is a naturally the very first wireless communication device that a person uses. While the baby does not really know what benefits it brings, the parents can move around the home with the assurance that no activity performed by the baby will be missed. It is a device that brings the baby and parents closer together.

Concern over liability and its litigation is always an issue with healthcare-related products as they are often required to comply with different sets of regulations governing the marketing and sales of such products imposed by different authorities. This is an important challenge that many healthcare device manufacturers face because some countries even have different state and provincial requirements. Many manufacturers may seek local partnership prior to importation to overcome these requirements.

In many metropolitan cities, people are exposed to pollutants, toxins, and electromagnetic fields on a regular basis. We may not be able to do a lot to contain environmental pollutions but attempts to reduce electromagnetic emission by electronic devices can reduce the effect of electromagnetic fields radiated from electronic appliances. Power control and proper device shielding would be effective ways to reduce such impacts.

Electromagnetic interference (EMI) is certainly an important issue both in terms of operational reliability and radiation safety to the baby's healthy growth. Of course, no wireless device can be made EMI-free, but laboratory testing to determine the design and operating characteristics of the device should be carried out. In this regard, imposing a minimum separating distance between the baby and the device can mitigate much of the EMI risk. Active power control that minimizes transmission power output can limit the amount of time the transmitter is active thereby reducing exposure to radiation by the user. The maximum transmission power should be suppressed within the Food & Drug Administration (2009) limit. The proximity also plays an important role in controlling EMI exposure. Figure 7.8 shows the field strength relative to distance. A substantial amount of energy is lost within the first 10–20 cm and drops to only 5% at a distance of about 20 cm from the transmitter. Therefore, placing the transmitter 20 cm or 8 in. away from the baby will drastically

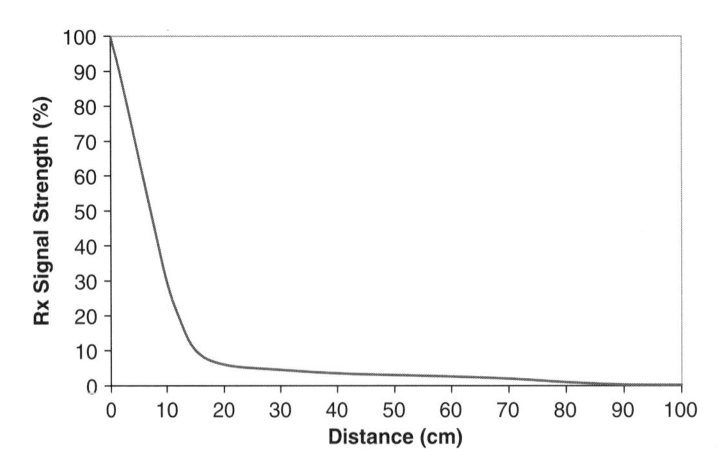

Figure 7.8 Field strength versus distance.

reduce the radiation associated with EMI since the field strength emission decreases rapidly as the distance increases. In Section 2.1.4, we discuss the concept of electromagnetic compatibility (EMC) as opposed to EMI, meaning that the device needs to be compatible with the surrounding EM environment so that it can still operate reliably when subject to interference, in addition to satisfying the emission limits of EMI that may affect the operation of other nearby devices. EMI "shielding" therefore becomes necessary. It is also worth noting that older equipment can degrade over time and become more susceptible to EMI.

A baby's skin is very delicate and sensitive. The choice of material and the housing must be very carefully designed to ensure optimal ergonomics and safety. Battery leakage is also a problem that can lead to serious consequences as toxic chemicals can reach the baby. To keep away from such risk, the battery compartment should be properly sealed in such a way that toxic chemicals would be contained within the monitor in the event of a battery leakage.

7.3.3 Marketing Myths, What Something Claims to Achieve

The idea about myths, perspectives, and inversions of marketing a product is perhaps best described with the following characteristics (Vargo and Lusch 2004):

- Intangibility: lacking the palpable or tactile quality of goods.
- Heterogeneity: the relative inability to standardize the output of services in comparison to goods.
- Inseparability of production and consumption: the simultaneous nature of service production and consumption compared with the sequential nature of production, purchase, and consumption that characterizes physical products.
- Perishability: the relative inability to inventory services as compared to goods.

Marketing is often used as a tool to deal with these (Zeithaml and Bitner 2000). What is said may or may not be true. The effectiveness of a given product may only be tested in a laboratory. Unfortunately for the consumer, this information is usually withheld, opening the door for manufacturers to exaggerate what their products can do. As with all consumer products, the fine print in the warranty will often contradict what the marketing people say about their products. This is particularly the case in a competitive market, like consumer healthcare products, where many manufacturers offer very similar products. Sometimes marketing materials can be deceptive. Take, for example this quote direct from the product box of a massage chair: "we've earned a sound reputation for producing durable, reliable, stable, industrial-grade healthcare products over three decades." Now compare that statement to what is specifically excluded in the warranty terms and conditions:

- wear and tear from moving parts
- commercial and industrial use.

So, is it made to be stable and reliable? Is it meant to be "industrial-grade"? Marketing statements are often misleading and contain no more than superficial gimmicks. Claims are sometimes made based on some studies. For example, "certain university studied … and confirmed …" These studies may be conducted in a well-controlled environment to support a claim by an expert in the field. Playing with psychology is often an effective marketing deception (Boush et al. 2009).

7.4 Telehealth in General Healthcare and Fitness

Medical technology is not always used for assistive remedy and it is also extremely important in providing a range of solutions for maintaining optimal health. There are many ways that we can keep ourselves healthy with technology, such as dietary monitoring and various massaging devices that apply reflexology to alleviate stress and tension. Further, there are exercise therapies such as yoga and martial arts that enhance circulation and flexibility while easing chronic pain. After all, a healthy lifestyle is about nutrition, exercise, and stress relief. Throughout this chapter, we have discussed how technology has been developed to help us maintain a healthy lifestyle with all these. To conclude the chapter, we look at ways telehealth and related technologies can be used to assist us with optimizing our health while exercising.

7.4.1 Technology Assisted Exercise

Being physically active and maintaining a healthy lifestyle benefits people of all ages. Moderate physical activity on a regular basis will help maintain optimal health and reduce health risks. Not all physical exercise needs to be taken in a gym, for example something like a half-hour walk can also keep us physically active. Earlier work by Tuomilehto et al. (2001) has shown that physical exercise can reduce the risk of developing diabetes for those at high risk. A wide range of exercise can be taken that fits an individual's schedule. And from walking and jogging to swimming and ball games, there is so much that technology can offer.

Technology can ensure exercise is taken at an appropriate and comfortable pace. For example, a simple measurement of the heart and respiratory rates can prevent any difficulty in breathing or fainting during or after exercise. For many who simply take a short walk after dinner, there is a step counter, also known as a pedometer, that counts your steps and determines the distance you have covered and the calories you have burnt while walking. A pedometer mainly works by sensing body motion and counts the number of footsteps. All pedometers count steps, although they may have different counting methods. These can be piezo-electric accelerometers, a coiled spring mechanism, and a hairspring mechanism. As illustrated in Figure 7.9, all these operate by compression and subsequent expansion. Each cycle translates to one step count. With knowledge of the user's usual stride length, the count can be multiplied by the nominal stride length to obtain the distance covered. The simplest pedometers only count your steps and display steps and distance. This can even be implemented on some mobile phones as a built-in feature that works by relaying data transmitted from the pedometer's sensor. When designing a pedometer, it is important to ensure that the count reset button is not easily depressed unintentionally and sufficient memory is available to store the number of steps for a specified number of days. Some even obtain geographical information from a global positioning system (GPS) for tracking continuous speed and distance;

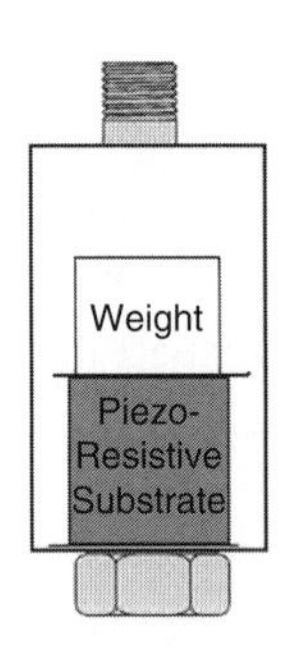

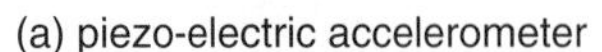

(a) piezo-electric accelerometer (b) coiled spring mechanism (c) hairspring mechanism

Figure 7.9 Fundamental components of a pedometer.

therefore speedometers and odometers like those found on vehicle dashboards are also available. The user can download and overlay the workout to a map and obtain information about the elevation and gradient of the hill that they climb. There is, however, a potential problem with tracking: walking under trees or tall buildings may temporarily disrupt the wireless communication link from the GPS.

GPS technology can also assist cyclists. A GPS unit can replace a conventional cycle computer to provide features such as route mapping, recording heart rate, and pulse data that can be downloaded to a computer for fitness analysis. The electronics can do just what the control panel of a gym exercise bike can offer.

7.4.2 In the Gym

There are many choices of fitness equipment in a typical gym: treadmills, exercise bikes (including elliptical cross trainers), rowing machine, boxing, weights, and bars. While how they operate and what potential benefit this equipment brings is not within the scope of this text, we intend to explore the technologies that support these devices.

Taking the treadmill as an example, it is perhaps the easiest to break down so reliability becomes a primary issue. The motor is the heart of a treadmill. The major cause of a treadmill motor failure is high walking belt friction created by a lack of lubricant. The cost of replacing a motor can run into hundreds of dollars, and preventive maintenance can certainly prolong the motor's life significantly. Condition based prognostic management can provide a solution to service the motor prior to an anticipated failure. Apart from keeping the treadmill up and running, this seemingly simple to operate equipment is fully packed with technology. Apart from controlling the speed of the walking belt, which also can cause a reliability issue since this is the part that wears out quickest owing to persistent rubbing between the shoe sole and its surface, a treadmill has a lot of features to offer. Upon completion of a workout, a set of statistics can be collected, as in Figure 7.10, where it shows the duration, amount of calories burnt computed from the effort of the workout, and the distance covered as a function of the number of revolutions the walking belt has completed throughout the workout. This information can then deduce the average speed and pace. Note also that there is an

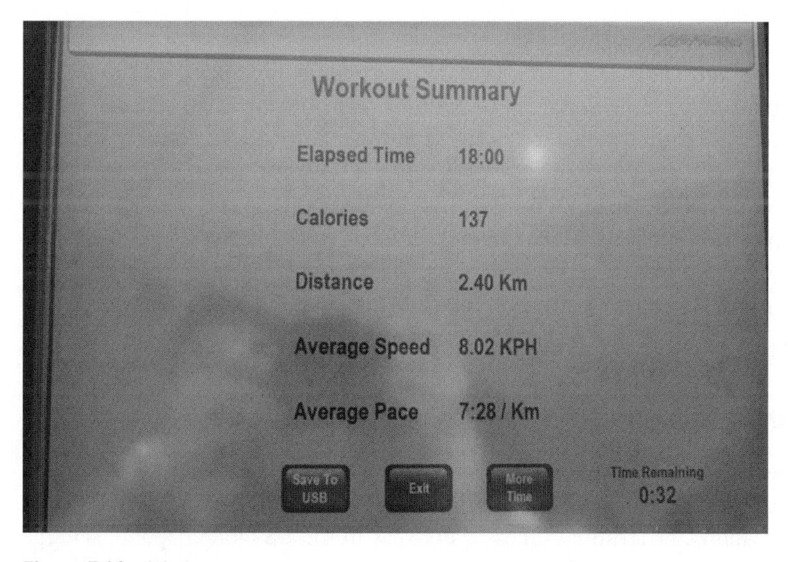

Figure 7.10 Workout summary.

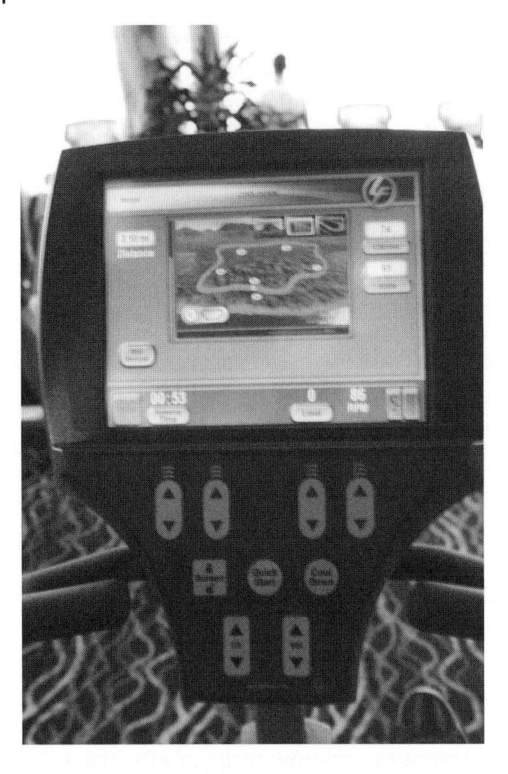

Figure 7.11 Simulated cycling path profile.

option to "Save to USB." The data can be stored for analysis to keep track of the user's fitness. It is also possible to automatically download the data via a wireless link.

One common feature found in most equipment is the handgrip-mounted heart rate monitor. This helps the user to calculate the calories they are burning during their workout and see whether they are working hard enough, or too hard. Other common features include simulated workout profile, such as the cycling path profile shown in Figure 7.11. A number of devices such as heart rate and blood pressure monitors can also be wirelessly linked to collect different body signs for health assessment.

Having covered what we commonly use in the gym, let us conclude by taking a brief look at the telehealth technology used in professional sports training. An array of small sensors can be attached to the user to collect real-time data about a sports training session. The performance can be analyzed and recorded by a computer. There are many types of sensors to meet different needs, for example boxing requires an accelerometer for gait and posture analysis and pressure sensors to track where the user has been hit and by what amount of force. Such biomechanical measurements can quantify functional performance during training. This can also be accomplished by video motion tracking. Another area is surface electromyography (EMG) analysis, which measures muscle contraction that initiates limb movement and so is used to quantify muscle activity and fatigue.

7.4.3 Continual Health Assessment

Telehealth is proven to be extremely helpful for those patients who need the most frequent contact, when literally millions of people receive home care because of acute illness, long-term health conditions, permanent disability, or terminal illness every year in the US alone. Upon discharge from the hospital to their homes, patients with chronic disease often require close surveillance throughout the rehabilitation process. Telehealth enables medical professionals to monitor the patient continuously and make real-time identifications and interventions in the care of their patients. Although telehealth is extremely important for those with special needs, its benefits also extend to normal healthy people. In addition to assisting us with daily exercise, technology can also help us maintain optimal health.

So far we have talked about using various meters to check body signs, but t we can do more than analyze these figures to ensure we are maintaining our health optimally. For example, we can keep track of what we eat and balance the nutrition intake. We can record the amount of food we eat and compare it with the amount of exercise we have undertaken in the day, to ensure there is a balance between energy consumed and energy used. Under the assumption that the food package provides true nutritional information about its contents, we can use such information to log the amount of food we have taken with a breakdown of individual components, such as salt and carbohydrate. An optimal meal can also be determined with reference to the FGP. A software-based FGP automated analysis system, such as that proposed in Muthukannan (1995), can be installed on a mobile phone for continuing analysis. This may be something nice to know for a healthy person but can be extremely helpful to those who need to control their diets for a variety of reasons. For example, this helps a patient with a kidney problem to ensure no excess sodium is ingested.

We have seen many healthcare applications that do not utilize traditional medical science and technology for healing and maintaining general health. Although medicine will certainly continue to be the mainstream cure for all of us, there are many other alternative ways that technology can be used for healthcare.

References

Barsoum, G., Perry, E.P., and Fraser, I.A. (1990). Postoperative nausea is relieved by acupressure. *Journal of the Royal Society of Medicine* 83 (2): 86–89.

Berecibar, M., Gandiaga, I., Villarreal, I. et al. (2016). Critical review of state of health estimation methods of Li-ion batteries for real applications. *Renewable and Sustainable Energy Reviews* 56: 572–587.

Bock, D. (2009). Acupressure points for stretching & releasing tension. *EC Martial Arts Blog* (21 February). http://www.fightingarts.com/reading/article.php?id=624 (accessed 20 January 2020).

Boush, D.M., Friestad, M., and Wright, P. (2009). *Deception in the Marketplace*. Routledge Academic.

Brandt, A. (2004). The game room: gaming for fitness. *PCWorld* (7 February). https://www.pcworld.com/article/114559/article.html (accessed 20 January 2020).

Bratman, S. (1997). *The Alternative Medicine Sourcebook*. Lowell House.

Business Insider (2020). Esports Ecosystem Report 2020: The key industry players and trends growing the esports market which is on track to surpass $1.5B by 2023. https://www.businessinsider.com/esports-ecosystem-market-report?r=US&IR=T (accessed 20 January 2020).

Chitturi, S. and Farrell, G.C. (2000). Herbal hepatotoxicity: an expanding but poorly defined problem. *Journal of Gastroenterology and Hepatology* 15 (10): 1093–1099.

Dobie, T.G. and May, J.G. (1994). Cognitive-behavioral management of motion sickness. *Aviation, Space, and Environmental Medicine* 65 (10 Pt 2): C1–C2.

Ezquerra, N. and Mullick, R. (1996). An approach to 3D pose determination. *ACM Transactions on Graphics* 15 (2): 99–120.

Food & Drug Administration (2009). Electromagnetic compatibility (EMC). https://www.fda.gov/radiation-emitting-products/radiation-safety/electromagnetic-compatibility-emc (accessed 20 January 2020).

Fong, B. and Li, C.K. (2011). Methods for assessing product reliability: looking for enhancements by adopting condition-based monitoring. *IEEE Consumer Electronics Magazine* 1 (1): 43–48.

Fong, B., Fong, A.C.M., Li, C.K. et al. (2012). A study on the reliability optimization of LED-lit backlight units in mobile devices. *IEEE/OSA Journal of Display Technology* 9 (3): 131–138.

Gerogianni, G., Babatsikou, F., Polikandrioti, M., and Grapsa, E. (2019). Management of anxiety and depression in haemodialysis patients: the role of non-pharmacological methods. *International Urology and Nephrology* 51 (1): 113–118.

Grand View Research (2019). Complementary and alternative medicine market worth $210.81 billion by 2026. https://www.grandviewresearch.com/press-release/global-alternative-complementary-medicine-therapies-market (accessed 20 January 2020).

Horstmanshoff, H.F.J., Stol, M., and Van Tilburg, C.R. (2004). *Magic and Rationality in Ancient Near Eastern and Graeco-Roman Medicine*. Brill Publishers.

Hoshiai, K., Fujie, S., and Kobayashi, T. (2009). Upper-body contour extraction using face and body shape variance information. *Lecture Notes in Computer Science* 5414 LNCS: 862–873.

Kasraee, B., Masouye, I., and Piguet, V. (2009). PlayStation palmar hidradenitis. *British Journal of Dermatology* 160 (4): 892–894.

Kausar, S., Multani, M.K., Zahoor, B., and Nazeer, A. (2017). Augmented reality based self-treatment using acupressure. In: *13th International Conference on Emerging Technologies (ICET)*, 1–5. IEEE.

Liu, T.Y., Yang, H.Y., Kuai, L., and Gao, M. (2007). Problems in traditional acupoint electric characteristic detection and conception of new acupoint detection method. *Zhongguo Zhen Jiu* 27 (1): 23–25.

Lu, G.D. and Needham, J. (1980). *Celestial Lancets: A History and Rationale of Acupuncture and Moxa*. New York: RoutledgeCurzon Publishers.

Maa, S.H., Sun, M.F., Hsu, K.H. et al. (2003). Effect of acupuncture or acupressure on quality of life of patients with chronic obstructive asthma: a pilot study. *Journal of Alternative & Complementary Medicine* 9 (5): 659–670.

Marshall, V.W. and Altpeter, M. (2005). Cultivating social work leadership in health promotion and aging: strategies for active aging interventions. *Health & Social Work* 30 (2): 135–144.

Miller, K.E. and Muth, E.R. (2004). Efficacy of acupressure and acustimulation bands for the prevention of motion sickness. *Aviation, Space, and Environmental Medicine* 75 (3): 227–234.

Mordor Intelligence (2020). Wireless healthcare market - growth, trends, and forecast (2020–2025). https://www.mordorintelligence.com/industry-reports/global-wireless-healthcare-market-industry (accessed 20 January 2020).

Muthukannan, J. (1995). The food guide pyramid: approach to an automated analysis of foods. *Journal of the American Dietetic Association* 95 (9): A49.

Nakao, M., Nomura, S., Shimosawa, T. et al. (1997). Clinical effects of blood pressure biofeedback treatment on hypertension by auto-shaping. *Psychosomatic Medicine* 59 (3): 331–338.

Nunn, J.F. (2002). *Ancient Egyptian Medicine*. Red River Books.

Rawlinson, G. (1956). *The History of Herodotus*. Tudor Publishing Company.

Riccio, G.E. and Stoffregen, T.A. (1991). An ecological theory of motion sickness and postural instability. *Ecological Psychology* 3 (3): 195–240.

Robertson, A. (2008). Can you really get fit with Wii exercise games? *MedicineNet* (30 May). http://www.medicinenet.com/script/main/art.asp?articlekey=90064 (accessed 20 January 2020).

Sheppard, A.L. and Wolffsohn, J.S. (2018). Digital eye strain: prevalence, measurement and amelioration. *BMJ Open Ophthalmology* 3 (1): e000146.
Shupak, A. and Gordon, C.R. (2006). Motion sickness: advances in pathogenesis, prediction, prevention, and treatment. *Aviation, Space, and Environmental Medicine* 77 (12): 1213–1223.
Stern, R.M., Jokerst, M.D., Muth, E.R., and Hollis, C. (2001). Acupressure relieves the symptoms of motion sickness and reduces abnormal gastric activity. *Alternative Therapies in Health and Medicine* 7 (4): 91.
Tuomilehto, J., Lindström, J., Eriksson, J.G. et al. (2001). Prevention of type 2 diabetes mellitus by changes in lifestyle among subjects with impaired glucose tolerance. *New England Journal of Medicine* 344 (18): 1343–1350.
Unschuld, P.U. (2003). *Huang Di Nei Jing Su Wen: Nature, Knowledge, Imagery in an Ancient Chinese Medical Text*. University of California Press.
NRCS. (2009). The PLANTS Database. http://plants.usda.gov (accessed 20 January 2020).
Vargo, S.L. and Lusch, R.F. (2004). The four service marketing myths: remnants of a goods-based, manufacturing model. *Journal of Service Research* 6 (4): 324–335.
Veith, I. (1972). *The Yellow Emperor's Classic of Internal Medicine*. University of California Press.
Welsh, S., Davis, C., and Shaw, A. (1992). Development of the food guide pyramid. *Nutrition Today* 27 (6): 12–23.
Weiss, R.F. (2000). *Herbal Medicine*, 2e. Beaconsfield: Beaconsfield Publishers.
Wilson, E.O. (2001). *The Diversity of Life*. Penguin Publishing.
Yuan, R. and Lin, Y. (2000). Traditional Chinese medicine: an approach to scientific proof and clinical validation. *Pharmacology & Therapeutics* 86 (2): 191–198.
Zeithaml, V.A. and Bitner, M.J. (2000). *Services Marketing: Integrating Customer Focus Across the Firm*, 2e. Boston: McGraw-Hill.

8

Digital Health for Community Care

The primary objective of any healthcare service is to make everyone safer, healthier, and live longer. Prevention often reduces the need for medical treatment. Telemedicine therefore seeks to provide advice and support to reduce the chance of illness or injury. Technology has evolved over centuries to provide people with a better future. This is exactly why we are interested in enhancing medical and healthcare technology for people everywhere by providing more efficient and affordable services with ease of access to as many people as possible.

Technology benefits both service providers and patients in many ways involving physicians, healthcare professionals, end users, engineers, equipment manufacturers, authorities, care centers, clinics, and hospitals. Healthcare services are no longer limited to certain locations such, as clinics and hospitals as communications technology is able to bring many of these services away from clinics and hospitals to users on the move or at home. Caring for the community is about assisting disabled people, taking care of children and older people, healing the sick or injured, and supporting vulnerable individuals.

In the first seven chapters, we look at many different types of wireless communication technologies being used in many different applications that cover the entire human body. In this chapter, we look at various aspects of healthcare that are used in caring for people under different circumstances.

8.1 Telecare

Through technical advances in telecommunications, many people who require special attention can live alone with the assurance that help is always available and they are taken good care of. Although telecare focuses more on subsequent responses to a situation rather than actively preventing an event from happening, caregivers can easily find out their whereabouts and attend to them whenever the need arises. Sometimes, telecare can even save the caregiver's visit because help can simply be provided remotely. Telecare puts the two otherwise contradictory attributes of independence and monitoring together in a mutually supporting way. Very simply, people can enjoy the freedom of being left alone while knowing that assistance is always there if or when needed.

Customization is a key feature of telecare. Necessary tools are provided to a user based on their individual needs. Telecare can be as simple as an alarm to call for emergency assistance or a sophisticated system that monitors the user's health condition. It can be an assistive network of devices for various routine tasks or an automated personal assistant that reminds the users of different things, such as taking their medication or switching off the gas stove after cooking. The list of what

Telemedicine Technologies: Information Technologies in Medicine and Digital Health, Second Edition.
Bernard Fong, A.C.M. Fong, and C.K. Li.
© 2020 John Wiley & Sons Ltd. Published 2020 by John Wiley & Sons Ltd.

telecare can do goes on. Telecare also includes a communication link that connects the user to a clinician or response center for alerts, health monitoring of vital signs, and personal advice. As a brief summary, telecare involves using telecommunications technology for health monitoring and to provide on-demand caring support.

The advantages brought to both users and caregivers by telecare are evident. The range of features telecare offers means that it uses many different technologies. We look at the building blocks of telecare in this section by first introducing the term "telehealth," which is defined by HealthIT.gov (2019) as a means to improve access to quality healthcare.

8.1.1 Telehealth

Telehealth is widely considered as a subset of telecare with the specific purpose of body vital sign monitoring (Li et al. 2006). It brings technology and advancing clinical practice together to collect patient information for monitoring with continuous feedback as well as scheduling for appointments. General health assessment is one key feature promoted by telehealth that uses a wide range of devices covering virtually all parts of the human body. Figure 8.1 shows a number of commonly used ones and they can all be connected as a small mobile healthcare center. Telehealth provides comprehensive coverage for the entire human body and enables patients to perform their own tests that automatically update their electronic patient record (EPR). This is particularly helpful for patients while waiting for their medical consultation. With the range of possibilities shown in Figure 8.1, basic parameters such as body temperature, body mass index (BMI), oxygen saturation, and heart rate can all be swiftly obtained and automatically made available to the doctor during the consultation.

Another key feature of telehealth is that automated health surveys about a patient's current state can be sent back to hospital and also update the EPR. Intake of any pharmacy medicine bought over the counter by the patient without a prescription can also be recorded. The device shown in Figure 8.2 provides a touchscreen for conducting medical surveys with a barcode reader that scans the barcode of any medicine that the patient may have taken. By linking to the pharmacy's database, detailed information about the medicine can be known.

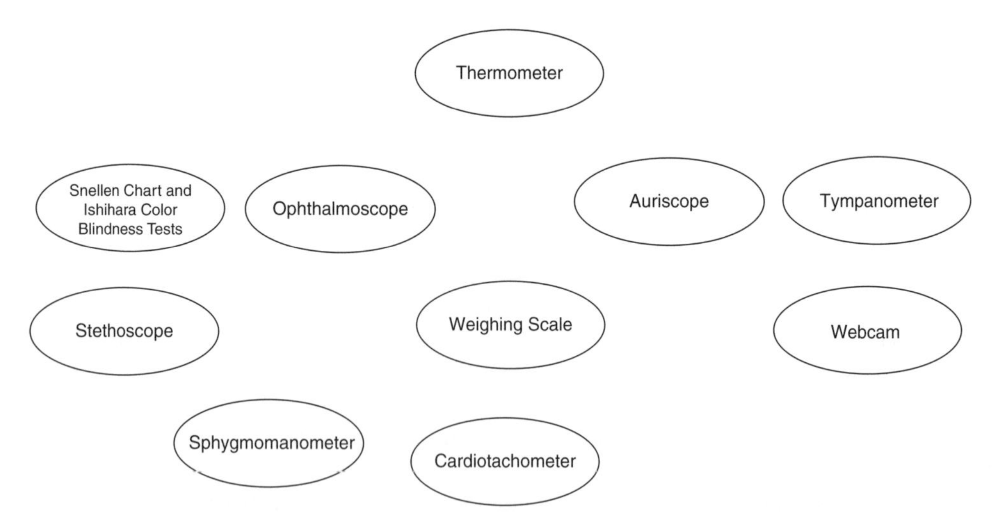

Figure 8.1 Collection of telehealth devices covering everything from simple consumer healthcare to delicate medical devices.

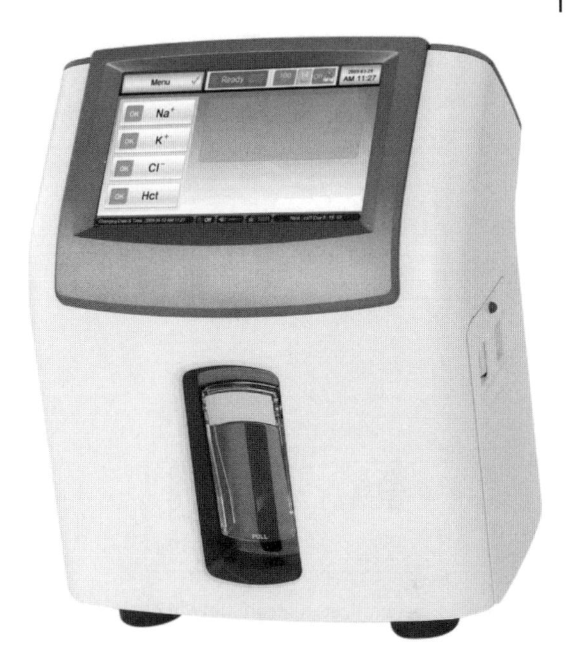

Figure 8.2 Pharmacy kiosk.

8.1.2 Equipment

Owing to the diversity of applications offered, there are many different types of equipment for telehealth covering telecommunications, physical assessment, diagnosis, cameras, and sensors. All telehealth systems rely on a good wireless communication network for data delivery. Other equipment involved will depend on the specific applications, examples include:

- Cardiology: stethoscopes, cardiac ultrasound, and electrocardiography (ECG) monitors.
- Radiology: probes, magnetic resonance imaging (MRI), and x-ray scanners.
- Ophthalmology: retinal cameras, ophthalmoscopes, pachymeters, and ophthalmometers.
- Otology, laryngology, and rhinology: otoscope, endoscope, laryngoscope, and rhinoscope.
- Dermatology: dermatoscope and autoclaves.

To make telehealth more accessible, telehealth services are often delivered through low cost monitoring devices. Another major part of telehealth equipment is a computer server that captures all data through health assessment and is also used to update the EPR system.

A telecare network can be generalized as Figure 8.3, where remote care is provided to end users by various entities via a telemedicine network. Here, the term "telemedicine network" is used with clear distinction from "telecare network," since the same network can be used for other medical applications sharing the same communication system. Primary providers would have all the medical equipment listed above, adequate for providing all types of services. A "response center," usually a regional hospital with expertise in many areas, is in charge of clinical support and advisory related matters to all "request centers," including end-users and rural/mobile stations. Technical support is provided by people who look after system maintenance and requires mainly diagnosis, network management, and monitoring tools to ensure network availability and data integrity. The telehealth network, which is a complex communication system by itself, as shown in the logical diagram linking various entities together in Figure 8.4, consists of equipment for data acquisition, storage, and transmission of all medical data across a multipoint-to-multipoint network infrastructure. Database maintenance to ensure EPRs are properly maintained and data security assured is

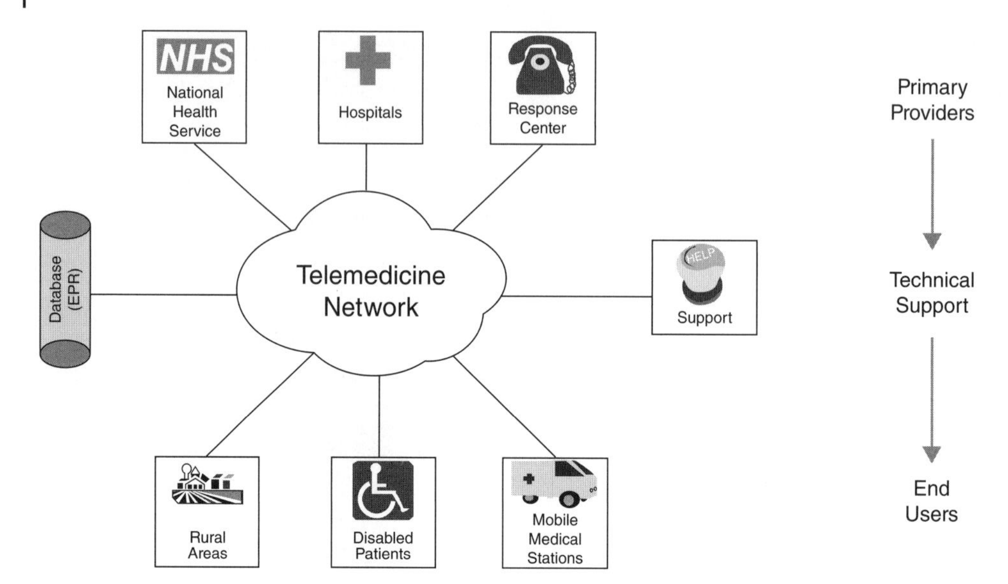

Figure 8.3 Generalized telecare network.

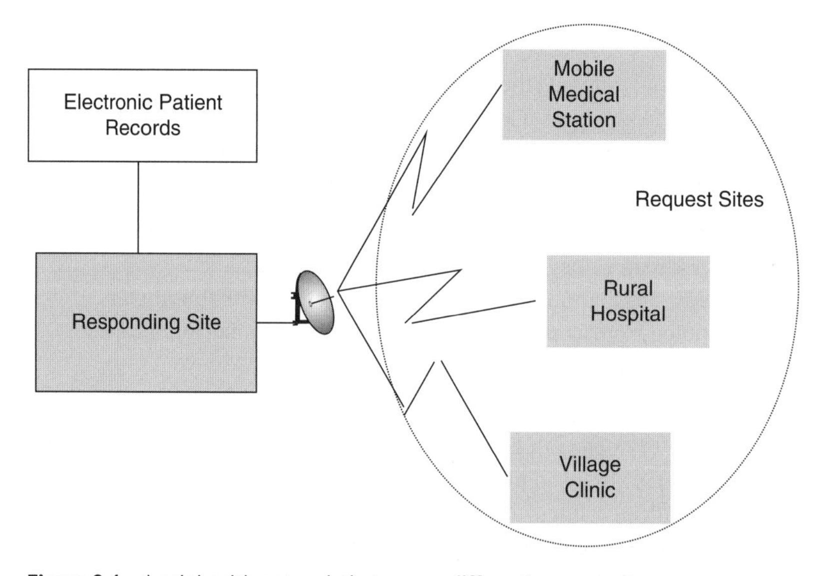

Figure 8.4 A telehealth network that serves different request sites.

taken care of by technicians. A wide range of biosensors and remote patient monitoring devices will be installed, either temporarily or on a permanent basis, at the end user's site or at mobile support centers. Although there are too many different types of equipment involved for providing support to different applications, in Sections 8.3 and 8.4 we look at one case study each in serving rural areas and in the care of older people. From there, we can learn more about the specific requirements for different telecare applications.

8.1.3 Sensory Therapy

Here, we are looking at senses of the human body, not the biosensors that are attached to healthcare devices. Telecare is also available to those who suffer from sensory and cognitive impairments. It offers activities to stimulate the five senses (vision, hearing, taste, smell, and touch). Multimedia technology allows interactive applications to be built for different kinds of therapies to heal or simply to enjoy a relaxing environment. Multimedia technology facilitates interactivity through audio/visual (AV) and haptic sensing (see Section 11.1 for details) for the eye, ear, and hand. In order to support these multimedia services, a system needs to have a vast amount of bandwidth to provide sufficient quality of service assurance (Vergados 2007). Data traffic requirements, indoor or outdoor propagation characteristics, and network structure need to be taken into consideration when designing a multimedia healthcare system.

This leads to the interesting topic of music therapy that subsequently became popular. It is primarily applied in relieving pain perception to improve a patient's physiological and cognitive state (Standley and Prickett 1994). It is also reported that music has an effect on cardiac output, heart, and respiratory rate, blood pressure and circulation, and electrical conductance of tissues. Scott (2019) even reports a positive impact on cancer treatment. In addition to healing, music therapy is also used in stress relief. As different individuals have different tastes in music, not all music is suitable for use in therapy. Also, the effectiveness of the same therapy applied to two different persons under identical conditions may differ (Darrow et al. 2001). Finding what might work for an individual is difficult and will involve some trial and error. Irregular physiological response may produce adverse electroencephalographic (EEG) patterns. Generally speaking, music with a slow rhythm (slower than the nominal heart rate of about 70 bpm) tends to be more effective, whereas faster music often serves as an effective stimulus.

8.1.4 Are we Ready?

Telecare is by no means a new area in healthcare services. eHealth Europe (2009) reports that the Spanish Fundación Andaluza de Servicios Sociales (FASS) of the Junta de Andalucía, being the 100 000th telecare deployment, supports the senior citizens of the southern Spain areas with supporting stations situated in Malaga and Seville. It involves a number of entities under the coordination of the Ministry for Equality and Social Welfare of the Autonomous Government of Andalucía. More issues exist when more entities are involved.

To promote telecare, there are certain fundamental issues that need to be addressed. First, for sparsely populated areas where no more than a few dozen residents live, there may be no existing network infrastructure available. Other basic supporting resources may also be scarce. Another issue, in the context of deployment considerations in continental Europe, is the lack of standardization such that current international standards may not support EPRs in certain languages. Some widely accepted international health data standards – such as HL7, DICOM, and SNOMED, and American's Health Insurance Portability and Accountability Act (HIPAA) – are widely used clinical data standards throughout the world. These cover regulatory requirements, privacy rules, standards, and recommendations for implementation. However, these standards are based on the English language and cannot be applied directly to other languages without some kind of translation. Little incentive exists for practitioners to manually convert patient data into English unless mechanisms are in place for direct entry in the specific language. For example,

FASS does have support for information in Spanish. Nationwide implementation is reasonably straightforward, provided the entire system is developed in one single language. However, any attempt to cover countries across Europe may require multilingual support.

But linguistic issues are not the only question we need to ask ourselves. There is also the overarching issue of who is in charge: who will be responsible in the event of a systems failure? Answering these questions will make us know whether we are ready for telecare, and this is what we examine in the next section.

8.1.5 Liability

Telecare works on the basis that people in different locations are involved in serving end-users. The wider telecare covers, the more complex the system will be. Local authorities can supervise everything within a city. When coverage involves areas overseen by different authorities, there may be issues such as state versus provincial ownership open for negotiation. To provide comprehensive telecare support, the following entities each with its own organizational structure may be involved:

- Hospitals and clinics: providing advice and treatment.
- Pharmacies: supplying medication and other medical resources.
- Government agencies: policies and administrations.
- Medical schools, public, and corporate research centers: research and development.
- Equipment manufacturers: including but not limited to medical devices and sensors, telecommunications, computers, data storage, etc.
- Telecommunication service providers: provide and maintain communication links to connect various entities together for secure and reliable data transfer between them.
- Health insurance companies: claims and payouts.
- Patients and end-users: citizens on both temporary and long-term care and/or monitoring.

Legal issues may also arise in the event of a system failure or when no attention is paid to an event. Failures can occur almost anywhere throughout the telecare system. For example, failure to attend to a situation may be caused by sensor or equipment failure, or a network outage or a cable being damaged by workers when maintenance work is carried out. Legal disputes may arise in telecare practices and who to be held responsible in the event of a mishap can be a problematic and complicated issue to address.

8.2 Safeguarding Senior Citizens and the Aging Population

The natural aging process and health degradation makes emergency response particularly challenging for older people (Veenema 2018). Aging and health degradation have a substantial impact on a person's vulnerability when coping with a disaster (Adams et al. 2018). Quarantelli (2008) discusses a number of individual as well as group behavioral factors in the disaster context, and although Lemyre et al. (2009) conducted a comprehensive study on the psychosocial considerations, research on other health-related factors is still lacking. Recent advances in digital health technologies open up new opportunities for safeguarding the aging population through preventive care and health monitoring. Understanding the problems associated with health degradation due to aging can improve safety and wellness through providing them with the appropriate assistive care solution (Hussain et al. 2015).

An aging population is an increasing problem in most developed nations where a higher proportion of citizens are older people. The current trend of population aging will certainly lead to a shortage of caregivers as well as funding for elderly care in the next one to two decades. As a direct result, deterioration in quality of elderly care can be expected unless something is done in the near future. Although the Nobel Prize in Physiology or Medicine 2009 was awarded to three scientists for their contribution to our understanding of the effects on cell aging (Nobel Prize 2009), present genetic engineering technologies may still be far from adequate in stopping or even reversing the body aging process. Before this can be achieved, telecare is changing the way people cope with healthy aging. Perhaps in the distant future, methodologies in aging reversal might make telecare obsolete.

According to the Office for National Statistics, over 24% of people living in the UK will be aged 65 or older by 2042, a substantial increase from 18% in 2016.

(National Statistics 2018). The screenshot in Figure 8.5 shows that the aging population trend in the UK is obviously on the increase. The screenshot in Figure 8.5 shows that the aging population trend in the UK is obviously on the increase. The situation in Americas is not much better. The US Administration on Aging puts the estimate that 20% of the US population will be aged 65 or over by the year 2030, rising steadily from 12.4% back in 2006. Statistics Canada also shows a very similar picture with an increase of 65+ to 23.4% for the next 25 years from 13.7% in 2006. The chart in Figure 8.6 shows the same alarming yet consistent trend among G8 nations, with Japan being projected as the country most affected. The financial burden on supporting national healthcare services will certainly increase in the foreseeable future (Kelliher and Parry 2015). Given the severity of population aging and its impact on society, we must explore how technology can relieve its adverse impacts.

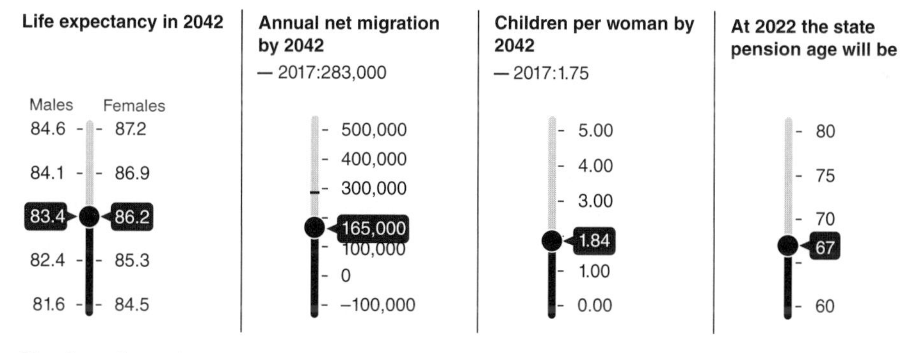

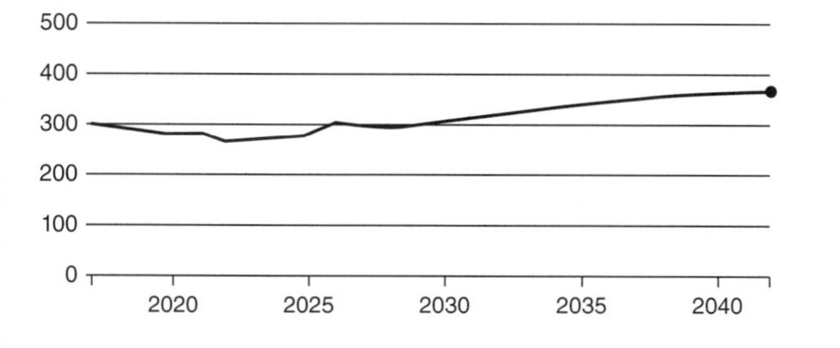

Figure 8.5 Screenshot of statistics on aging.

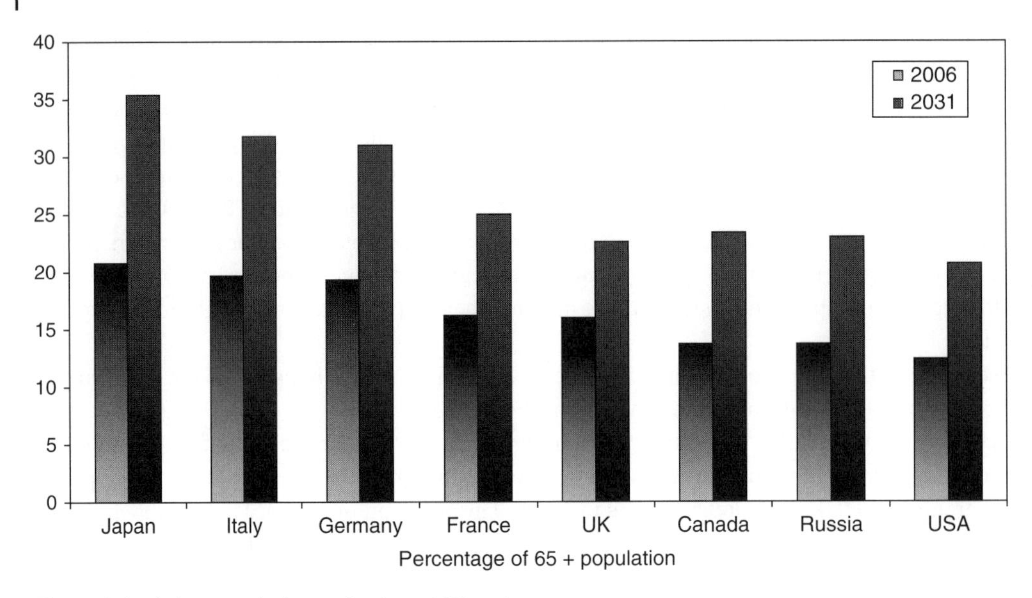

Figure 8.6 Aging population projection of G8 nations.

8.2.1 Telecare for Senior Citizens

Telecare, although not intended to provide a preventive solution for senior citizens, can improve efficiency and cost effectiveness to serve the older population. In this section, we take a look at an example where an information and communication technology (ICT) solution is implemented. This enables caregivers to remotely monitor the well-being of older people. To help senior citizens with their daily tasks and make them feel safer while leaving home, a system that serves as an electronic guard by utilizing wireless communications technology can help older users stay connected. A generic wearable device that provides advance alerts in anticipation of potential dangers and to remind the user of certain routine tasks is offered. The system can be customized to suit individual needs based on budget and circumstances. For example, a user at higher risk of falling can be equipped with accelerometers that automatically detect a fall and send an alert to a caregiver for immediate attention; a user suffering from any form of cognitive impairment can be reminded of various tasks, such as flushing the toilet after use, washing one's hands, and taking prescribed medicine at certain times.

Based on an off-the-shelf mobile phone, it provides an inexpensive means of helping older people with self-care and information gathering by making use of a wearable device and existing wireless communication systems. To the caregivers, ranging from rest home to offsite remote support, they can easily receive updates on a user's conditions and receive warnings of emergency situations, making remote monitoring more efficient. To the older users, they can be assured that they are well looked after. In the event of an accident, an emergency response will be offered. Their health conditions will be monitored. A guard is readily available 24/7 to remind them of various tasks. They can also obtain advice on demand. Their safety and well-being can therefore be assured.

To provide comprehensive telecare services linking older users to their caregivers, the system consists of two separate modules. Each operates independently and is linked together through a backbone 3G cellular wireless network. 3G is selected for the sake of greater rural coverage in many countries, where the same operating principle applies in locations where 4G/5G coverage is available. Readers are encouraged to refer to Section 2.2.6 for key differences between

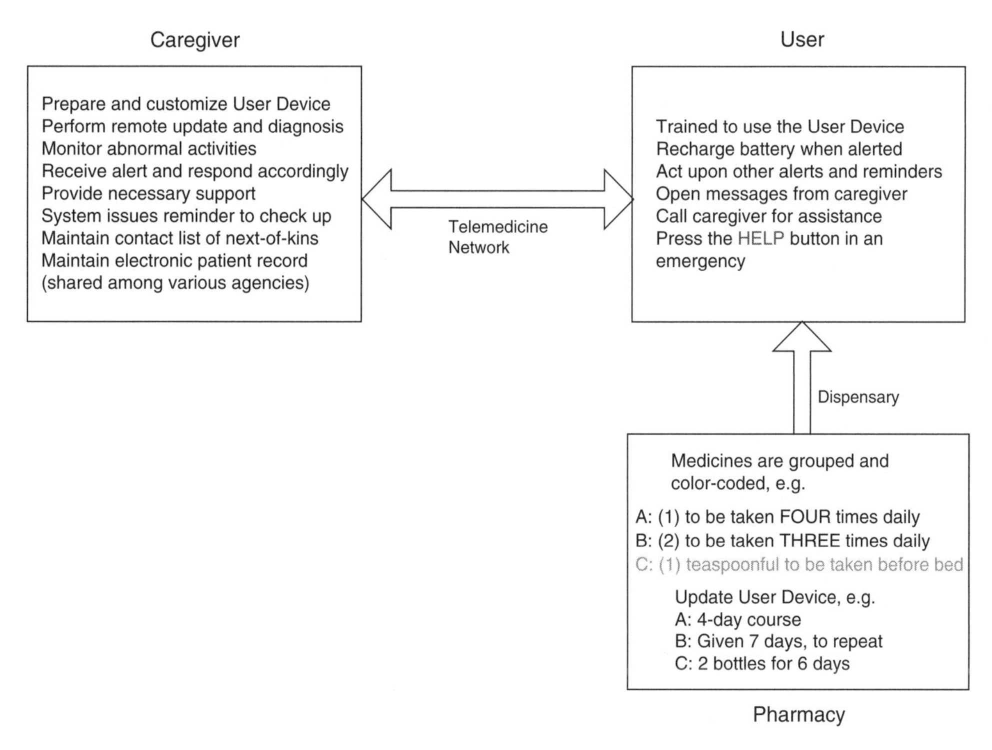

Figure 8.7 Telehealth for older patient care.

different generations of cellular networks. A system block diagram is shown in Figure 8.7, where the caregiver's side is responsible for tasks such as customization of end-user device and acting as a response center. Whereas the end-user's side, namely the older user's home, can be as simple as just a pre-programmed mobile phone, to a sophisticated system with comprehensive features that serves users with special needs such as those with chronic diseases.

The system layout is reasonably simple where the user only needs to undergo informal training that introduces the device's features and how to respond to various types of alerts and messages. They also need to learn how to seek help in certain situations. More on the benefits to the users is described in an example below.

The system relies on cooperation from pharmacies when preparing drugs such that a color coded system can be developed. Medications will be packed with appropriate color bags. To serve this purpose, one solution that demands minimal effort would be to color print labels, so that the drug's name can be printed using a specific color according to its quantity and frequency of intake. Medication information can also be updated to the user device by a Bluetooth link. The information can be embedded in the prescription so that the device can remind the user of when to take their medication.

At the caregiver's side, a number of supporting tasks need to be maintained. One of the design motives is to minimize the efforts of older users so that most maintenance related matters are dealt with by the support center. Prior to delivery, the user device will be fully programmed and configured for its specific functions based on an individual user's needs. A number of separate modules (software or accompanied attachment devices) will be installed during the configuration process.

Once the device is prepared, the remaining tasks will be reasonably similar to the regular duties of caregivers such as nurses and social workers. The system is designed to assist them with a range of these tasks, including reminder of activities, remote checkup in lieu of site visits on certain occasions, and automatic alerts to certain situations. Some necessary support such as consultation can be provided remotely, for example capturing data for analysis or archiving, recovery of progress tracking, monitoring ECG of users either diagnosed with cardiopathy or classified as high risk allowing any abnormal activities to be identified. To minimize the response time in the event of an accident, the system is designed to remotely detect situations such as a fall that can immediately trigger an alarm. This feature is particularly helpful in nursing homes, where older people may wander around unsupervised.

This system's modular design can be easily tailored for users with different needs and budgets. Implementation can be as simple as a digital assistant that provides a convenient communication link to the caregiver and serves as a reminder for a variety of tasks. Modules can be fitted into the system for the permanent monitoring of certain parts or be installed on a temporary basis to serve certain immediate needs, such as rehabilitation or illness.

A system functional diagram is shown in Figure 8.8. It is capable of serving a user with dementia who is recovering at home after an operation. This particular system consists of both permanently installed sensors and equipment that is installed for post-surgical rehabilitation. This example shows the following permanently available features with a central control console enabling smart home technology:

- thermometer to regulate ambient temperature.
- smoke detector for fire hazard.
- kitchen sensors to ensure safe use of stove and reminds user of activation.
- medication console to ensure prescribed medicines are taken on time.
- first aid kit with reminders for replenishment and alert for expiry.

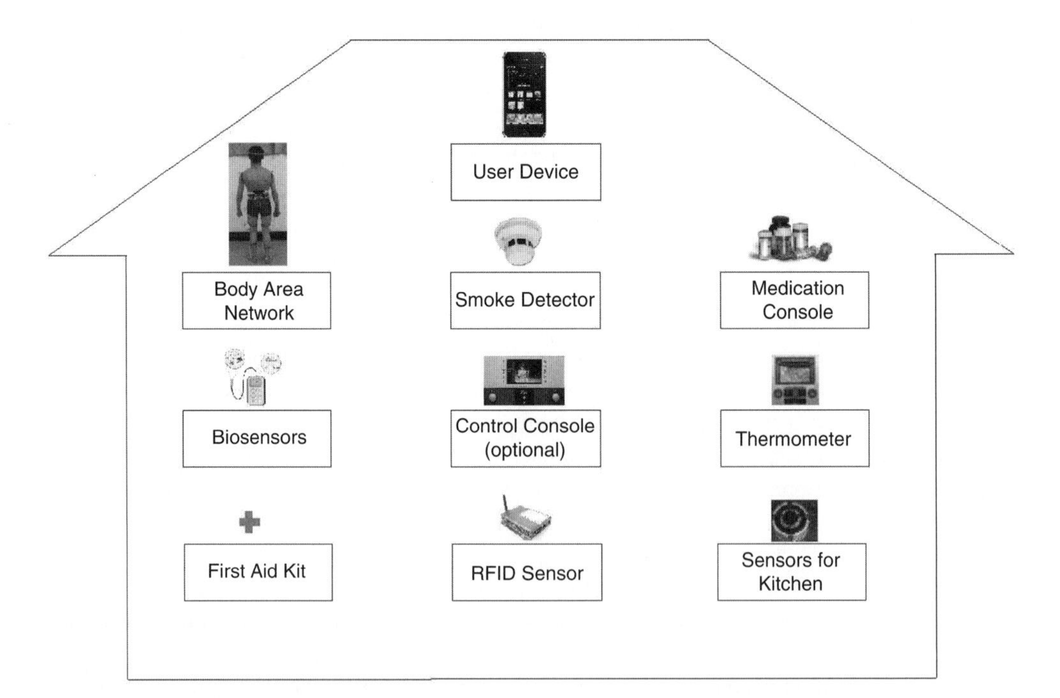

Figure 8.8 Assistive home setup for living alone, comprehensive services ensure that older patients are well looked after and supported while enjoying an independent life.

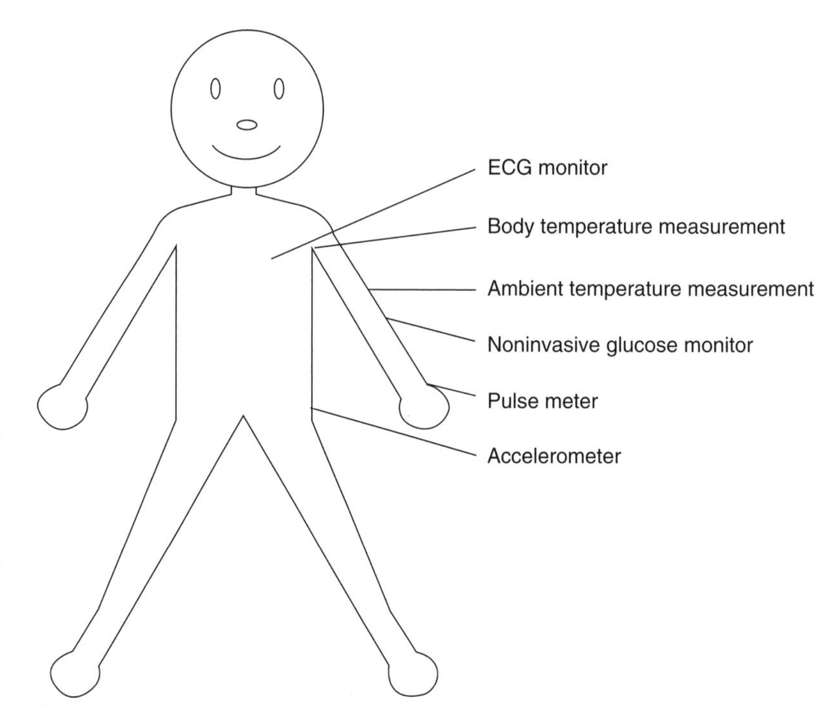

Figure 8.9 Body area network biosensors.

In addition to these, one key design feature is the incorporation of smart clothing technology (see Section 9.1.3 for details), where a body area network, as shown in Figure 8.9, is set up with biosensors and accelerometers to detect movement and activities so that various signs related to the user's health state can be collected. Different features can be added based on individual needs. In this example, sensors embedded in clothing can monitor a range of signs including ECG, temperature, blood sugar level, and pulse rate. Also, should the user fall, the response center will be automatically alerted.

Radio frequency identification (RFID) readers can be installed for a variety of features. For example, when used in the medication console users can be tracked for medication being taken and when repeat or replenishment should be sought. A reader that is installed at the door can remind users to bring the keys and to lock the door. The reader, of course, can also be programmed to automatically lock the door after sensing the user leaving home.

At each user's end, there will be a standalone device that stimulates users with various control methods. Audio command, particularly for users with dementia, may require filtering and synthesis to make the speech recognizable. Mobility is also an important consideration as the current system is primarily designed for users remaining at home. The intention of this system is to serve as a companion that can be easily carried. One feature of the device is a communication module that makes the user feel as though they are being cared for while away from home.

This system incurs minimal setup costs to the older user as the device can be created by customizing an existing mobile phone. Specific software applications are installed to support a wide range of tasks at the user's end. The device will be pre-programmed for the user depending on individual needs. The Java-enabled portable device is used to serve as a personalized assistant and as a monitor. With Java programming, virtually any modern mobile phone can be supported. One suitable mobile phone that can be used as an assistant to older users by running apps especially designed for older users can be bought from the open market for as little as $70. While one of the main design objectives is to minimize user interaction so that most tasks are performed automatically, primitive regular operations like battery recharging need to be taken care of by the user. Also, since most mobile phones currently make use of micro-SD memory cards, it is possible to roll out feature enhancements and to provide software updates by swapping the installed card for re-programming. Other functions such as continual diagnosis of both the user's well-being and the system's condition can be performed remotely.

To illustrate how flexible the device can be, we refer to the device in Figure 8.10, where the following functions are supported by a smartphone with touchscreen user interface. Although a touchscreen makes it easy for use by older people, this is not a basic requirement to enjoy basic telecare services brought to them by wireless technologies. This particular device has the following features:

- All basic functions of a mobile phone, plus a message filtering capability that sorts messages from the caregivers into its default mailbox.
- A help button for real-time support is also provided.
- Checking the user's health, analysis by captured vital signs such as heart and respiratory rates, blood glucose levels, etc.

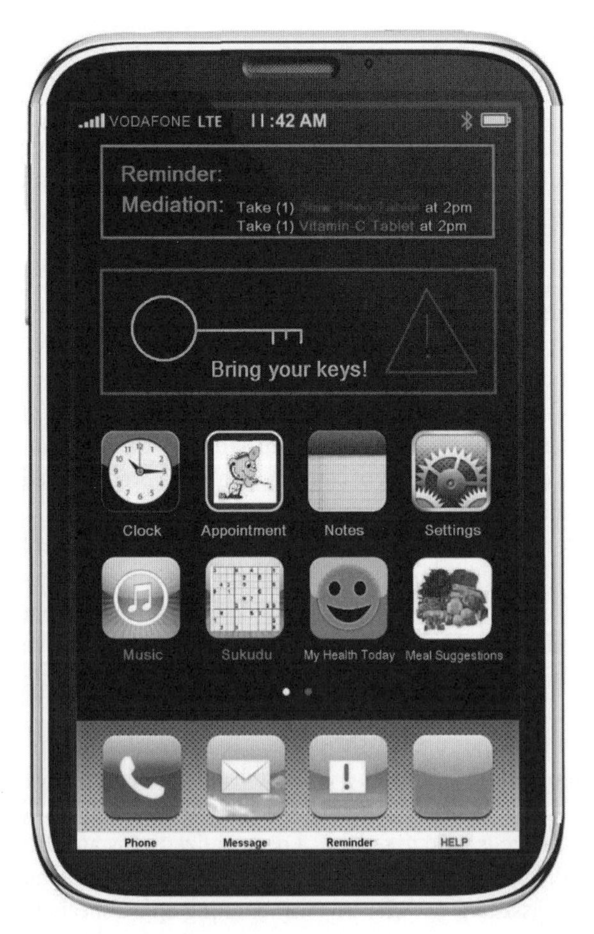

Figure 8.10 Assistive care smartphone.

- Suggest what to eat for the next meal, based on nutritional balance and any existing medical conditions, can be linked to an online ordering system for home delivery, similar to that of some paid TV subscription set top boxes.
- Link to a home medication console where medicine is stored, the color coded system described above is implemented to assist with ensuring appropriate time and quantity of each medicine taken. An RFID system would keep track of medicine taken and remaining stock.
- The RFID sensor featured in Figure 8.8 is intended to serve as a door guard at the main entrance of the user's home. Its main purpose is to ensure that the user has not forgotten their door keys before leaving home. If the user unlocks the door without removing the keys, an alarm will be generated. Similarly, the system will also remind the user to lock the door securely before leaving. An automatic locking mechanism can also be engaged.
- Entertainment features are also available. In this illustration (Figure 8.10), these two options are darker than the others, since they are auxiliary functions that are less important than healthcare-related support functions. Music also serves as an effective tool to calm the user down when agitated; memory games that stimulate the user can keep the user engaged in some brain-training activities.

The flexibility of Java allows many other services to be included with appropriate networking sensors. These can include fall detection, stroke detection, body temperature monitoring, prognostics of implanted medical devices, etc. Built upon open-source software that is not restricted by licensing issues, applications serving different needs can be developed for cross-platform compatibility. The same app can run on both iOS and Android smartphones.

User-friendliness is an important design consideration since most senior citizens are not familiar with technology. Another major function is to collect information about users' health conditions such as blood pressure, body temperature and SpO_2 readings, medication and nutritional intake, and fall history. Such clinical information will be analyzed on a regular basis for monitoring purposes. In addition, the clinical information can be connected to and shared with healthcare facilities (e.g. general practitioners or hospitals) using an existing wireless network. This feature is particularly suitable for older adults with cognitive impairment users who are recovering at home after hospitalization (after hip fracture surgery, for example) while still under close surveillance by hospital staff. In addition, this feature can help reduce demands in hospital resources as well as travel time for older patients.

8.2.2 The User Interface

Routine activities of senior citizens can be supported by a multisensory telecare system as an electronic guard. Older users with special needs such as memory loss and cognitive impairment sufferers can greatly benefit from technological advancements in human–computer interaction (HCI) and wireless communications. A wearable therapeutic device provides general assistance, health monitoring, calling for emergency assistance, alerts, and reminders, and can provide dementia sufferers with peace of mind. This solution also links care providers and older people, particularly those living alone, so that they can stay in touch.

The HCI interface, which determines how user-friendly a device is, must be very carefully designed. In particular, attention must be paid to ensure older users will find it easy to operate. HCI involves consideration of the following:

- language
- engineering feasibility and cost effectiveness

- mechanical reliability and durability
- precision
- ergonomics and human factors
- cognitive psychology and sociology
- ethnography.

There are virtually infinite choices of implementation methods, including keypad, computer mouse, touchscreen, navigation menus, etc. In dealing with user interface design, we must mention Shneiderman's "Eight Golden Rules of Dialog Design" (Shneiderman 2005):

1. Strive for consistency
   - Consistent sequences of actions should be required in similar situations.
   - Consistent styling with color, layout, capitalization, and fonts should be used throughout.
   - Identical terminology should be used in prompts, menus, and help screens.
2. Enable frequent users to use shortcuts
   - Abbreviations, special keys, hidden commands, and macros can be assigned to increase the efficiency of interaction.
3. Offer informative feedback
   - The system should respond in some way for every user action so that the user knows the input has been collected.
4. Design dialogs to yield closure
   - Sequences of actions should be organized into groups with a beginning, middle, and end. Instructive feedback at the completion of a group of actions confirms command execution.
5. Offer error prevention and simple error handling
   - Forms to be organized in such a way that obvious errors will be disallowed, caution should be exercised to accept certain exceptions, e.g. telephone entry may include characters such as "+," "–," and brackets for area codes.
   - Instructions should be offered upon detection of an error and to offer simple, constructive, and specific instructions for correction.
   - Segment long forms and process each section separately such that any error will not cause total loss of information already entered.
6. Permit easy reversal of actions
   - Let users go back through menus.
7. Support internal locus of control
   - User override and manual intervention. Must ensure ease of information retrieval and avoid monotonous data entry sequences.
8. Reduce short-term memory load
   - Theory suggests that a typical person can store something between five and nine pieces of information for a short time. One can relieve short term memory load by designing screens with clearly perceptible options or using pull-down menus and icons that list every available option to avoid the need to memorize items.

As a final note, operational reliability depends heavily on the prevention of errors whenever possible. Necessary actions can be taken in user interface design in such a way that error occurrence is minimized by using methods such as organizing screens and menus functionally; – and designing screens to be distinctive thereby making it almost impossible for users to mistakenly carry out irreversible actions that may cause data loss or system malfunction. Understanding the behavior of target users is of particular importance when designing systems for older users, and this should include identifying mistakes users may make when using the device, for example pressing the wrong button or becoming confused between various apps. Exception handling would prevent unpredictable system response in the event of a user executing an invalid command.

8.2.3 Active Versus Responsive

As a reminder to the readers, telecare is not intended to prevent accidents from happening. For example, telecare systems do not have the ability to counterbalance the user in the event of falling. Such systems operate more reactively as certain rules are programmed to respond to different scenarios. There are telecare devices that are more responsive to assist with preventive care. Any such device helps stimulate users so that they are trained to keep themselves engaged in some kind of activities.

Although "prevention is better than cure" may be heard countless times, accidents do happen, despite of all best practices in place. Although technology can sometimes prevent an accident from happening, technical solutions are far more often passive than active. As in the case of modern motor vehicles where many safety features are built in to enhance safety, many of them only reduce the risk of an accident happening or minimize its impact: many of these technical features do not have the ability of stop an accident from happening. For example, parking distance control (PDC) automatically alerts the driver when coming close to a physical obstacle. However, it does not apply brakes to stop the vehicle from bumping into an obstacle. Collision can only be avoided if the driver stops the vehicle manually. Similarly, telecare technology is mainly responsive as many of the items featured in Figure 8.8. Very few systems have the capability of actively preventing an accident by proactively performing a task upon early detection of hazardous activities.

Most active telecare systems involve artificial intelligence. For example, detection and analysis of daily activities so that early signs of warning can be generated before something serious happens. This is particularly useful for older users, since what they see and what their inner ears sense may sometimes differ. In theory, any proactive system should address such differences and initiate corrective actions before something goes wrong. For example, a fall can be prevented if an imbalance is detected so that counterbalance can be activated prior to an actual fall. An older person may see the ground and sense how to move across it, but the actual action taken when taking a step forward may differ from that perceived. This happens because their visual reference may be distorted by poor eyesight. Counterbalancing such difference can be accomplished in a similar way as a cruise ship stabilizer that reduces the effect of the rocking motion. Stabilizers help the ship to stay straight and upright in waves and adverse weather conditions (Deer and Kemp 2006). These stabilizers function by extending wing-like flaps on both sides of the ship. The stabilizing mechanism will counteract the ship swaying by exerting a force, through distribution of its own weight, in the opposite direction, thus maintaining good balance. In the context of telecare, an equivalent system takes into account the user's life habits by constantly monitoring what the user does. The system can be "trained" to respond to any abnormalities to initiate responsive actions.

In addition to these existing solutions that provide assistive older people's care, there are also other implementation options such as using a set top box based way of providing monitoring and information for healthcare via the TV remote control, linking to information services but also directly to telemedicine and security equipment such as monitors and wireless cameras (Weinstein et al. 2018). Putting various technologies such as smart home, artificial intelligence, wireless sensing together opens up numerous opportunities for supporting older people's independent living by giving patients peace of mind while they enjoy their retirement.

8.2.4 Supporting Independent Living

Many retirees choose to live in rural communities for a combination of reasons, which brings its own challenges when seeking healthcare services. Issues such as vast travel distances between home and healthcare facilities, stringent privacy and security requirements related to safeguarding patient records, and the shortage of nursing home vacancies and healthcare providers are all contributing factors that drive the tremendous demand of supporting independent living through telemedicine. Smart and assistive technologies have emerged as the new paradigm for the older community in automation, quality of life, well-being, and peace of mind. Self-cognizant capability has become a requirement for the operation of homecare systems used in both independent living and nursing home settings (Fong et al. 2018). Round-the-clock care is made readily accessible through telemedicine by providing a wide range of services from tracking a patient's daily food intake and activities to receiving remote consultation without leaving home.

Real-time interactions between patients and physicians imply that a simple photograph allows diagnosis such as teledermatology and teleophthalmology to be provided without a visit to the clinic. One scenario that demonstrates the importance of telemedicine for senior citizens is disease management, from infectious disease prevention to chronic disease management. Before a patient leaves home, one can check on the intended destination so that risks associated with any location of interest such as commutable diseases and severe air pollutions can be avoided. A map showing cases of pandemic outbreak can allow the patient to avoid high risk areas, as in the case of Figure 8.11, which shows confirmed cases of measles in the region.

By using appropriate sensors according to an individual patient's health condition, monitoring a patient's health conditions like cardiac patients allows immediate response to be made when heart rate or blood pressure change is detected. Other risks, such as fall, can also be automatically detected by using accelerometers so that immediate assistance can again be provided. All these

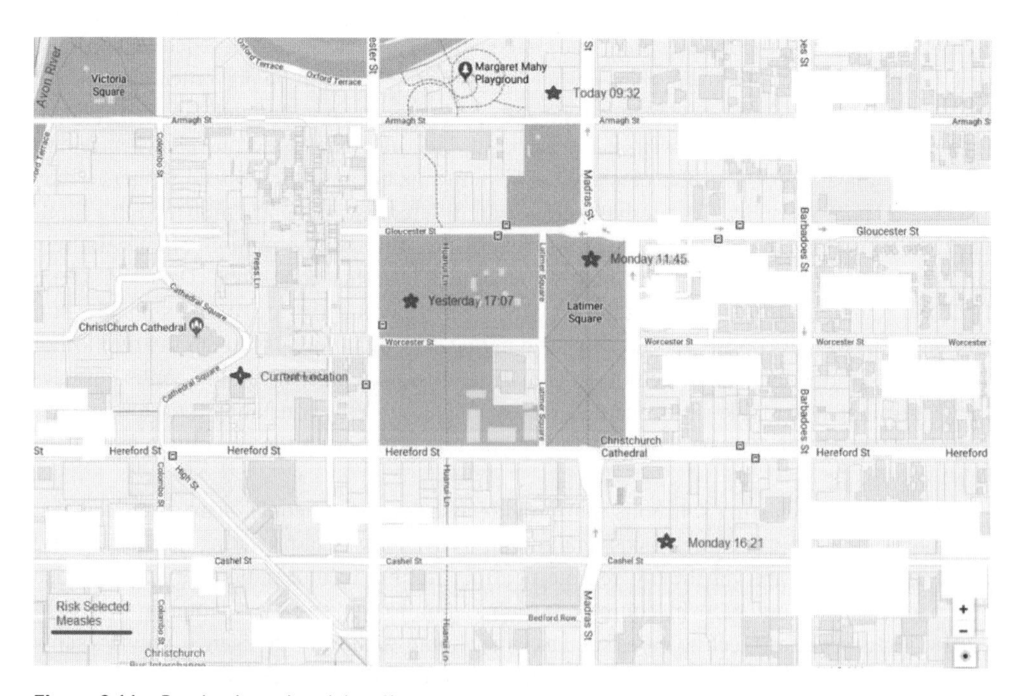

Figure 8.11 Pandemic outbreak locality map.

technologies ensure senior citizens that they are always well looked after while living alone and that help is readily available.

8.3 Telemedicine in Physiotherapy

Physiotherapy, or physical therapy, is widely practiced in relieving deterioration in movement due to aging (geriatric), injury (cardiopulmonary and orthopedic), or disease (neurological). As physical movement involves areas such as limbs and the back, it commonly relates to biomechanics of joints and spinal manipulation. As such, sensors may need to be small and able to detect minute movements in 3D space.

8.3.1 Movement Detection

As recovery and progress monitoring involves detection of movements, there are two major methods, namely sensors and video analysis. Many sensory systems are infrared-based, which means that movement of infrared emitting sources, such as human bodies, is tracked. Others involve mechanical switches and sensors, such as accelerometers and vibration sensing. Movement of different parts of the body may require different mechanisms. For example, spinal curvature (Chow et al. 2007) has very different requirements when compared to knee position (Brinker et al. 1999). Regardless of which technology is used, there is always a tradeoff between coverage area and precision.

Video sensing, as illustrated in Figure 8.12, can easily track the movement of the entire body inside a confined area. The coverage area depends on camera placement and lens focal length. In this example, six cameras, as shown in Figure 8.13, are installed. All cameras are connected to the computer, either with cable or wireless, so that the image captured from each camera at a given point in time can be compared and analyzed. By comparing the images acquired by all cameras with those of adjacent frames, movement can be tracked. The camera in Figure 8.13 has a photographic lens mounted, just as those used in single-lens reflex (SLR) cameras. The longer the lens's focal length, the more detail is captured with a greater closeup view so that greater precision is yielded. However, the angle of coverage is also reduced. Conversely, a wide-angle lens

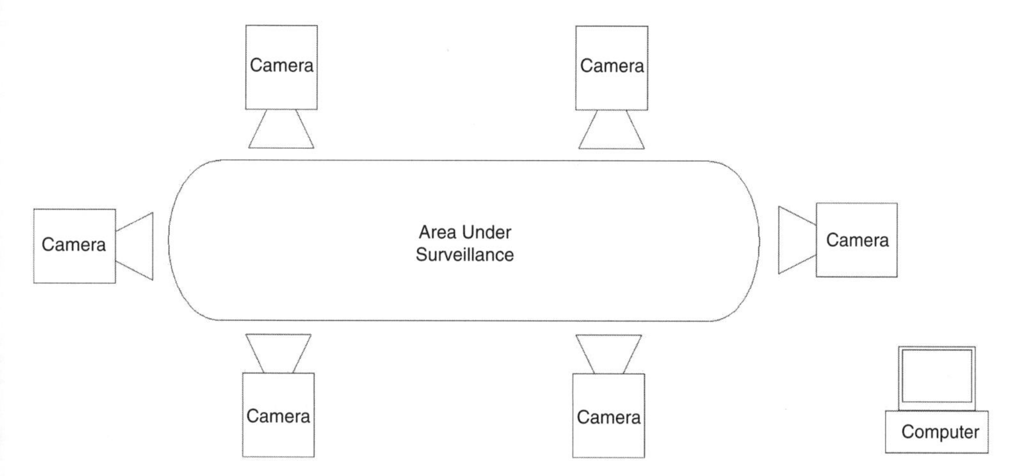

Figure 8.12 Video motion sensing network.

Figure 8.13 Motion tracking camera.

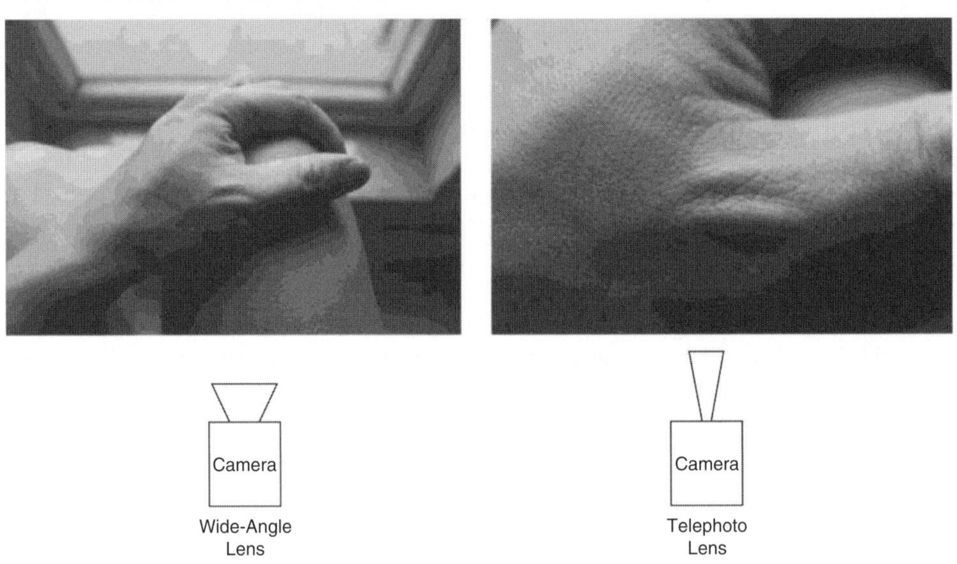

Figure 8.14 Lens focal length versus coverage angle.

provides wider coverage at the expense of less detail and precision. Such a tradeoff is illustrated in Figure 8.14.

A widely used alternative is the accelerometer network. By placing a number of accelerometers on the subject's body, such as the dummy shown in Figure 8.15, when the subject moves, each accelerometer will sense the movement in each of the three dimensions. Although the example in Figure 8.15 connects each accelerometer together using wires, additional circuitry can be installed with Zigbee communication capabilities. For details on the standards governing communications between medical devices, please refer to Chapter 2. An accelerometer is low cost and simple to fit.

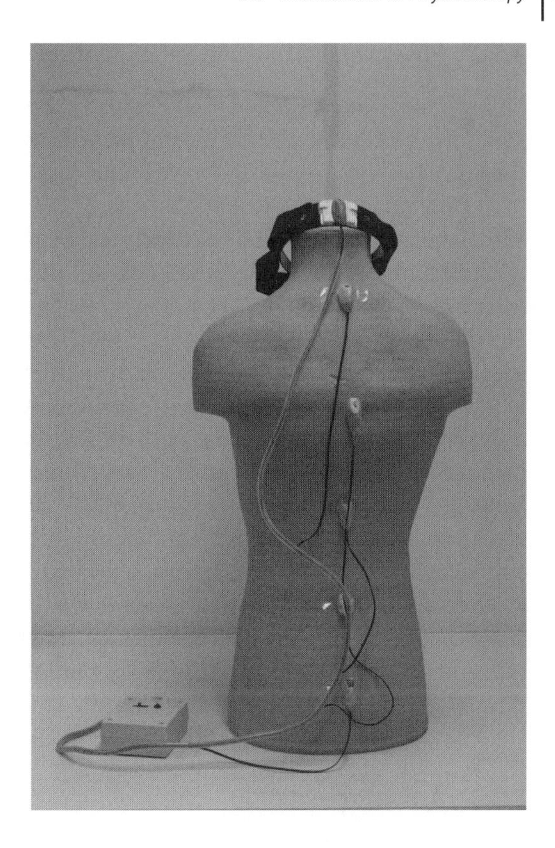

Figure 8.15 Installation of accelerometers on a dummy.

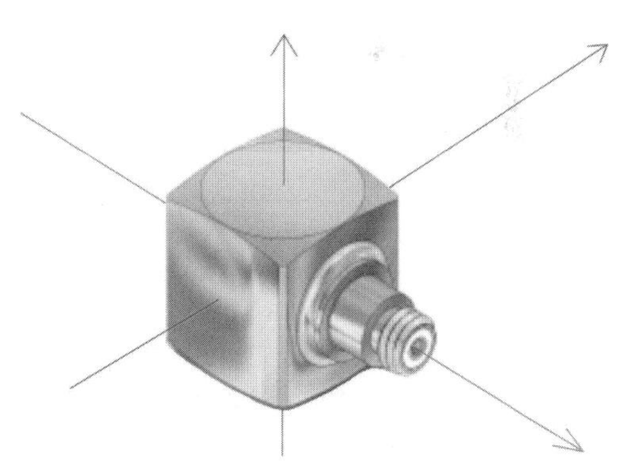

Figure 8.16 An accelerometer senses 3D movement.

The one depicted in Figure 8.16 is capable of detecting 3D movement. When fitted, any movement can be sensed, and sudden acceleration (change in speed and or direction) that may indicate a fall can trigger a remote alarm.

All these rely on technologies for detecting the magnitude, orientation, direction, and speed of movement. One major drawback of using an accelerometer for fall detection is that it operates by measuring its acceleration relative to freefall due to gravity. Therefore, an accelerometer will not produce an output when it undergoes freefall. To combat this problem, an accelerometer should be installed at an offset angle that produces a relative movement with respect to its vertical axis while falling downward.

8.3.2 Physical Medicine and Rehabilitation

Physical medicine and rehabilitation, also known as physiatry, is aimed at regaining functional abilities in an effort to combat the impact of disability. It deals with the recovery of muscles, bones, tissues, and nervous systems. Prognosis for various neuromuscular disorders can be accomplished by nerve conduction studies (NCS) and needle electromyography (EMG). As NCS involve electrical stimulation to peripheral nerves, these can be conducted remotely so that the patient does not have to travel to the clinic for diagnosis. This is particularly suited for spinal cord monitoring and has been used in studies on the impacts of schoolbags on children's backs (Chow et al. 2006). We look at a case study involving prevention of spinal injury by studying the effects of weight distribution of schoolbags on children in a case study in Section 7.2.2. Before this, we continue to look at how technology advances in telemedicine can bring relief to patients with physical impairments.

Palliative care and rehabilitation have long been considered as two important parts of comprehensive medical care for patients with advanced disease (Santiago-Palma and Payne 2001). Santiago-Palma and Payne also suggest that physical function and independence are, as in the case of older people, important attributes for both patients and caregivers. Palliative care involves psychological and spiritual support as a means of relieving distressing symptoms. Regulations governing the application of palliative care may differ in different countries from country to country. For example, the US requires certification by two physicians for a terminally ill patient whose remaining life expectancy is less than six months to be eligible for enrolment (Lamba and Mosenthal 2012). No such regulation exists in most other countries as of early 2020.

8.3.3 Active Prevention

Although telecare does not normally deal with prevention, technology does provide a mechanism for active prevention. For example, a patient who exercises after knee arthroscopy may need to restrict the amount of movement to prevent causing further injury in the event of overstretching. Necessary actions such as controlled passive stretching, hold–relax, repeated contractions, and assisted active exercises may be necessary for the recovering limb and free active exercise for unaffected areas to reduce edema. This does not consolidate and cause joint stiffness. Patellar tracking would become necessary for ensuring speedy recovery (Brunet et al. 2004). The appropriate installation of accelerometers would detect early signs of movement that may cause contractures and deformities, essentially serving as a splint that can dynamically track movement, instead of using a traditional static splint that immobilizes the entire limb. A limited range of movement is therefore allowed without the risk of overstretching. There is, however, a small catch with sensor placement, since sensors must be installed without nerve compression. Also, the force exerted on the sensors may be reduced by padding for bony prominences or areas where the bones protrude slightly below the skin. In situations where sensors are affixed to the patient's skin, skin impedance is often measured, as illustrated in Figure 8.17 (Thakur et al. 2019). In addition to the small catch discussed, we also need to remember two important aspects: first, the sensors themselves should only detect movement specific to the limb, but not the vibration that may be caused when the patient walks. To compensate for vibration, a series of tests under different use conditions may be necessary for the associated electronic components so that the sensors can continue to capture reliable readings as the subject moves (Fong and Li 2011). Another consideration is the wireless transmission of captured data, ensuring data integrity is vital when responding to a sudden situation and also to ensure no critical event is missed. A mechanism for ensuring a continual communication link is available may be necessary for each sensor and its networking

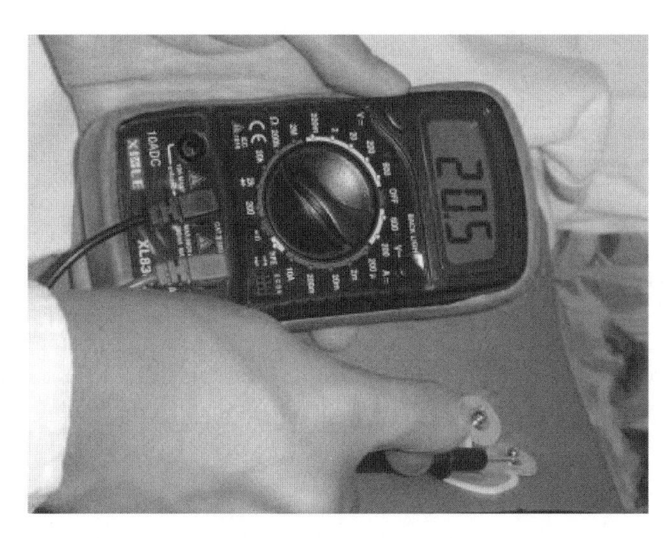

Figure 8.17 Skin impedance measurement.

device. For example, a polling system that sequentially checks the readiness of each sensor would ensure that all sensors are in range. A controller must be preprogrammed to detect early signs of a possible risk. This may involve the implementation of "fuzzy logic," an "intelligent" problem solving algorithm installed in an embedded system. The key feature of fuzzy logic is the ability to derive a decision based on equivocal and incomplete information. In this context, the algorithm is capable of detecting an alarming situation prior to its occurrence, based on subtle abnormal signs.

Unlike simple embedded system controllers that execute basic responses based on a number of predefined parameters (such as in the simple case of wireless insulin control system in Figure 8.18, where a simple controller regulates the glucose level by feedback from the glucose meter that controls the amount of insulin pumped solely upon the meter's reading), in a fuzzy logic controller, which takes multi-valued logic with values not just "0"s and "1"s. The algorithm relies on picking

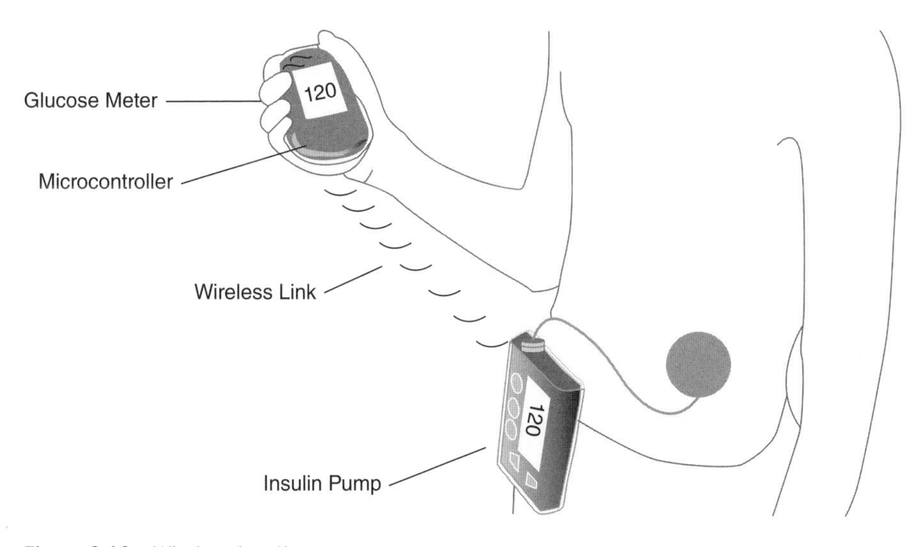

Figure 8.18 Wireless insulin pump.

up the rate of change of reading captured from the sensors connected to it such that it responds to the detected changes. For example, an array of accelerometers installed to detect the fall of an older patient may rely on successive readings that exhibit a significant change in movement relative to the regular pattern being read while walking steadily. The sudden significant change in reading within a relatively short period of time may be many times greater than that of what normal activities will generate. These readings do not have to follow any logical pattern in order to be identified as the detection of a fall so that it is not necessary to tweak the reading into any logical description.

Fuzzy logic implementation involves defining the control criteria and parameters. In the example of fall detection the parameters would be readings obtained from individual accelerometers. What is the normal range of reading when the patient undertakes normal activities? Are all sensors experiencing the same readings? What are the input and output relationships? Does the simultaneous detection of sudden acceleration downward indicate a fall or the patient intentionally bending down? The rule-based nature entails a series of expressions:

IF X AND Y THEN Z

that collectively define the output response for the given set of input conditions, namely the X's and Y's of each expression in the series. The simultaneous occurrence of X and Y would trigger the corresponding predefined action Z.

Remember, one objective of implementing fuzzy logic is not only to detect the occurrence of an event but also to proactively warn of the risk of an event. So, an output should be generated to indicate the risk when something is detected prior to its happening. An imbalance that may lead to a fall should therefore be detected and a warning issued prior to an actual fall. This will be triggered by a set of abnormal readings, such as the situation where X suddenly rises while Y descends. Any possible preventive actions can therefore be activated. Other expressions may include consequences of post-event action such as automatically alerting a response center after a fall.

8.4 Healthcare Access for Rural Areas

The problems associated with providing healthcare services in rural areas are very different from those in urban areas. Rural residents face a unique combination of factors that create disparities in healthcare. Lack of recognition by legislators and the isolation of living in remote rural mean that inhabitants are often overlooked by policymakers, or the money necessary to improve matters is deemed not worth spending. In isolated areas, where residents are more likely to be either self-employed or retired, people are far less likely to enjoy employer-provided healthcare coverage.

Funding is one major issue in any national healthcare system (Roemer 1993). Any vast project in extending healthcare services must therefore produce observable return on investment (ROI). Providing healthcare services to rural areas can be a significant challenge because of the population density that makes support very expensive. To demonstrate the connotation of serving rural areas, let's look at a case study from the United States. First, the financial incentive is an issue for operators. Of the millions of people living in rural areas, only a minority are Medicare beneficiaries. Medicare margins are particularly lower with small hospitals. Justification for the establishment of any adequately equipped hospitals would be difficult from a financial point of view. Driving down the cost of providing healthcare services would increase profit margin for service providers. This can be accomplished by advances in healthcare technologies and simplifying processes and formalities, and could mean that medical services could be extended to rural areas more efficiently and cheaply.

Another major problem arises from accident recovery and the prolonged delay between an accident and its response. Many of these delays are related to increased travel distances in rural areas and personnel distribution across response centers. In response to these problems in the 1990s, the US government's Telemedicine Report to Congress (Kantor and Irving 1997) stated: "Telemedicine also has the potential to improve the delivery of health care in America by bringing a wider range of services such as radiology, mental health services, and dermatology to underserved communities and individuals in both urban and rural areas," acknowledging the importance of providing healthcare services to rural areas through telemedicine.

Telecare is particularly suitable for rural areas where people can live alone with the assurance that they are well looked after. It has the following key features:

- Bring medical and healthcare technology to people everywhere.
- Provide more efficient and affordable services with easy access.
- Healthcare services are no longer limited to certain locations such as clinics and hospitals.
- To assist disabled people, take care of children and older people, heal the sick or injured, and to support vulnerable individuals.

There are, however, certain prerequisites that need to be dealt with. First, supporting infrastructure that provides coverage to the areas of concern must be available, for example an existing wireless network with sufficient bandwidth that can support all healthcare services. As telecare involves people in various locations, liability issues must be sorted out before providing any remote services. In this context, we may need to ask questions like who is responsible for overseeing the process, what if a mishap leads to insurance related issues, what happens if something fails and who would be held liable, etc. All these decisions and liabilities need to be thoroughly contemplated.

One main deployment consideration is whether existing infrastructure, if any, can support the desired services in terms of providing adequate resources and geographical coverage. In vast areas and low population densities, there may be no support at all; small settlements may have only very primitive telecommunication networks such as the plain old telephone system (POTS) available for nothing more than voice calls. Serving the farming community may be even more challenging, because the houses can be several miles apart. Even a small local clinic with the most basic equipment can be difficult since there are perhaps only a dozen of people within its proximity. Providing wireless telecare services is extremely difficult because of excessive signal loss. Going back to the fundamental issue of existing infrastructure again, the lack of adequate networking resources is even more acute in developing countries. Cloud computing has become increasingly popular in recent years. It may change the way IT infrastructure advances This is likely going to be particularly helpful for developing countries (Cleverley 2009). According to the definition on *wiki*, the "cloud" is a metaphor for the Internet, such that can support virtually any kind of services, including a range of healthcare services. Cloud computing, originally developed as a platform for supporting various common business applications online that are accessed from a Web browser, aims at making over the ICT model from a largely static connection between applications and hardware, and discrete expansion dictated by physical equipment limitations to an integrated computing platform capable of more granular scalability and flexibility. Such deployment freedom allows a wide range of multimedia telecare services to operate via different entities and different modes of application delivery. The basic cloud computing conceptual model can provide a number of telecare services, as shown in Figure 8.19.

Let us take a look at a case study in Grainger County, Tennessee, a rural area with approximately 20 000 residents without a hospital. This area is geographically isolated with limited road access caused by the Clinch Mountain and lakes. The project Rural Health Care Through Telemedicine:

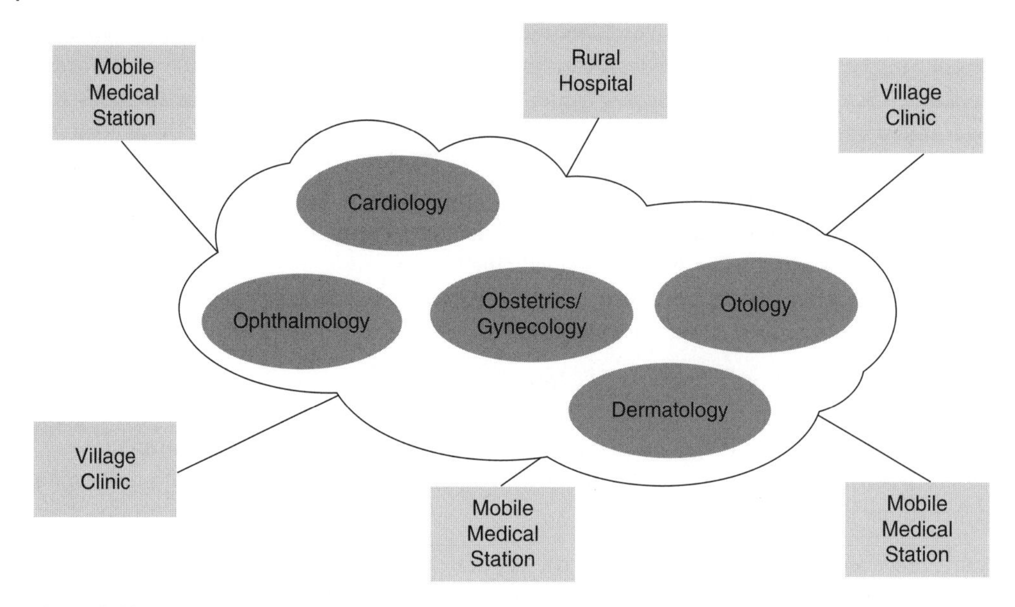

Figure 8.19 Telecare network.

An Interdisciplinary Approach has been implemented by the US Office of Rural Health Policy Rural Telemedicine Grant Program and the University of Tennessee with the main objective of improving access to healthcare services and to reduce the isolation of service providers in the county. Each of the county's four clinics had an interactive audiovisual telemedicine system installed and clinicians were trained to use the system for patient consultations. This supported a primary care physician in one of the rural clinics to examine a patient with remote support by specialist physicians at the university's medical center some distance away. For emergency health services, two ECG units each capable of transmitting 12- lead ECG data from a patient to both the local clinic and the remote university medical center were connected via a mobile phone network. For other nonemergency consultations, patients could communicate with service providers using a video phone connected to the POTS network. This system provided a mechanism for basic healthcare services for residents in a geographically isolated area. This was only possible given the necessary funding that supports initial deployment. To implement similar systems in a rural area, financial feasibility is most likely a major constraint. The readiness of an adequate existing network infrastructure and interoperability standards for necessary supporting software may also be issues that need to be addressed.

8.5 Healthcare Technology and the Environment

The industrial revolution changed the landscape of manufacturing and mining in America around the dawn of the nineteenth century. Fossil fuel burning and toxic gas discharge have significantly increased that in turn have created health-related issues such as air pollution and acid rain. Although there is no doubt that industrialization has a direct negative impact on people's health, the trend of industrialization has continued to spreads eastward into Asia in the postwar era. For example, the highly unsanitary business of battery manufacturing saw its shift from the US to Japan around the 1970s then into China about two decades later, and so the health hazards associated with industrialization have shifted gradually from developed countries to developing countries, which appear to be more willing to trade health deprivation for monetary profits. There are close relationships between healthcare and the environment, as well as the technologies behind them. We intend to conclude this chapter by taking a look at why healthcare is so closely linked to the environment, how healthcare technology plays a role in environmental protection, and the kind of environments healthcare technology is bringing to us. Healthcare technology has many implications for the environment, including everything from pollution of biological waste to radiation that may be hazardous. Conversely, environmental impacts can affect healthcare and the technologies related to it. For example, regulatory constraints may prohibit the use of certain materials; the environmental impact of the spread of disease has also caused great concern over the centuries.

8.5.1 A Long History

The links between healthcare and the environment have been deliberated for centuries. The first reported plague pandemic was probably that of 541 CE and originated in Egypt. Better known as The Plague of Justinian, it affected much of the Eastern Roman Empire (Little 2008). It is widely believed that bubonic plague first made its way to Europe via grain ships that had housed an immense rodent population. Believed to have wiped out around half of Europe's population by the year 590 (Maugh 2002), the plague continued to roam the world for another century before it subdued. Next came the "black death," which haunted much of the world around the middle of the fourteenth century. It was probably the best-known example of the close tie between healthcare technology and the environment. Some 600 years ago, there were thought to be three types of plagues responsible for wiping out an estimated half of Europe's people. Kelly (2005) suggests that the culprit was most likely a viral hemorrhagic fever spread by rodents. It was suggested that fleas that carried the plague originated from Asia and rats carried them into Europe via merchant vessels (Bramanti et al. 2016). When symptoms started to appear, a victim typically had a remaining life expectancy of about a week. Without any defense or knowledge of the cause of the pestilence, physicians were unable to provide any cure so those who got infected were abandoned. The disease proliferated very vigorously as victims communicated it to anyone who came near them.

Any effort in containing a disease must involve knowledge about its spread. Any mechanism that can combat the disease requires information to be gathered. Any such "technology" was not available during the outbreak. In fact, the causes of plague were not discovered until the nineteenth century. The plague, originally attached to rodents, was believed to be transmitted to humans by fleas. A flea, carrying ingested plague-infected blood from its host, i.e. the rodent, could live for as much as a month away from that host before finding its way to a new host, i.e. a human being. The plague therefore spread as the flea sucked blood from the human body when injected into that victim some of the blood already within it. Early telemedicine found its presence when people realized that the spread of plague could have been contained after a certain period of isolation. Ships suspected to have carried the plague were therefore quarantined and identified with a flag. They were only allowed to dock when the plague was thought to have vanished after the quarantine period. The primitive telemedicine health information communication technology shown in Figure 8.20 is a good example of early telemedicine deployment. In this example, information about the environment and the situation inside the ship was sent out for investigation in a remote location. The ship will communicate with the control center so that rodents and fleas that carry the plague cannot board the shore before the ship's environment can be assured to be safe.

Technology, if available at the time, would have helped in many ways. First, a telemedicine system would have helped diagnose and quarantine those who were infected, providing information

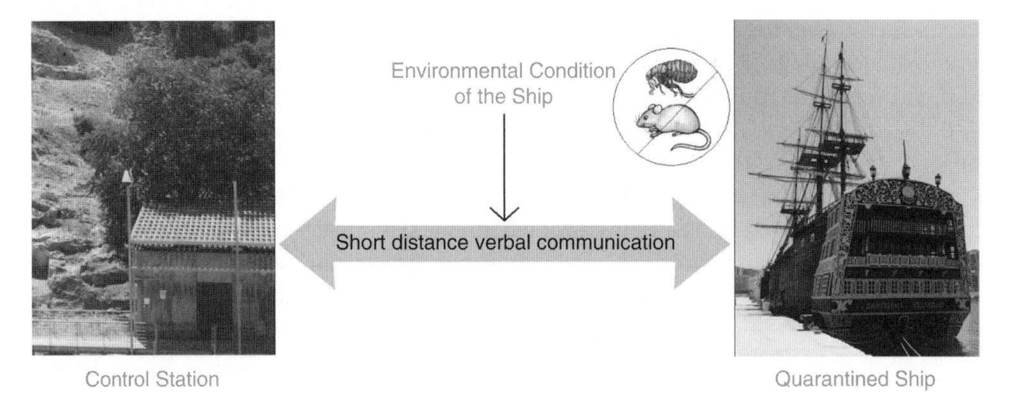

Figure 8.20 An ancient telemedicine system.

on treatment and therefore a better chance of survival. Clusters of those infected could be linked together for information sharing. Also, the plague's spread pattern could have been analyzed, thereby reducing the risk of spreading further by containing it. The study of plague could provide some insights into other types of pandemics. Although we know by now that plagues can be controlled by antibiotics such as streptomycin, gentamicin, or tetracycline (Massachusetts Institute of Technology 2008), some 2000 deaths from plague are still reported around the world each year. Telemedicine could provide a measure for combating plagues with mobile medical monitoring devices that continuously assess the health of people in different areas. Such monitoring systems will improve the ability of health authorities in different areas to react to and predict disease outbreak and its epidemiological spread due to different bacteria or viruses.

To assess the disease spread pattern and analyze the environmental impact on disease outbreak, computation modeling is regarded as the most appropriate method (Bloch 2009). A computation model can be generated by the collection of information based on spatial and temporal information of the occurrence of an infection, for example a progression model that analyzes epidemic spread in homogeneous and heterogeneous networks (Zhang et al. 2011). Such models are developed for computing the spread of effects such as environments, climate, and population movements between countries and regions in various scenarios. Information about each reported case is collected for computing the dynamic history of an outbreak to connecting clusters by adding the appropriate demographic and population mobility information. This would expand to a spatially and chronologically structured stochastic disease model that simulates the spread of epidemics from a suspected origin across the globe. For example, the 2009 A(H1N1) swine flu was believed to have originated from Mexico (Centers for Disease Control and Prevention 2009). To predict its spread a computational model would have commenced by using the first set of data for simulation with a cluster of infections that occurred in March 2009. As the disease spread, the predicted infected areas would have expanded outward from its origin. Initially, data was collected at the level of individual countries. Spots throughout the world appeared, owing to rapid population movement resulting from air travel. Further cases of outbreak about the location and time were then appended to construct a more comprehensive model over time.

There are, however, uncertainties involved in modeling disease spread. In the modern world, the seemingly random movement of airplanes that bring infected people across the world may carry the disease in a haphazard pattern such that spread phenomena with physics-based knowledge can no longer be used. As with historical events that date back to the Plague of Justinian, the environment plays a vital role in manipulating disease spread. The relationship between environment and disease could also be seen in the spread of diseases as changes to the ecosystems caused by environmental pollutions encouraged pathogen growth (Briggs 2003). Water contamination, poor sanitation, and poor hygiene are all contributing factors to rapid disease spread. Environmental protection therefore remains an important factor in disease control.

8.5.2 Energy Conservation and Safety

Energy conservation is always perceived as closely linked to environmental protection because of the general belief that the consumption of nonrenewable sources impacts the environment. Designing a medical device that is energy efficient is one important step toward maximizing ROI and product reliability. Particularly important for mobile medical devices, energy efficiency improves the cost effectiveness of a device and prolongs battery life. Safety related to the use of medical devices is vitally important since a device failure may lead to fatality. Safety assurance may entail:

- Identifying potential hazards during operation.
- Quantifying damage potential, e.g. through computational modeling or prognostics techniques.
- Evaluating all necessary safety measures.
- Taking remedial measures for reducing and controlling risks.
- User training to ensure proper use.

Protective housing is always a vital measure to ensure operational safety. However, there may be a tradeoff between efficiency and the level of protection related to the material used. For example, although a metal housing may provide protection against physical impact and electromagnetic interference, metal is generally not a suitable material for medical devices used in telemedicine applications, because many of them transmit and receive data through wireless media. With wireless medical diagnosis and monitoring devices, metal is generally not suitable, because of its conductive properties that reflect electromagnetic energy at their surfaces except at extremely high frequencies. Electromagnetic energy penetrates a distance into the metal that increases with wavelength, known as the "skin depth." Lower frequency electromagnetic waves can therefore propagate through metal with a certain amount of attenuation if the wall is thin enough, while higher frequency electromagnetic waves, because they do not significantly penetrate into the metal, are reflected like a mirror. Such properties can also be useful since the conductive housing of the device then effectively shields the internal electronic circuitry from higher frequency electromagnetic interference that could adversely affect device operation.

Careful design consideration is necessary for energy conservation and transmission efficiency if the transmitting antenna is also within the conductive housing. Telemetry, technology that allows the remote measurement and reporting of information, must be performed with lower carrier frequencies since the housing effectively acts as a low-pass filter. This reduces the effective data rate that can be supported by the wireless link and increases the necessary transmitting power for the implantable device when sending captured data to a remote device. Such additional transmitting power requirements for the implantable device result in limiting the range over which it can operate.

In these transmitting devices, particularly for implantable devices, efficiency and orientation may impact power consumption. An antenna for such application should be coupled with a reflective plate to increase the gain of an antenna in a selected direction. This normally points away from the patient's body. Electromagnetic waves radiated by the antenna tend to attenuate when encountering obstructions such as tissue and water. An antenna that is designed with a selected transmission direction is known as a "directional antenna." Its main advantage is to enhance the power of the antenna in the selected direction, thereby increasing the transmission distance. Such types of antenna usually have a reflective plate on one side to increase the directionality of the antenna and increase the gain of the antenna. The reflective plate reflects radiated signals for transmission and receiving and hence increases the directional radiation gain of the antenna.

In addition to antenna efficiency, control of the specific absorption rate (SAR) also needs to be optimized for power saving. Transmitted devices are required to meet certain regulatory requirements for maximum SAR levels in some countries. Such regulations are aimed at imposing appropriate limits for users of wireless devices from the perspective of energy absorption into body tissues. SAR is a description of the time t derivative of the incremental energy dU deposited in an incremental mass dM contained in a volume element dV of density ρ, written as:

$$SAR = \frac{d\left[{}^{dU}/_{dM}\right]}{dt} = \frac{E^2}{\rho} \tag{8.1}$$

SAR, measured in watts per kilogram (or equivalently milliwatts per gram mW/g), is a measure that estimates the amount of radio frequency power absorbed in a unit mass of body tissue. The SAR restriction varies from around 1.6 to 2.0 mW/g depending on legislation. Also, SAR limits can be different for different regions of the human body. Compliance with the applicable maximum SAR limits is usually obtained under specific environmental and operational conditions.

Practical SAR values can deviate from anticipated measurement results for a number of reasons:

- frequency or energy of the incident radiation relative to the composition of the tissue mass being measured.
- radiation intensity of the device and the proximity of the device to the tissue.
- any nearby reflecting surfaces and their orientation.
- transmission power of the device to establish and maintain communication.
- orientation (polarity) of the field vectors relative to the tissue.

All these can be managed during the design stage. However, in order to comply with certain standards the carrier frequency may be fixed. Active control of the transmission power output can optimize battery life. To improve transmission efficiency, it is also necessary to prevent antenna detuning, which can occur because of proximity to objects, owing to electromagnetic capacitive or inductive coupling. Shielding by use of appropriate housing or active control of the antenna can combat these problems.

8.5.3 Medical Radiation: Risks, Myths, and Misperceptions

Ever since the discovery of X-ray for medical imaging that relies on the different rates of energy absorption by different tissue (and bone) types, radiation exposures from diagnostic medical examinations has been considered for over a century (Filler 2010). As discussed in Section 4.2.1.2 earlier, the effectiveness of X-ray radiography is governed by the intensity of radiation dosage. The amount of radiation that may cause health problems needs to be thoroughly investigated. Villforth (1985) suggests that, in America, human exposure to ionizing radiation is almost all related to medical diagnostic radiology, which suggests that radiation from the ambient environment could just be as high as from medical radiography. This is a question that does have grounds for dispute because "gamma rays" from disintegrating nuclei of radioactive substances that naturally exist discharge even more energy than X-rays do (Als-Nielsen and McMorrow 2001).

The therapeutic use of radiation naturally involves higher exposures. Its associated risk is assessed by a physician before examination. Standardized radiation dose estimates can be given for a number of typical diagnostic medical procedures, yet the dosage can be very different depending on individual circumstances. Each patient's metabolism and the type of examination are important considerations when determining the dosage. The exposures are widely considered comparable to those that are routinely generated from natural radiation in the surrounding environment. Obviously, some energy of the X-ray is absorbed within the body since bones and tissues block some of the radiation that in turn forms the radiograph showing up as shadows on the film. As a consequence, some cells may die prematurely – although, the amount of cell damage is quite minimal. Such damaged cells do not actually pose any risk since they are naturally replaced. However, the health risk possibly comes because some of the cells may not die but instead sustain genetic damage. Such damage can, in rare cases, result in the cell becoming cancerous.

The dosage varies, depending on applications and diagnosis areas. For example, the typical dosage of dental X-ray is around one-third of that of a chest X-ray. Computed axial tomography (CAT) scans, also known as CT scans, subject the patient to the X-ray scanner for less than 30 minutes to complete a full body scan. Some CAT scanners use up to 300 X-ray scanners taking 300 pictures each. This generates some 90 000 X-ray slices, or tomograms, to form the overall picture. The amount of radiation received in a CAT scan is usually around 10 mSv. This is equivalent to about 60 medical X-ray doses. Note, incidentally, that this is approximately twice the recommended maximum radiation dose for a pregnant woman. CAT should therefore be avoided for pregnant women.

As described in Section 4.2.2, positron emission tomography (PET) is a nuclear medicine imaging technique that relies on the circulation of an injected radioactive substance that emits "positrons" (high speed electrons) and "gamma rays" (highly energetic ionized radiation produced by subatomic particle interactions. PET relies on detecting pairs of gamma rays emitted indirectly by a radioactive tracer, and the scanner reads gamma rays like the CAT scanner reads X-rays. As the radioactive tracer travels through certain parts of the body, PET is capable of producing more detailed images of a specific organ. However, the emission of gamma rays may pose health hazards. With a typical dose somewhat higher than that of a CAT scan, the use of PET should be precluded unless it is absolutely necessary.

Sources of radioactivity in the ambient environment include the colorless, odorless radioactive noble gas radon (Rn^{222}), itself a product of the natural radioactive decay chain of uranium (U^{238}) found in soil and rocks around the world (Adams et al. 1964). Both radon and uranium emit gamma rays. Owing to uranium's enormously long half-life of billions of years ("half-life" is a term that corresponds to the time period in which half of the atoms decay into another element, e.g. from uranium into radon), both of these radioactive substances will retain their presence at the same concentrations, thus the amount of radioactivity caused will remain the same (Agency for Toxic Substances and Disease Registry 2012). Exposure to excessive concentrations of radon only poses a health risk in low elevation indoor enclosures such as basements, and is known to increase the risk of developing lung cancer (National Cancer Institute 2020). Radon and its floating radioactive products, such as polonium (Po^{218}) and lead (Pb^{214}), can be absorbed in a human body through inhalation. Heavy metal particles therefore accumulate inside the body as radon over decades. Along with other gases such as oxygen and carbon monoxide, radon readily dissolves in the blood and circulates throughout the body. So, radon is sucked in along with air whenever we breathe. It can also leave the body by exhalation through the lungs or sweating through the skin. Its seriousness is reported by the Environment Protection Agency (2016), which reports that over 20 000 people die in the US alone because of radon induced lung cancer. An average person receives a higher radiation dose from radon at home than from anywhere else with other natural and manmade radiation

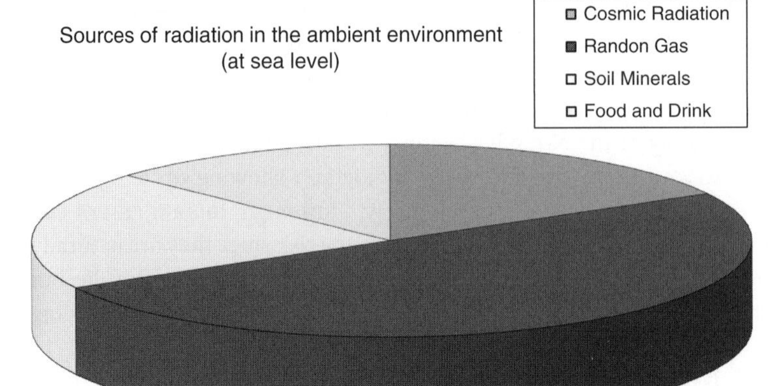

Figure 8.21 Radiation sources.

sources combined. From the composition shown in Figure 8.21, we can see that about half of all natural radiation sources come from radon.

In the natural world, any inhaled radon atom that decays before it has a chance to leave the body will form heavy metal particles that accumulate in the lungs and tracheobronchial tree, predominantly in bifurcations. Subsequent radioactive decay of the accumulated heavy metal may emit sufficient energy to damage surrounding epithelial cells. If trapped in the bloodstream, there is also a small risk of causing leukemia or sickle cell anemia, owing to radioactive residues left by radon decay.

Cosmic radiation originating from outer space and the sun consists of energetic charged particles such as protons and helium ions, and is known to affect air travelers. The biological damage caused by subatomic particles is widely believed to be more serious compared to X-rays or gamma rays. The intensity of cosmic radiation depends on altitude, latitude, and solar activities, according to the US Federal Aviation Administration (FAA 2014). At cruising altitude of around 33 000 ft. (10 000 m), an airplane is subject to cosmic radiation of some 100 times of that at sea level. The cosmic radiation intensity generally increases as we fly away from the equator toward the poles because of the diminishing shielding effect of the earth's magnetic field. On average, a few hundreds of flying hours per year would absorb a comparable amount of radiation dosage by an average person on the ground (Lewis et al. 2000).

Energy emitted by radiation, both X-ray and radioactivity, can carry sufficient energy to trigger genomic changes to the cell's DNA, including mutation and transformation. The consequential effect of genetic mutations and chromosome aberrations may cause birth defects in future generations if the defective gene is carried. Another potential problem is chemical radicals that can be created inside the cell.

Radiation risk to fetuses is higher than to children as the excessive energy can damage fragile embryonic cells. Children are more susceptible to radioactive emissions due the combined effect of their rapidly dividing cells and higher breathing rates; the latter translates to breathing in more radioactive radon gas. A single x-ray dose to a pregnant woman in the first six weeks of pregnancy can lead to as much as a 50% increase in cancer and leukemia risk to the unborn baby. Carcinogens cause random damage to the chromosomes and DNA molecules inside the cell's nucleus. While such damage normally destroys the cell completely, there is still a risk that a partially damaged cell can survive and reproduce with its defects sustained. Such cells can then proliferate in a cancerous behavior that ultimately develops into a cancer tumor.

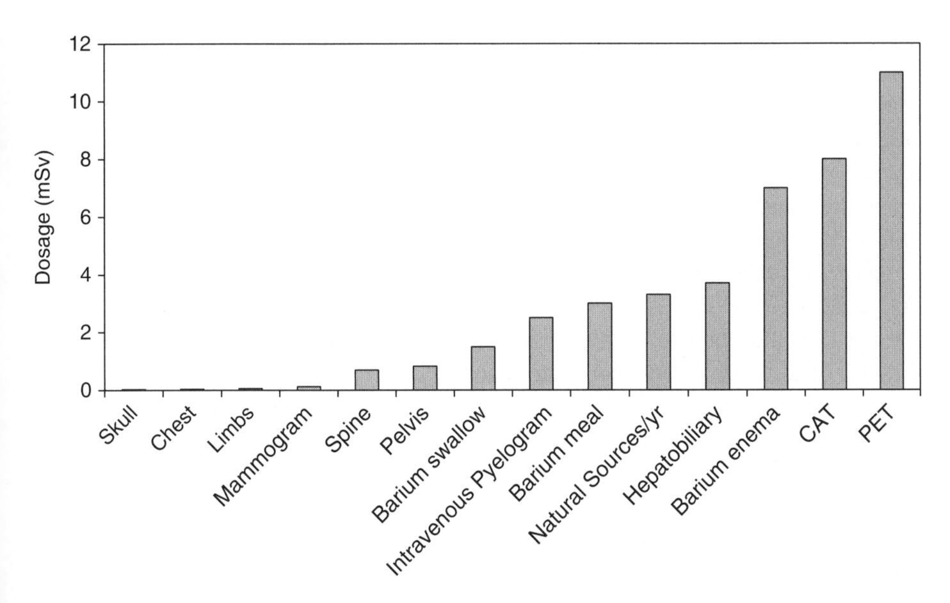

Figure 8.22 X-ray dosage.

So, how much is too much? Quantitatively describing the amount of radiation (from medical diagnosis and the natural environment alike) can sometimes be confusing because different standards and units exist. The millirem, mrem, millirad, and mrad are all identical measurement units. Also, 1 mSv is equivalent to 100 millirem. To understand how much one unit of mSv is, we generate a chart that shows the typical dose of X-ray based on figures given by Wall and Hart (1997) and UNSCEAR (2017) in Figure 8.22. This chart shows us that the amount of dosage from a few X-ray examinations combined would be very insignificant compared to what an average person is subjected to from natural radiation annually. Cumulative exposure from CAT scans may slightly increase the risk of cancer (Reinberg 2009).

References

Adams, J.A.S., Louder, W.M., Phair, G., and Gottfried, D. (1964). *The Natural Radiation Environment*. US Department of Energy Nuclear Testing Archive, Accession Number: NV0053511, Document Number: 57452.

Adams, C., Ide, T., Barnett, J., and Detges, A. (2018). Sampling bias in climate–conflict research. *Nature Climate Change* 8 (3): 200–203.

Agency for Toxic Substances and Disease Registry (ATSDR) (2012). *Toxicological Profile for Radon*. Atlanta, GA. https://www.atsdr.cdc.gov/ToxProfiles/tp.asp?id=407&tid=71 (accessed 20 January 2020): Public Health Service.

Als-Nielsen, J. and McMorrow, D. (2001). *Elements of Modern X-ray Physics*. Wiley.

Bloch, C. (2009). NIH Awards Grants. *Federal Telemedicine News* (8 September). http://telemedicinenews.blogspot.com/2009/09/nih-awards-grants.html (accessed 20 January 2020).

Bramanti, B., Stenseth, N.C., Walløe, L., and Lei, X. (2016). Plague: A disease which changed the path of human civilization. *Advances in Experimental Medicine and Biology* 918: 1–26.

Briggs, D. (2003). Environmental pollution and the global burden of disease. *British Medical Bulletin* 68 (1): 1–24.

Brinker, M.R., Garcia, R., Barrack, R.L. et al. (1999). An analysis of sports knee evaluation instruments. *American Journal of Knee Surgery* 12 (1): 15–24.

Brunet, M.E., Brinker, M.R., Cook, S.D. et al. (2004). *Disease and History*, 2e. Sutton Publishing.

Centers for Disease Control and Prevention (2009). Outbreak of Swine-Origin Influenza A (H1N1) Virus Infection: Mexico, March–April 2009. *MMWR* (30 April). http://www.cdc.gov/mmwr/preview/mmwrhtml/mm58d0430a2.htm (accessed 20 January 2020).

Chow, D.H., Kwok, M.L., Cheng, J.C. et al. (2006). The effect of backpack weight on the standing posture and balance of schoolgirls with adolescent idiopathic scoliosis and normal controls. *Gait & Posture* 24 (2): 173–181.

Chow, D.H., Leung, D.S., and Holmes, A.D. (2007). The effects of load carriage and bracing on the balance of schoolgirls with adolescent idiopathic scoliosis. *European Spine Journal* 16 (9): 1351–1358.

Cleverley, M. (2009). How ICT advances might help developing nations. *Communications of the ACM* 52 (9): 30–32.

Darrow, A.A., Johnson, C.M., Ghetti, C.M., and Achey, C.A. (2001). An analysis of music therapy student practicum behaviors and their relationship to clinical effectiveness: an exploratory investigation. *Journal of Music Therapy* 38 (4): 307–320.

Deer, J.C.B. and Kemp, P. (2006). *The Oxford Companion to Ships and the Sea*, 2e. Oxford University Press.

eHealth Europe (2009). FASS up to 100,000 telecare deployments. http://www.ehealtheurope.net/News/4985/fass_up_to_100000_telecare_deployments

Environment Protection Agency (2016). Assessment of risks from radon in homes (EPA 402/K-12/002 | 2016 |): a citizen's guide to radon: the guide to protecting yourself and your family from radon. US Environmental Protection Agency. https://www.epa.gov/sites/production/files/2016-12/documents/2016_a_citizens_guide_to_radon.pdf (accessed 20 January 2020).

FAA (2014). Advisory Circular 120-61B In-Flight Radiation Exposure. https://www.faa.gov/regulations_policies/advisory_circulars/index.cfm/go/document.information/documentID/1026386 (accessed 20 January 2020).

Filler, A.G. (2010). The history, development and impact of computed imaging in neurological diagnosis and neurosurgery: CT, MRI, and DTI. *Internet Journal of Neurosurgery* 7 (1): 1–69.

Fong, B. and Li, C.K. (2011). Methods for assessing product reliability: looking for enhancements by adopting condition-based monitoring. *IEEE Consumer Electronics Magazine* 1 (1): 43–48.

Fong, A.C.M., Fong, B., and Hong, G. (2018). Short-range tracking using smart clothing sensors: a case study of using low power wireless sensors for patient tracking in a nursing home setting. In: *2018 IEEE 3rd International Conference on Communication and Information Systems (ICCIS)*, 169–172. IEEE.

Hussain, A., Wenbi, R., da Silva, A.L. et al. (2015). Health and emergency-care platform for the elderly and disabled people in the smart city. *Journal of Systems and Software* 110: 253–263.

HealthIT.gov (2019). Telemedicine and telehealth. https://www.healthit.gov/topic/health-it-initiatives/telemedicine-and-telehealth (accessed 20 January 2020).

Kantor, M., & Irving, L. (1997). Telemedicine Report to Congress. US Department of Commerce in conjunction with the Department of Health and Human Services. http://www.ntia.doc.gov/reports/telemed/index.htm (accessed 20 January 2020).

Kelliher, C. and Parry, E. (2015). Change in healthcare: the impact on NHS managers. *Journal of Organizational Change Management* 28 (4): 591–602.

Kelly, J. (2005). *The Great Mortality: An Intimate History of the Black Death, the Most Devastating Plague of All Time*. HarperCollins Publishers.

Lamba, S. and Mosenthal, A.C. (2012). Hospice and palliative medicine: a novel subspecialty of emergency medicine. *Journal of Emergency Medicine* 43 (5): 849–853.

Lemyre, L., Gibson, S., Zlepnig, J. et al. (2009). Emergency preparedness for higher risk populations: psychosocial considerations. *Radiation Protection Dosimetry* 134 (3–4): 207–214.

Lewis, B.J., Bennett, L.G.I., and Green, A.R. (2000). Cosmic radiation exposure on Canadian-based commercial airline routes. *Radiation Protection Dosimetry* 87 (4): 299–301.

Li, J., Wilson, L.S., Qiao, R.Y. et al. (2006). Development of a broadband telehealth system for critical care: process and lessons learned. *Telemedicine Journal & e-Health* 12 (5): 552–560.

Little, L.K. (2008). *Plague and the End of Antiquity: The Pandemic of 541–750*. Cambridge University Press.

Massachusetts Institute of Technology (2008). Bacterial "battle for survival" leads to new antibiotic. *ScienceDaily* (27 February). www.sciencedaily.com /releases/2008/02/080226115618.htm (accessed 20 January 2020).

Maugh, T. H. (2002). An empire's epidemic: scientists use DNA in search for answers to 6th century plague. *Los Angeles Times* (6 May). https://www.latimes.com/archives/la-xpm-2002-may-06-sci-plague6-story.html (accessed 20 January 2020).

National Cancer Institute (2020). Radon and cancer. https://www.cancer.gov/about-cancer/causes-prevention/risk/substances/radon/radon-fact-sheet (accessed 20 January 2020).

National Statistics (2018). Population estimates for the UK, England and Wales, Scotland and Northern Ireland: mid-2018. https://www.ons.gov.uk/peoplepopulationandcommunity/populationandmigration/populationestimates/bulletins/annualmidyearpopulationestimates/mid2018 (accessed 20 January 2020).

Nobel Prize (2009). The Nobel Prize in Physiology or Medicine 2009. http://nobelprize.org/nobel_prizes/medicine/laureates/2009 (accessed 20 January 2020).

Quarantelli, E.L. (2008). Conventional beliefs and counterintuitive realities. *Social Research: An International Quarterly* 75 (3): 873–904.

Reinberg, S. (2009), As CT radiation accumulates, cancer risk may rise. *US News and World Report* (31 March). http://health.usnews.com/articles/health/healthday/2009/03/31/as-ct-radiation-accumulates-cancer-risk-may-rise.html (accessed 20 January 2020).

Roemer, M.I. (1993). *National Health Systems of the World: The Issues*. Oxford University Press.

Santiago-Palma, J. and Payne, R. (2001). Palliative care and rehabilitation. *Cancer* 924: 1049–1052.

Shneiderman, B. (2005). *Designing the User Interface: Strategies for Effective. Human-Computer Interaction*. Reading, MA: Addison-Wesley Publishers.

Standley, J.M. and Prickett, C.A. (1994). *Research in Music Therapy: A Tradition of Excellence*. The National Association for Music Therapy, Inc.

Thakur, R., Jin, A., Nair, A., and Fridman, G.Y. (2019). Nerve cuff electrode pressure estimation via electrical impedance measurement. *Journal of Neural Engineering* 16 (6): 064003. https://doi.org/10.1088/1741-2552/ab486f.

UNSCEAR (2017). *Sources and Effects of Ionizing Radiation*. New York: United Nations Scientific Committee on the Effects of Atomic Radiation.

Veenema, T.G. (ed.) (2018). *Disaster Nursing and Emergency Preparedness*. Springer Publishing Company.

Vergados, D.D. (2007). Simulation and modeling bandwidth control in wireless healthcare information systems. *Simulation* 83 (4): 347–364.

Villforth, J.C. (1985). Medical radiation protection: a long view. *American Journal of Roentgenology* 145 (6): 1114–1118.

Wall, B.F. and Hart, D. (1997). Revised radiation doses for typical X-ray examinations: report on a recent review of doses to patients from medical X-ray examinations in the UK by NRPB: National Radiological Protection Board. *The British Journal of Radiology* 70 (833): 437–439.

Weinstein, R.S., Krupinski, E.A., and Doarn, C.R. (2018). Clinical examination component of telemedicine, telehealth, mhealth, and connected health medical practices. *Medical Clinics* 102 (3): 533–544.

Zhang, H., Small, M., and Fu, X. (2011). Staged progression model for epidemic spread on homogeneous and heterogeneous networks. *Journal of Systems Science and Complexity* 24 (4): 619.

9

Wearable Healthcare

Advances in consumer wearable devices over the past decade open up numerous possibilities in telemedicine services. In particular, the miniaturization of components such as sensors and batteries allows many healthcare applications to be incorporated in a single wearable device (Tricoli et al. 2017). Many mobile healthcare systems have been miniaturized to tiny multifunction wearable devices, and in this chapter we focus on opportunities in providing comprehensive and preventive care in telemedicine through wearable technologies. These devices range from consumer health monitors for general fitness tracking to noninvasive medical devices that provide round-the-clock support for chronic disease management. The main distinction between these two types of wearable device is the measurement accuracy for the purpose of medical diagnosis and assertions, which is described in more detail in Section 9.2.

9.1 From Mobile to Wearable

Many consumer electronics devices have shrunk in size over recent years as a direct result of system-on-chip (SoC) and battery advances (Alioto and Shahghasemi 2017). The shrink in size not only makes consumer healthcare as well as medical devices ultraportable but also opens up numerous opportunities for implantable medical sensing systems and invasive diagnosis (Naranjo-Hernández et al. 2019). In fact, many of these devices and systems are connected for purposes like control and health monitoring, such as those found in automatic insulin pumps and cardiac monitors. Like most consumer electronics products such as the transition from CRT (cathode ray tube) televisions of the last century to modern Internet-enabled OLED (organic light-emitting diode) televisions as well as analog mobile phones of the 1980s to smartphones in recent years, these devices have been made much smaller and lighter while offering far more features than before; it is not surprising that the same trend applies to healthcare devices and systems. How comfortable is a device being worn or even implanted into a patient all depends on its size while not compromising on safety and performance.

9.1.1 Size Matters

Mobile devices from the last decade have evolved to tiny wearable devices that patients can wear with them day and night without affecting their daily activities. Telemedicine plays a vital role in connecting these devices. As an integral part of an Internet of things (IoT) healthcare system, more comprehensive and efficient health services can be supported through a smart city framework, as in Figure 9.1, (Dimitrov 2016). When viewed by a patient, the patient has comprehensive

Telemedicine Technologies: Information Technologies in Medicine and Digital Health, Second Edition.
Bernard Fong, A.C.M. Fong, and C.K. Li.
© 2020 John Wiley & Sons Ltd. Published 2020 by John Wiley & Sons Ltd.

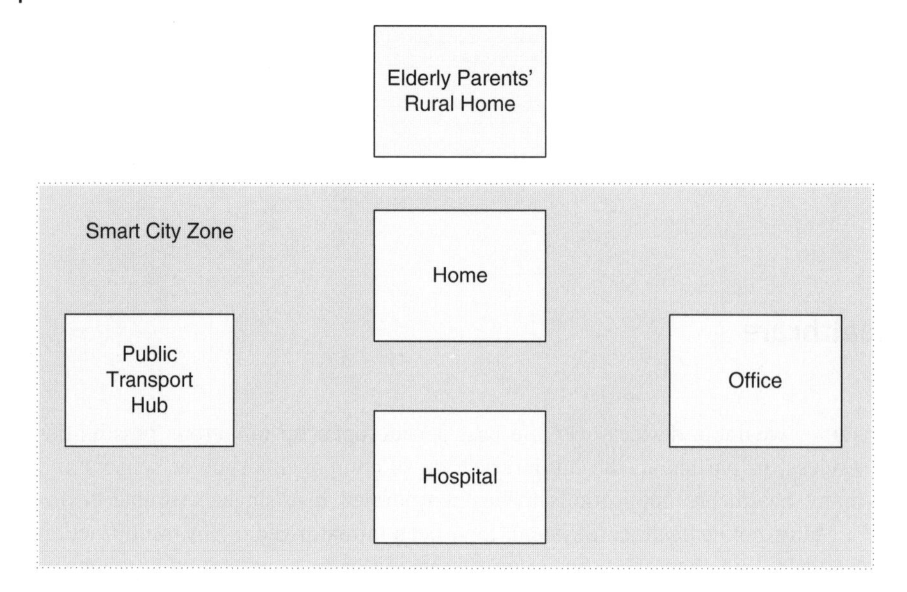

Figure 9.1 Telemedicine in a smart city framework.

coverage throughout the city, being linked to the hospital at all times across an IoT backbone that supports not only health monitoring but also other aspects of life to enhance efficiency (Mohanty et al. 2016). As an example, a patient with chronic obstructive pulmonary disease (COPD) can obtain real-time information on air pollution severity, and an optimal commuting route can be customized to avoid the most heavily polluted areas (Pérez-Roman et al. 2020). Additionally, the patient can also check out the elderly parents' well-being remotely through smart home integration, as described in Section 8.2.1.

The relationship between aging and chronic disease is a growing challenge in many metropolitan cities across the world (Dantas et al. 2019). The idea of extending connectivity beyond the smart city zone facilitates independent living while maintaining optimal health and increases social participation by connecting to caregivers, friends, and family members. IoT in smart city telemedicine is set to provide solutions for older people's healthcare and lifestyle management in addition to providing efficient healthcare services to residents within the smart city zone. Telemedicine enables personalized and preventive care services to be provided by hospitals and clinics through utilizing IoT with both fixed and wearable sensors for both disease management and diagnosis.

Telemedicine services keep track of parameters like food intake, daily activities, and geographical location using numerous applications designed for everything from general dietary monitoring and chronic disease management to safeguarding patients with cognitive impairments. Acquisition as well as subsequent analysis and processing of such information are all carried out at the device level, i.e. the device should be capable of performing critical functions such as anomaly warning. Any information to be sent across the telemedicine network should be already analyzed for initiating assistance or storage in electronic patient record (EPR) databases. This is because sending any data out to a remote location, such as a response center or data storage facility, is subject to uncontrollable factors and the data can simply be lost or corrupted, as discussed in Section 2.4. As in the example of a smartwatch, small wearable devices are capable of performing many functions with a variety of built-in sensors. Small devices and systems can function in different forms, like being embedded in clothing, shoes and accessories.

Table 9.1 Measurement from a controlled experiment.

Subject 1:	Female	32 y/o	5′ 5″	103 lb				
Time	Pulse (times/min)	Hemoglobin (g/L)	Blood Glucose (mmol/l)	Blood Flow (ml/min)	SaO_2 concentration	Finger Temperature (Celsius)	Humidity (%)	Room Temperature (Celsius)
09 : 52	57	151.5	6.6	180	98	29.6	84.6	26.1
11 : 56	79	149.8	17.5	203	99	30.5	83.4	26.9
Subject 2	Male	28 y/o	5′ 10″	139 lb				
Time	Pulse (times/min)	Hemoglobin (g/l)	Blood Glucose (mmol/l)	Blood Flow (ml/min)	SaO_2 concentration	Finger Temperature (Celsius)	Humidity (%)	Room Temperature (Celsius)
09 : 45	82	157	7.4	292	97	34.1	99	24.2
11 : 17	84	156.5	11.9	285	96	32.5	92.9	23.3

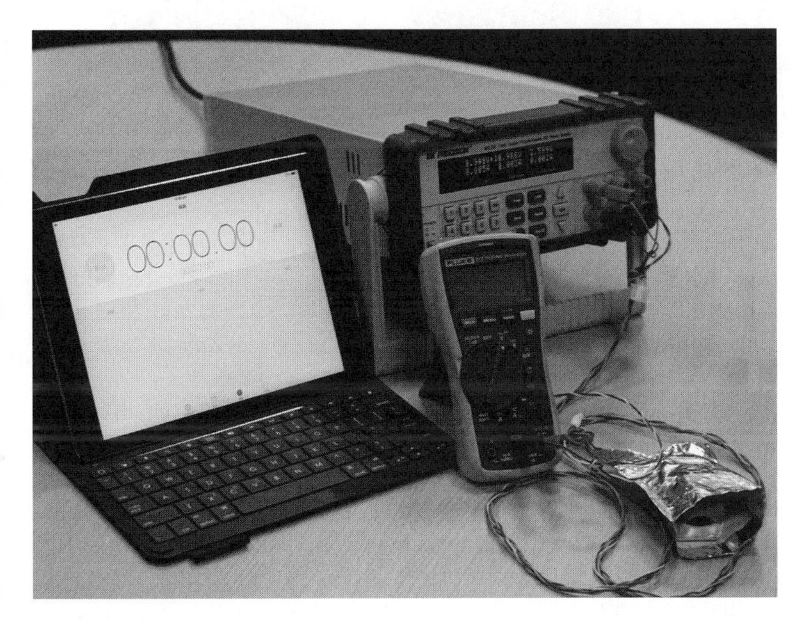

Figure 9.2 Experimental setup for noninvasive glucose monitor prototyping; reading that reflects the variation of blood glucose level before and after meal is listed in Table 9.1 with two test subjects.

In Section 10.5.2, we introduce the design of a noninvasive glucose monitor in the form of a wristband. Here we take a look at a case study that carries out a simple experiment for the purpose of prototyping, as shown in Figure 9.2. The key components of measuring the rate of infrared absorption due to glucose concentration is set up with a clip that houses a pair of infrared light-emitting diodes (LEDs) with 950 nm wavelength and photodiode in a light-seal bag. The test subject places their index finger into the clip and the amount of light absorption is measured by comparing the measured power of the light through the finger versus the output power of the LED.

A working prototype must balance the battery size and how long the device can operate per charge. This concept is exactly the same as our smartphones where the battery capacity, measured in milliampere hours (mAh), determines how many hours we can use our phone after its battery is

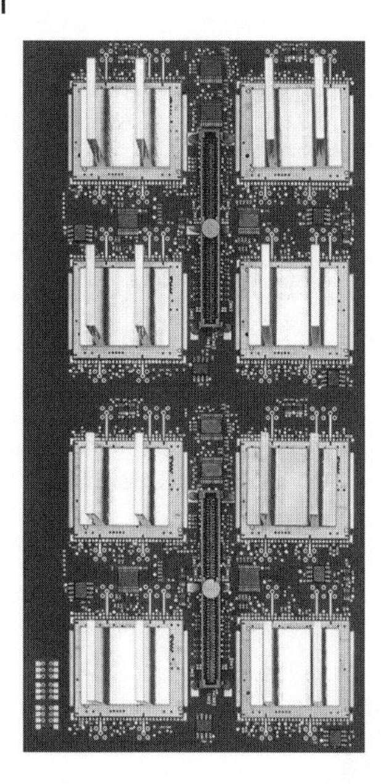

Figure 9.3 An MIMO antenna (left) performs better than a microstrip antenna but is more costly to manufacture than a simple patch of copper on a PCB.

fully charged. For example, a smartphone with a battery of capacity 3000 mAh that draws a constant electrical current of 150 mA while operating will last 3000/150 = 20 hours when the battery is fully discharged from 100 to 0%. Obviously, a battery with a larger capacity can power the device for longer at the expense of additional size and weight. For illustration purposes, we examine a device that takes either common AA or AAA size batteries, whose capacities are 2400 and 1000 mAh, respectively. In this particular example, using an AA battery makes the devices operate longer but will also add size and weight at the same time.

Another important component is the antenna module. The same principle applies when balancing between portability and operational performance. An MIMO (multiple input, multiple output) antenna array with better reception sensitivity and interference mitigation will be physically larger than a single microstrip antenna, as shown in Figure 9.3 (Chi et al. 2016). So, optimizing operational performance and portability is an important design consideration. In contrast to a single microstrip antenna, the MIMO antenna in Figure 9.3a features an array of eight antennas in a 2 × 4 matrix that are individually fed units, t. They collectively work together as a single antenna mounted on a printed circuit board (PCB). On the reverse side of the PCB is a simple copper backplane that serves as a reflector.

9.1.2 Continuous Versus Continual Monitoring

As battery capacity is fixed for the type and size being installed, how long a device can last per charge largely depends on how often it is being used, i.e. for health monitors, how often a reading is taken and how often captured data is transmitted. Ideally, readings should be taken less frequently to reduce battery usage while ensuring the risk of missing a critical event is minimized.

Taking the glucose monitor as an example, continual reading of 2–3 times/day is usually adequate for type 2 diabetes patients, whereas more frequent readings are needed for type 1 diabetes patients (Cariou et al. 2017). In both cases, continual reading of no more than once every hour is normally sufficient for the majority of diabetes patients (Beck et al. 2017).

Before going further into the context of telemedicine in digital health, let us remind ourselves that the fundamental difference between continuous and continual measurement is that continuous measurement takes nonstop readings, such as in the case of sampling a continuous stream of electrocardiographic ECG signals without any break in between the line that represents the captured ECG, whereas continual measurement repeats readings with diabetes monitoring where glucose measurement is done several times a day.

How often a reading should be taken depends largely on the type of disease being managed. While we note that diabetes management typically entails continual monitoring of glucose levels several times per day, examples like ECG anomaly detection and cardiac arrhythmias in monitoring chronic renal failure patients can require continuous monitoring (Steinberg et al. 2017). Multiple sensors may be needed for certain types of patients like cardiac monitoring for epileptic seizures. Additional fall detection with accelerometers can also be added to enhance management efficiency (Cogan et al. 2017). As a general rule, we should consider battery life versus risk of a misdetection of an abnormal event when determining the frequency of measurement.

9.1.3 Wearable Monitoring for Everyone

Wearable health monitors not only become increasingly popular in patient management but also help improve the safety of others. For example, embedded sensors worn by occupational drivers can make the roads safer by detecting health risks such as seizures and drowsiness that can be coupled with automotive electronics to enforce the autonomous stopping of a vehicle in the event of driver blackout. A driver can be fitted with appropriate sensors to monitor health risks that could become hazardous while behind the wheel, as shown in Figure 9.4 where the test subject has a range of sensors attached (Fong et al. 2016). We look at the example of occupational drivers as they are often prone to a range of chronic diseases such as diabetes and cardiovascular disease, owing to their work environment (Egger et al. 2017).

This particular example illustrates a simple wearable heart rate monitor with pulse counters for the health hazard detection of a driver, as this cardiac scenario is of particular importance since it indicates the driver's health status while static, as in the case of driving behind the wheel. The main challenge of obtaining an accurate reading is caused by a patient tending to produce a higher than normal reading while they are being measured in a clinic, owing to anxiety. The ability to measure the patient at ease can ensure a more representative set of readings being taken (Man et al. 2017).

The heart rate can be measured at any location of a human body where an artery is near enough to the surface so that a pulse can be detected. For convenience, various arteries can be used for measurement, depending on situation and application:

- radial (wrist)
- carotid (neck)
- femoral (groin)
- brachial (elbow)
- princeps pollicis (thumb).

Small wireless pulse counters can be embedded in clothing such as cuffs and wristbands. All these counters can be hooked up to either the driver's smartphone or an in-vehicle console via Bluetooth communication links, to store the readings of each location. Readings can therefore be taken without the driver paying attention to the measurement process so that a normal resting heart rate can be obtained.

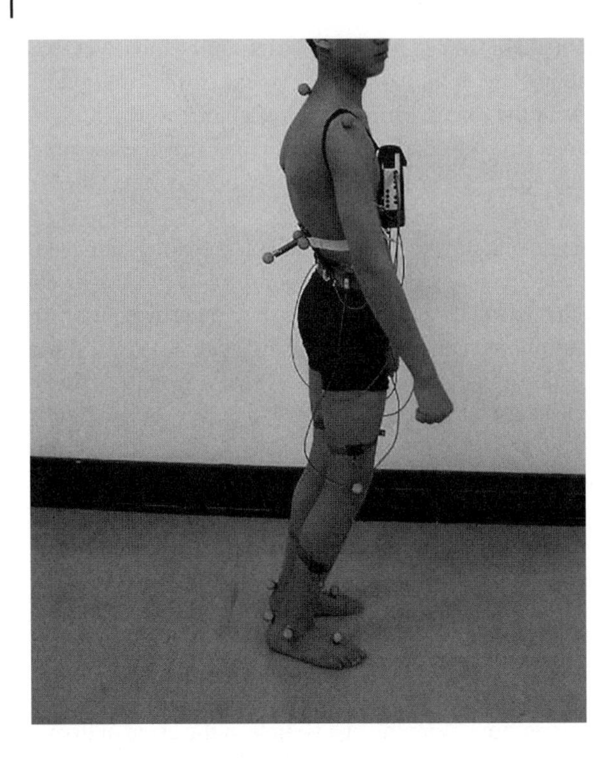

Figure 9.4 Test subject with wearable sensors, markers, and electromyographic (EMG) electrode placements. In this particular experiment, cardiac and EMG measurements are carried out while driving on a simulator.

Changes in resting heart rate can affect the measurement process. Variations can be due to illness or fatigue and influenced by substances such as caffeine and prescribed medications, and so appropriate adjustment should be made to avoid any false alarm or missed detection of anomaly. Another important factor that needs to be understood is prescribed medication being taken by the driver that could impair driving ability such as medicine which causes drowsiness.

Wearable health monitoring is becoming increasingly popular. It is capable of enhancing the safety not only of the monitored subject but also of other people surrounding the subject. Wearable monitors enhance the health and safety of people of all ages, like the baby monitoring system featured in Figure 9.5. It consists of two low cost sensors that are effective in cot death prevention and

Figure 9.5 A simple baby monitoring system.

alert for change of nappies while the baby sleeps, with such simple wearable monitoring system in place the parents can carry on with their daily activities while their baby sleeps next door knowing that an alarm for attention will go off on their smartphones as an when needed.

9.2 Medical Devices Versus Consumer Electronics Gadgets

Wearable electronics can take various different forms like embedded clothing and smartwatches. Numerous consumer electronics in the market are advertised to provide numerous health monitoring features, such as the apps shown in Figure 9.6. With so much being promised what they can actually do in terms of health monitoring remains largely unanswered, because the reported results generally are not backed by any clinical proof. In contrast, medical devices have to provide a sufficiently accurate reading for the purposes of diagnosis and treatment, and so differ from consumer healthcare devices in that they are not generally meant for everyday use (as we discuss in the next section).

9.2.1 Definition of Medical Devices

The World Health Organization (WHO) provides the following definition on its website (https://www.who.int/medical_devices/full_deffinition/en/):

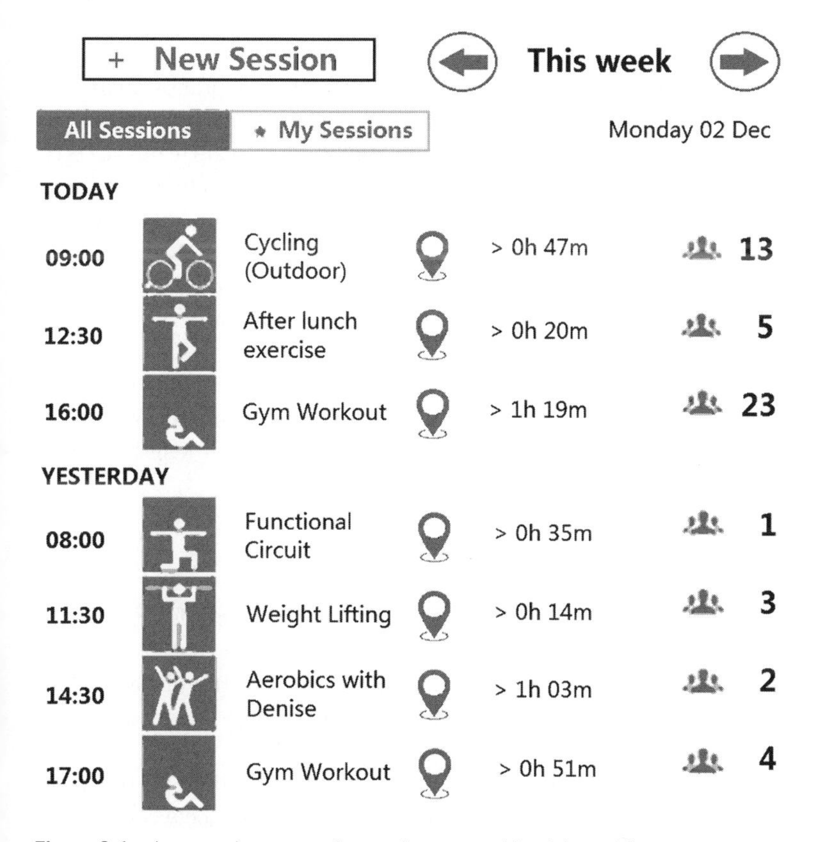

Figure 9.6 A smartphone app that assists general health tracking.

Medical Device – Full Definition

"Medical device" means any instrument, apparatus, implement, machine, appliance, implant, reagent for in vitro use, software, material or other similar or related article, intended by the manufacturer to be used, alone or in combination, for human beings, for one or more of the specific medical purpose(s) of:

- diagnosis, prevention, monitoring, treatment or alleviation of disease,
- diagnosis, monitoring, treatment, alleviation of or compensation for an injury,
- investigation, replacement, modification, or support of the anatomy or of a physiological process,
- supporting or sustaining life,
- control of conception,
- disinfection of medical devices
- providing information by means of in vitro examination of specimens derived from the human body;

and does not achieve its primary intended action by pharmacological, immunological or metabolic means, in or on the human body, but which may be assisted in its intended function by such means.

Note: Products which may be considered to be medical devices in some jurisdictions but not in others include:

- disinfection substances,
- aids for persons with disabilities,
- devices incorporating animal and/or human tissues,
- devices for.in-vitro fertilization or assisted reproduction technologies

We shall elaborate from the above definition and see what it means to medical professionals. Traditionally, medical devices are regulated by the Federal Food, Drug, and Cosmetic Act (FD&C Act) in the United States for many decades. The Center for Devices and Radiological Health (CDRH) within the US Food and Drug Administration (FDA) is currently in charge of regulation for medical devices. While the FDA defines medical devices fairly similarly as the WHO, one notable difference is that the FDA's definition covers "within or on the body of man or other animals," meaning that devices used in animals are also regulated in the same way. Any medical device to be used in the US is required to undergo General Controls and Premarket Approval (PMA) as well as subsequent post-marketing regulatory controls (Jarow and Baxley 2015).

Prior to launching a new medical device in the US, it is necessary to first obtain clearance to the market by 510(k) (FDA 2014) or approval to market through PMA. Post-marketing regulatory controls entail Device Listing, Medical Device Reporting (MDR), Establishment Registration, as well as Quality System Compliance Inspection.

9.2.2 Device Classification

The classifications by the FDA are assigned according to the risk a given medical device presents to a user, and the level of regulatory control is divided into three categories based on risk that the device can potentially pose, from Class I having the lowest risk up to Class III with the highest risk. All classes of devices are subject to General Controls according to the baseline requirements of the FD&C Act.

Device classification is generally based on the intended use as well as indications for the use of a given device. While we have started this section by making a clear distinction between medical devices versus consumer healthcare devices that do not require regulatory approval with a general intention of providing indicative readings, the "indications for use" concerning FDA's device classification has a different meaning, which is documented in FDA 2014), which actually indicates the type of intended usage, as shown in this example here:

> A wearable optical glucose monitor has an *intended use* of *carrying out non-invasive glucose concentration measurement*. More specialized indication for use is appended as a subset of intended use as *for taking continual measurement on the skin*.

It is vitally important to determine the classification of a given device prior to product launch. There are also cases where exemptions can be granted. As a starting point, finding the classification of a given device can be done by searching through the FDA's online Product Classification Database, as illustrated in Figure 9.7.

Similar regulatory requirements concerning medical devices are in place in many countries. For example, the Medicines and Healthcare products Regulatory Agency (MHRA) in the UK oversees the medical device product registration process, where the main difference between the US and UK classification is that Class II is further divided into *a* and *b*. It is also important to note that, as of the time of writing shortly before Brexit, a valid CE marking (which requires a Clinical Evaluation

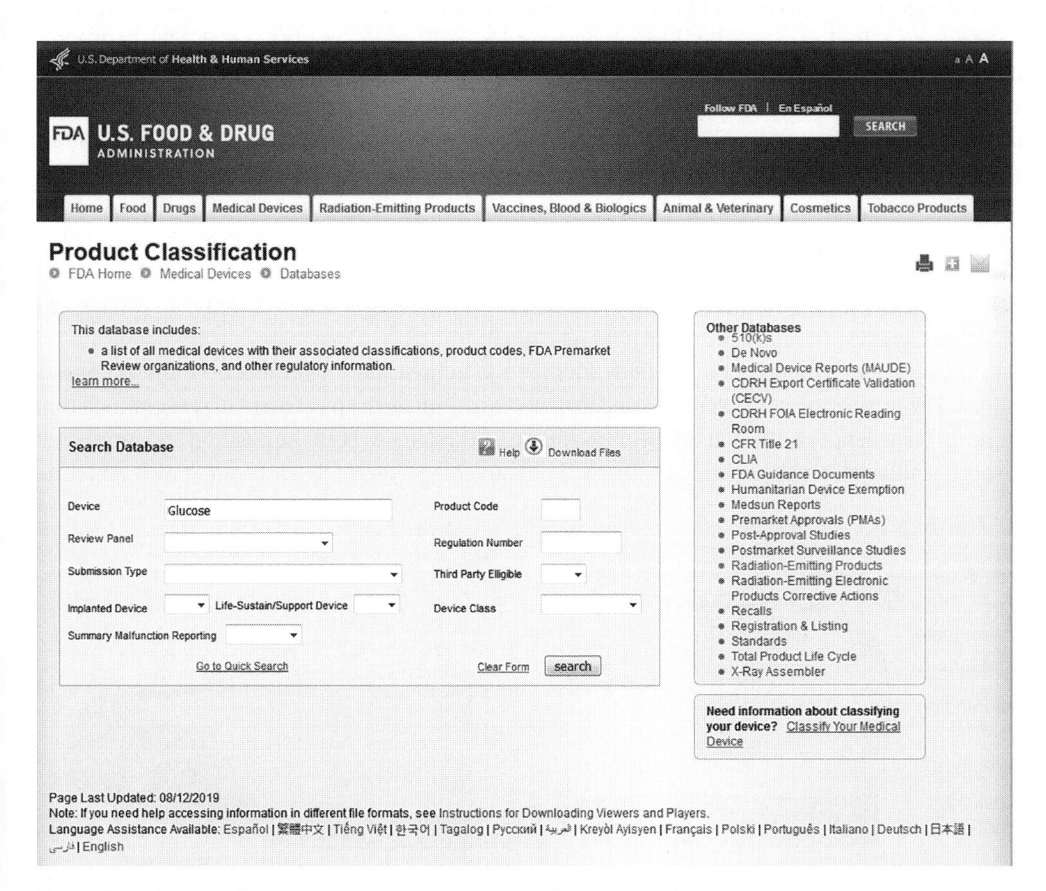

Figure 9.7 Product Classification Database for determining the device class.

Report, or CER) is mandatory for all medical devices marketed in Great Britain as a confirmation of conformity with the European Union's medical device regulations. The European Union has the separate Medical Devices Regulation (MDR) and *In Vitro* Diagnostics Regulation (IVDR) that govern medical devices marketed across the EU.

Having made the clear distinction between medical devices subject to strict regulatory compliance and general consumer healthcare devices, our discussion now concentrates on the latter as the specific details related to medical devices are not within the scope of this book and further reading on text specifically devoted to medical devices is recommended.

9.3 Connectivity

9.3.1 Deployment Options

Various consumer healthcare devices make use of built-in accelerometers to estimate the number of steps taken during a walk. Start pain management has been an area where monitoring can be accomplished through sensors inside the shoe, as shown in Figure 9.8. Wearable healthcare monitors a wide range of health parameters from head to toe. Examples include the smart helmet shown in Figure 9.9. This helmet not only tracks health and ambient environment parameters but also serves as a driving safety aid through detecting hazards in the road (Fong and Siu 2012), for example traffic congestion or a physical obstacle in the road. In addition, crash sensors will trigger an automatic emergency alert in compliance with the European eCall initiative (Lego et al. 2020). Throughout this book we have looked at a number of similar examples about how these different types of wearable health monitors form an important block of telemedicine system in digital health. We now go deeper into the deployment options by looking at the practical design of the smart shoe that monitors start pain using a magnetometer for gait analysis. These sensors not only analyze fatigue, steps for calories computation, and posture for injury prevention but also control an actuator for pumping the air cushion to create the optimal shape. All captured data can be sent to a smartphone via Bluetooth for analysis and storage.

In summary, the design rationale in both the shoe and helmet share a number of common considerations. The most important aspect related to data acquisition is the placement of sensors such that data accuracy will not be affected by user movement. This relates to both the position and number of sensors installed within the confined space without any impact on safety and ergonomics. Antenna

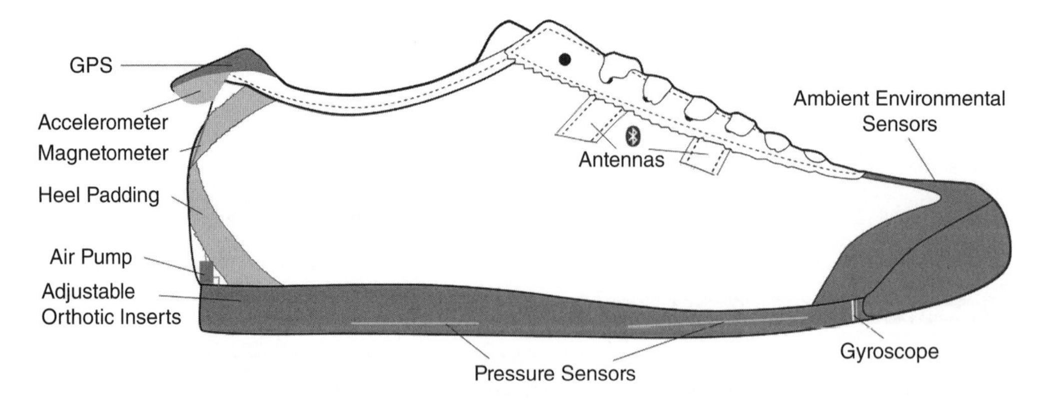

Figure 9.8 Smart shoe for start pain management.

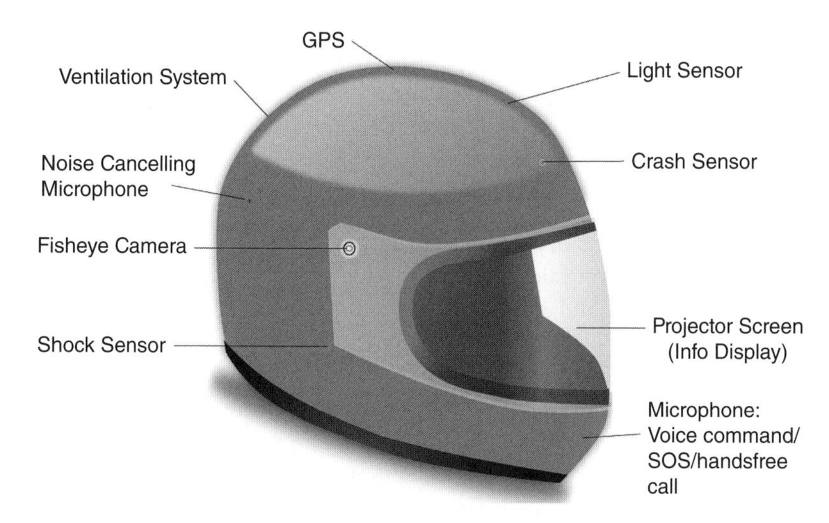

Figure 9.9 Helmet with integrated sensors for both health monitoring and safety enhancement.

placement is also an important consideration as this affects both power consumption as well as the reliability of a radio link to ensure connection availability.

9.3.2 Connectivity for Quality Monitoring

Connectivity, being an important part of telemedicine, serves as an important platform for quality monitoring in frameworks such as the JCI (Joint Commission International) accreditation for hospitals, clinical laboratories, etc. Safety enhancements such as fall prevention and fire risk can be managed through a simple sensing network. Wearable sensors have the capability of detecting falls and location tracking in the event of an emergency evacuation, which we take a closer look at using telemedicine for assisting JCI compliance through a case study of managing an emergency situation in the next section.

9.4 Enhancing Caring Efficiency

As we have learned in this chapter, wearable connected sensors and devices provide important functionalities in telemedicine. We extend our discussion on smart hospitals begun in Section 3.4 and wrap up this chapter on wearable healthcare through looking at a case study of the use of telehealth for enhancing nursing home care. With the number of senior citizens increasing steadily, there is a growth in demand for nursing homes (Lakdawalla et al. 2003). Nursing homes, also known as convalescent homes, are a type of residential care that provides around-the-clock nursing care for older or disabled people according to the definition from the US National Caregivers Library by Fidia Advisors, LLC.

The vulnerability of residents in nursing homes in response to any one from a growing list of disasters with a natural or manmade trigger need to be thoroughly investigated for the safety of residents (Pelling 2012). Addressing the needs for responding to a disaster when assisting older residents who may have physical or cognitive impairments is a vitally important aspect of enhancing the safety of senior citizens (Gajos et al. 2014), which the use of wearable healthcare solutions can assist with. For example, Tängman et al. (2010) suggest that nursing home residents are particularly vulnerable to falls while responding to an emergency situation such that connected accelerometers used in conjunction with a location tracker can automatically alert the caregiver to the exact location of the patient in the event the patient falls.

9.4.1 Mobility Assistance

Mergner and Lippi (2018) studied the effects of posture and movement disturbance among older patients often requiring walking aids. Walking aids assist older patients with balance and mobility at the expense of increasing loads on arms and elevated energy expenditure and heart rate (Mello et al. 2018). In an effort to enhance the mobility of older patients, a biomechanical study on the length of crutches and their effects on residents who are not suffering from any kind of disease or disability need to be analyzed based on earlier work by Shoup et al. (1974). Many older people who rely on crutches suffer from lumbar hyperlordosis, a common postural position where the curve of the back is accentuated, which can lead to muscle pain or spasms (Walicka-Cupryś et al. 2018). Patients are therefore prescribed with a pair of crutches to improve their balance (Yagi et al. 2017). Wearable sensors come in when analyzing the relationship between the changes of body alignment while using crutches so that the impact on mobility can be optimized for crutch usage.

Senior citizens with adequate physical abilities sometimes use a walking stick to provide extra support. Pang et al. (2007) propose the use of the Gross Motor Functional Classification scale to indicate the need for specially designed canes to assist patients categorized as level 3 after a stroke. One of the objectives is to study the effects of involuntary muscle contraction and spasm on older people with reduced mobility (Rooks et al. 2017). Wearable sensors appropriately placed for gait monitoring along the back can be configured as in the example in Figure 9.10 to assist the analysis of spasticity which is a velocity dependent stretch reflex of muscles in conjunction with muscle weakness, imbalance, as well as impaired selective motor control (Marinelli et al. 2016). The gait

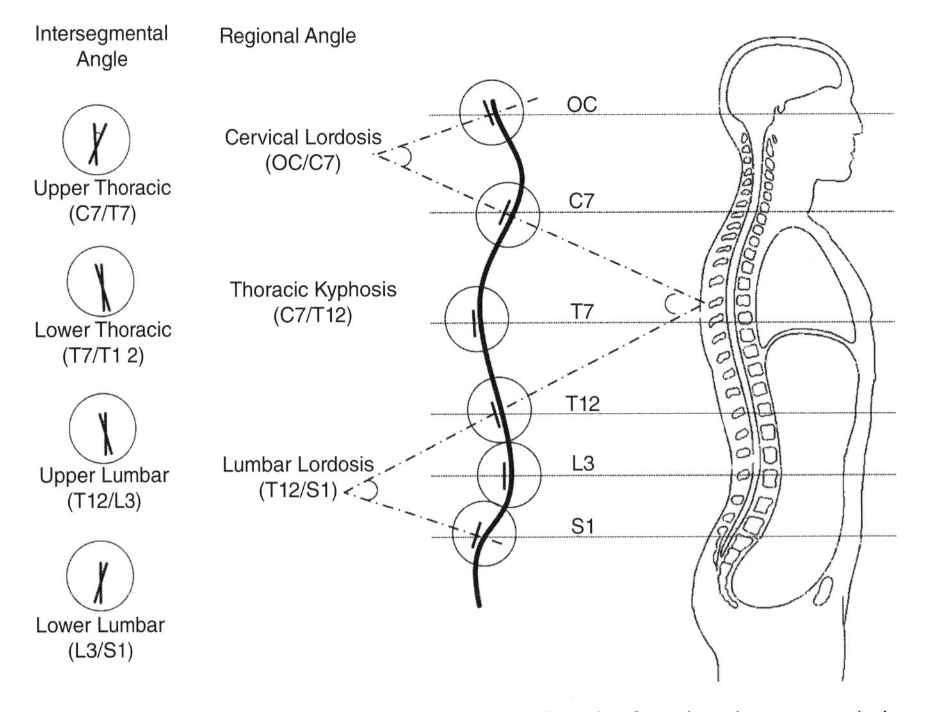

Figure 9.10 Wearable sensing network placed along the spine for gait and posture analysis.

and posture of older patients may have crouch knee, jump knee, stiff knee, and recurvatum knee that require assistance when evacuating (Marsden 2018). Additional sensors placed around the knee can enhance the overall understanding of the effects of using crutches.

9.4.2 Preparation for an Emergency Situation: A Case Study of a Nursing Home

A high concentration of senior citizens with reduced cognitive and/or motor capabilities makes nursing home patients more vulnerable to injuries (Donner and Rodríguez 2008). While a study by Chamberlain et al. (2017) reveals that the shortage of support staff poses significant challenges to emergency management in Canada, it also seeks to take an in-depth look into the resource issues in a local context so that enhancement solutions can be derived to cater for the specific needs in nursing homes around the world. The ability of a given member of staff can care for multiple patients is closely associated with their knowledge, skills, and accessible technology (Yager et al. 2011).

The model shown in Figure 9.11 summarizes the people involved, namely older residents and caregivers, who have different mobility health conditions, characteristics, skills, and abilities that affect how an individual responds to an emergency. These interactions link between the three

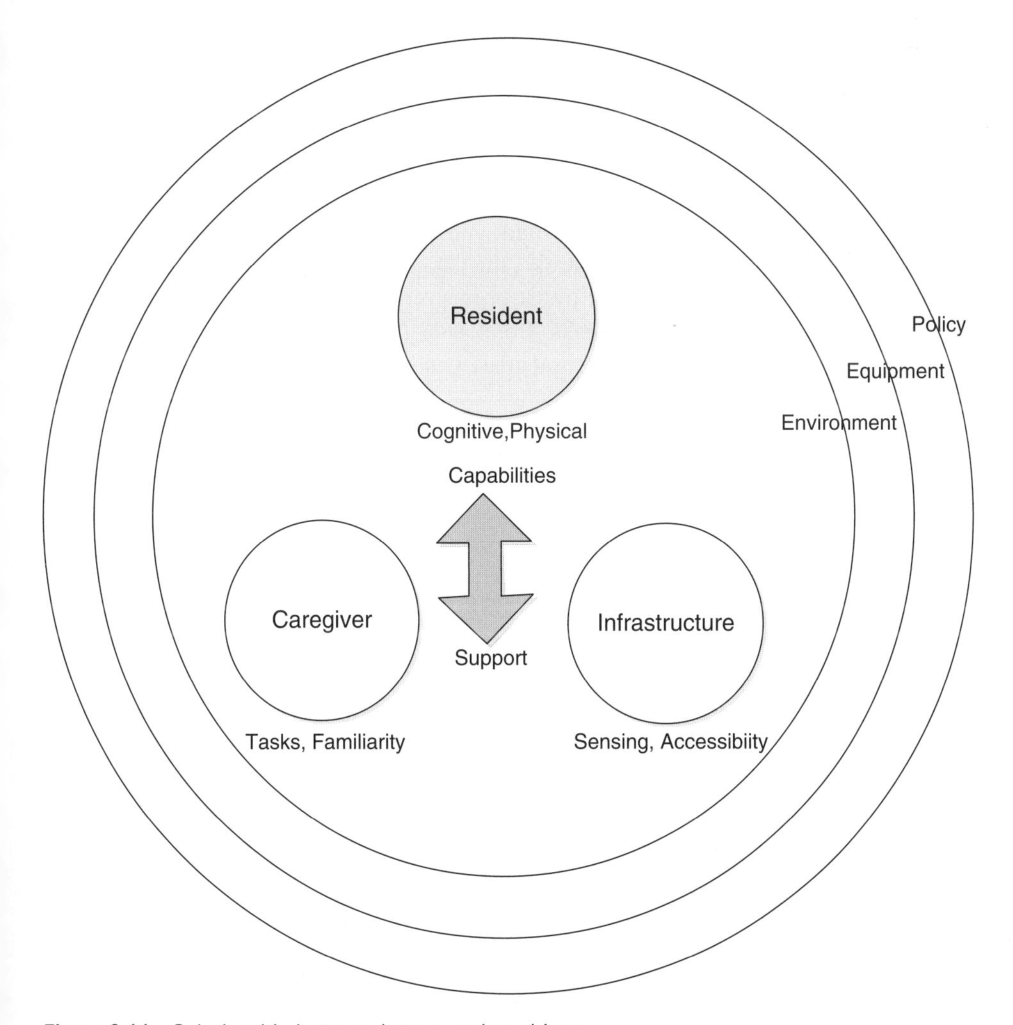

Figure 9.11 Relationship between human and machines.

entities in the model that in turn describe the capabilities of individual older people and what kind of technical solutions can assist them in coping with an emergency.

Nursing home clients often have difficulty responding to an emergency environment in situations such as fire, flooding, and earthquake (Sternberg 2003). Kholshchevnikov et al. (2012) report that one major issue associated with the emergency evacuation of senior citizens in the event of a fire is mobility in relation to the use of crutches, which means that nursing homes should assess the vulnerability of their residents in the event of an emergency. Rowland et al. (2007) emphasize the need to evaluate how nursing home staff cope with an emergency within a very short timeframe as well as being able to facilitate an imminent evacuation. The process of emergency preparedness involves the use of different types of wearable devices, sensors, and systems.

The exact nature of an emergency could have a severe impact on the ability of a patient to safely evacuate a building (Pierce and West 2017). Current methods for emergency management for cohorts of evacuees with limited mobility are heavily dependent on manual processes by caregivers where human factors play a substantial role in ensuring the safety of patients (Gaba 2000). Díaz et al. (2016) suggest that technology could play a key role in improving the evacuation of nursing homes during emergencies.

9.5 Wearable Physiotherapy

To improve our health, we can always try to maintain a healthier lifestyle. However, this does not include suffering from unavoidable accidents. After any accidents due to either overstretching our body (in work or in sport) or external events, consultation with a physiotherapist can be unavoidable. In other occasion, for example stroke, the patient also needs the intensive help of a professional therapist. In this case, their mobility can be recovered as long as treatment is performed within the golden period of the first three weeks (Horak et al. 2015).

Many patients, particularly those living in rural areas, can face difficulties in utilizing such services, owing to one or more of the following reasons:

- Shortage of professional physiotherapists in their area.
- Many patients cannot afford the service.
- Service providers are too far away from the patient.
- The mobility of the patient is limited by injury.

In these situations, telemedicine would be extremely beneficial, and yet it is seldom considered. seldom addressed. For example, patients often use transcutaneous electrical nerve stimulation (TENS) or electrical muscle stimulation (EMS) devices after muscular injuries, as these devices are commonly available on the market, have simple instructions, and are easy to use. However, to garner the most benefit from these devices, users need to know the exact dosage for the correct level of stimulation (Gibson et al. 2019). Also, analyzing the rate of recovery requires a substantial amount of information (Wu et al. 2018). With connected devices and interfaces such as Bluetooth, wearable TENS or EMS devices with built-in sensors would allow the monitoring of the stimulation strength of the initial treatment as set by the user and uploaded via an app, and qualitative information such as the user's subjective feeling and comment can also be collected for analyzing treatment effectiveness.

With artificial intelligence (AI) algorithms coupled with a domain expert engine, the user's next therapy needs and specifications can be downloaded for the next treatment session (Cypko and

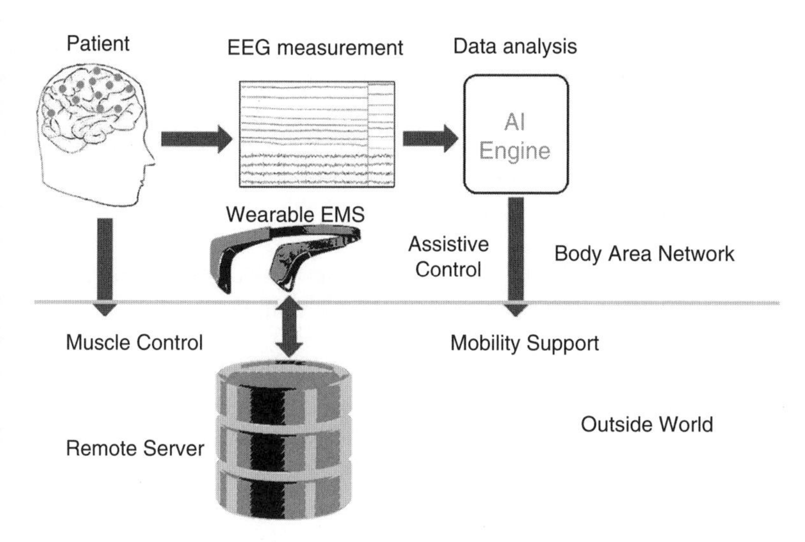

Figure 9.12 Wearable electrical muscle stimulation for stroke recovery.

Stoehr 2019). With this iterative feedback of advice and treatment, the fine-tuning of subsequent treatment can be achieved through appropriate adjustments (Amatya et al. 2018). This will enhance patient experience as well as provide useful feedback to other patients with similar situations.

Taking a look at post-stroke therapy as a case study, intensive care by therapists within the first three weeks is vital for rehabilitation (Cunningham et al. 2016). A wearable device based on exoskeleton devices with integration of electroencephalographic (EEG) monitoring shown in Figure 9.12 assists the patient to move their limbs. The EEG signals allows treatment program to be customized for a specific patient (Bundy et al. 2017). This can also be performed remotely via the normal telemedicine communication channels. After the patient shows improvement and potential for establishing the reconnection of relevant nerves as well as some mobility, the exoskeleton device can serve as an aid to help the patient undertake more exercise and training until the patient's mobility is fully recovered.

References

Alioto, M. and Shahghasemi, M. (2017). The Internet of things on its edge: trends toward its tipping point. *IEEE Consumer Electronics Magazine* 7 (1): 77–87.

Amatya, B., Young, J., and Khan, F. (2018). Non-pharmacological interventions for chronic pain in multiple sclerosis. *Cochrane Database of Systematic Reviews* (12): CD012622. https://doi.org/10.1002/14651858.CD012622.pub2.

Beck, R.W., Riddlesworth, T., Ruedy, K. et al. (2017). Effect of continuous glucose monitoring on glycemic control in adults with type 1 diabetes using insulin injections: the DIAMOND randomized clinical trial. *JAMA* 317 (4): 371–378.

Bundy, D.T., Souders, L., Baranyai, K. et al. (2017). Contralesional brain–computer interface control of a powered exoskeleton for motor recovery in chronic stroke survivors. *Stroke* 48 (7): 1908–1915.

Cariou, B., Fontaine, P., Eschwege, E. et al. (2017). Influence of organizational context on nursing home staff burnout: A cross-sectional survey of care aides in Western Canada. *International Journal of Nursing Studies* 71: 60–69.

Chamberlain, S.A., Gruneir, A., Hoben, M. et al. (2017). Influence of organizational context on nursing home staff burnout: a cross-sectional survey of care aides in Western Canada. *International Journal of Nursing Studies* 71: 60–69.

Chi, H.R., Tsang, K.F., Chui, K.T. et al. (2016). Interference-mitigated ZigBee-based advanced metering infrastructure. *IEEE Transactions on Industrial Informatics* 12 (2): 672–684.

Cogan, D., Birjandtalab, J., Nourani, M. et al. (2017). Multi-biosignal analysis for epileptic seizure monitoring. *International Journal of Neural Systems* 27 (01): 1650031.

Cunningham, P., Turton, A.J., Van Wijck, F., and Van Vliet, P. (2016). Task-specific reach-to-grasp training after stroke: development and description of a home-based intervention. *Clinical Rehabilitation* 30 (8): 731–740.

Cypko, M.A. and Stoehr, M. (2019). Digital patient models based on Bayesian networks for clinical treatment decision support. *Minimally Invasive Therapy & Allied Technologies* 28 (2): 105–119.

Dantas, C., van Staalduinen, W., Jegundo, A. et al. (2019). Smart healthy age-friendly environments: policy recommendations of the thematic network SHAFE. *Translational Medicine* 19: 103.

Díaz, P., Carroll, J.M., and Aedo, I. (2016). Coproduction as an approach to technology-mediated citizen participation in emergency management. *Future Internet* 8 (3): 41.

Dimitrov, D.V. (2016). Medical internet of things and big data in healthcare. *Healthcare Informatics Research* 22 (3): 156–163.

Donner, W. and Rodríguez, H. (2008). Population composition, migration and inequality: the influence of demographic changes on disaster risk and vulnerability. *Social Forces* 87 (2): 1089–1114.

Egger, G., Binns, A., Rossner, S., and Sagner, M. (2017). *Lifestyle Medicine: Lifestyle, the Environment and Preventive Medicine in Health and Disease*. Academic Press.

FDA (2014). *The 510(k) Program: Evaluating Substantial Equivalence in Premarket Notification [510(k)]: Guidance for Industry and Food and Drug Administration Staff*. US Department of Health and Human Services, Food and Drug Administration.

Fong, B. and Siu, W.C. (2012). A detection system for assisting a driver when driving a vehicle. US Patent 8,174,375, filed 19 April 2012 and issued 26 March 2013.

Fong, A.C.M., Chan, C., Situ, L., and Fong, B. (2016). Wireless biosensing network for drivers' health monitoring. In: *2016 IEEE International Conference on Consumer Electronics*, 247–248. ICCE.

Gaba, D.M. (2000). Anaesthesiology as a model for patient safety in health care. *BMJ [British Medical Journal]* 320 (7237): 785.

Gajos, A., Kujawski, S., Gajos, M. et al. (2014). Effect of physical activity on cognitive functions in elderly. *Journal of Health Science* 4 (8): 91–100.

Gibson, W., Wand, B.M., Meads, C. et al. (2019). Transcutaneous electrical nerve stimulation (TENS) for chronic pain-an overview of Cochrane Reviews. *Cochrane Database of Systematic Reviews* (4): CD011890. https://doi.org/10.1002/14651858.CD011890.pub3.

Horak, F., King, L., and Mancini, M. (2015). Role of body-worn movement monitor technology for balance and gait rehabilitation. *Physical Therapy* 95 (3): 461–470.

Jarow, J.P. and Baxley, J.H. (2015). Medical devices: US medical device regulation. *Urologic Oncology: Seminars and Original Investigations* 33 (3): 128–132.

Kholshchevnikov, V.V., Samoshin, D., and Istratov, R. (2012). The problems of elderly people safe evacuation from senior citizen heath care buildings in case of fire. In: *Proceedings of 5th International Symposium "Human Behaviour in Fire"*, 587–593. Cambridge: Interscience Communications.

Lakdawalla, D., Goldman, D.P., Bhattacharya, J. et al. (2003). Forecasting the nursing home population. *Medical Care* 41 (1): 8–20.

Lego, T., Mladenow, A., and Strauss, C. (2020). Assessment of eCall's effects on the economy and automotive industry. In: *Data-Centric Business and Applications* (eds. N. Kryvinska and M. Greguš), 409–431. Springer.

Man, S., ter Haar, C.C., de Jongh, M.C. et al. (2017). Position of ST-deviation measurements relative to the J-point: impact for ischemia detection. *Journal of Electrocardiology* 50 (1): 82–89.

Marinelli, L., Mori, L., Canneva, S. et al. (2016). The effect of cannabinoids on the stretch reflex in multiple sclerosis spasticity. *International Clinical Psychopharmacology* 31 (4): 232–239.

Marsden, J.F. (2018). Management of walking disorders in neurorehabilitation. In: *Neurorehabilitation Therapy and Therapeutics* (eds. P.S. Nair, M. González-Fernández and J.N. Panicker), 105–118. Cambridge University Press.

Mergner, T. and Lippi, V. (2018). Integrating posture control in assistive robotic devices to support standing balance. In: *International Symposium on Wearable Robotics*, 321–324. Springer.

Mello, J.L.C., Souza, D.M.T., Tamaki, C.M. et al. (2018). Application of an effective methodology for analysis of fragility and its components in the elderly. In: *Information Technology-New Generations* (ed. S. Latifi), 735–739. Cham: Springer.

Mohanty, S.P., Choppali, U., and Kougianos, E. (2016). Everything you wanted to know about smart cities: the Internet of things is the backbone. *IEEE Consumer Electronics Magazine* 5 (3): 60–70.

Naranjo-Hernández, D., Reina-Tosina, J., Buendía, R., and Min, M. (2019). Bioimpedance sensors: instrumentation, models, and applications. *Journal of Sensors* 2019 Article ID 5078209 doi:https://doi.org/10.1155/2019/5078209.

Pang, M.Y., Eng, J.J., and Miller, W.C. (2007). Determinants of satisfaction with community reintegration in older adults with chronic stroke: role of balance self-efficacy. *Physical Therapy* 87 (3): 282–291.

Pérez-Roman, E., Alvarado, M., and Barrett, M. (2020). Personalizing healthcare in smart cities. In: *Smart Cities in Application* (ed. S. McClellan), 3–18. Springer.

Pierce, J.R. and West, T.A. (2017). Mortality in evacuating nursing home residents. *Journal of the American Medical Directors Association* 18 (9): 803.

Pelling, M. (2012). *The Vulnerability of Cities: Natural Disasters and Social Resilience*. Routledge.

Rooks, D., Praestgaard, J., Hariry, S. et al. (2017). Treatment of sarcopenia with bimagrumab: results from a phase II, randomized, controlled, proof-of-concept study. *Journal of the American Geriatrics Society* 65 (9): 1988–1995.

Rowland, J.L., White, G.W., Fox, M.H., and Rooney, C. (2007). Emergency response training practices for people with disabilities: analysis of some current practices and recommendations for future training programs. *Journal of Disability Policy Studies* 17 (4): 216–222.

Steinberg, J.S., Varma, N., Cygankiewicz, I. et al. (2017). 2017 ISHNE-HRS expert consensus statement on ambulatory ECG and external cardiac monitoring/telemetry. *Heart Rhythm* 14 (7): 55–96.

Shoup, T.E., Fletcher, L.S., and Merrill, B.R. (1974). Biomechanics of crutch locomotion. *Journal of Biomechanics* 7 (1): 11–19.

Sternberg, E. (2003). Planning for resilience in hospital internal disaster. *Prehospital and Disaster Medicine* 18 (4): 291–299.

Tängman, S., Eriksson, S., Gustafson, Y., and Lundin-Olsson, L. (2010). Precipitating factors for falls among patients with dementia on a psychogeriatric ward. *International Psychogeriatrics* 22 (4): 641–649.

Tricoli, A., Nasiri, N., and De, S. (2017). Wearable and miniaturized sensor technologies for personalized and preventive medicine. *Advanced Functional Materials* 27 (15): 1605271.

Walicka-Cupryś, K., Wyszyńska, J., Podgórska-Bednarz, J., and Drzał-Grabiec, J. (2018). Concurrent validity of photogrammetric and inclinometric techniques based on assessment of anteroposterior spinal curvatures. *European Spine Journal* 27 (2): 497–507.

Wu, Y.N., Gravel, J., Chatiwala, N. et al. (2018). Effects of electrical stimulation in people with post-concussion syndromes: a pilot study. *Health* 10 (04): 381.

Yager, P.H., Lok, J., and Klig, J.E. (2011). Advances in simulation for pediatric critical care and emergency medicine. *Current Opinion in Pediatrics* 23 (3): 293–297.

Yagi, M., Ohne, H., Konomi, T. et al. (2017). Walking balance and compensatory gait mechanisms in surgically treated patients with adult spinal deformity. *The Spine Journal* 17 (3): 409.

10

Smart and Assistive Technologies

Over the past nine chapters, we have discussed how telemedicine and related digital health technologies assist various aspects of healthcare and medical practices. Most of the technologies have a long proven history. Data communications evolved from the first telephone designed by A.G. Bell and Elisha Gray formed the basis of many modern telemedicine systems deployed throughout the world today (Williams 2018). Technological advances and innovative breakthroughs are opening a wide range of possibilities in medical and healthcare services. *The Report of the National Advisory Group on Health Information Technology in England* published for the National Advisory Group on Health Information Technology in England (Wachter and Chair 2016) discusses the importance of using information technologies (IT) in the delivery of better and safer healthcare services through interdisciplinary research, planning, and management. Through integrated research involving medicine, data analytics, computational modeling, and statistics, it is possible to develop technical solutions in solving many healthcare challenges through smart and assistive technologies.

10.1 Affordability in Assistive Technologies

The idea of context home environmental interventions and assistive technology devices for older people's independent living was raised two decades ago by Mann et al. (1999). More recently, assistive home automation has been proposed with smart home control (Grossi et al. 2008). Consumer healthcare has grown substantially in terms of both features and availability over the past decade with faster communication links and smaller wearable devices.

10.1.1 Assistive Technology Becomes Affordable

Transmission of data, like the cost of using a megabyte (1 MB) of mobile data from the cellular network, has gone down substantially over recent years with the Internet of things (IoT) connecting a vast range of devices together both within a smart home environment and across the broader smart city environment (Zhu et al. 2017). In many countries the cost of data transmission has become so low that we can have many devices like our smartwatches, electrical appliances, as well as cars all connected for convenience like tracking our daily exercise and finding the best route to avoid traffic congestions; all these are easily accessible through connectivity. Connected devices bring a wide range of assistive technologies to us for both health and safety enhancement.

Telemedicine Technologies: Information Technologies in Medicine and Digital Health, Second Edition.
Bernard Fong, A.C.M. Fong, and C.K. Li.
© 2020 John Wiley & Sons Ltd. Published 2020 by John Wiley & Sons Ltd.

10.1.2 Connecting People and Machines

Video conferencing can be set up between friends and family members for networking, and with caregivers for advice and assistance. For example, a television set with a small webcam can facilitate real-time communication between different parties. Without keyboard or mouse, a user can get connected with a remote control or speech command. They can even participate in a range of videogames with people far away. Around the older user, doctors and other caregivers are well within reach.

All these are made possible by cheaper and faster telecommunications. The implication of elevated data capacity is that users can enhance mobility, fitness as well as social interactions. Wearable devices can provide health monitoring both at home and outside so long as they are within areas of cellular coverage. One of the key design challenge is to provide a comprehensive care solution package that is easy to use, so that all connected devices form an integral part of a user's daily life.

Users are connected to devices for countless activities, ranging from personal comfort to critical care. Devices are also interconnected so that comprehensive services can be supported. For example, we mentioned that a refrigerator can be connected to a microwave oven and recipes can be downloaded from the Internet. So, based on what is stored in the refrigerator, cooking instructions can be provided to the user. Other appliances such as food processors and coffeemakers can also be linked together to provide assistance for preparing a complete meal with ease.

Communication technology in a smart home environment not only benefits older residents but also their relatives. They can be assured that they are being kept informed of the older user's well-being by getting far more information than what a mobile phone can offer. When one is thousands of miles away from an elderly relative or working just a few blocks away, one can be assured that an alert will be received in case of an emergency and that help is always available.

10.1.3 Emotional Intelligence: Remaining Happy and Healthy

The importance of remaining happy is to live life to the fullest, and this goes for older people as well as the young. After all, the vast majority of older people have contributed decades of hard work to their communities in different capacities. Quantitative assessment of how well a smart home system performs can be easily measured for the communication network and sensor network. There are parameters such as bit error rate (BER), latency, data loss, and indeed what we have covered in Chapter 2. What about happiness, self-confidence, and self-esteem? What we have discussed so far only takes care of the user's physical well-being. What about technology that deals with emotional issues like loneliness and fear?

For those who live alone, a "talking machine" can help, a dummy that initiates a conversation when someone approaches, gives a news briefing about what is happening around the world, and suggests to go out for a meal. Body language and habitual behavior can provide information about the user's psychological well-being. With speech recognition, social interactions can be made possible (Chen et al. 2007). Emotional intelligence applications can take actions according to the user's mood, for example when the user is bored the system can suggest some entertainment. For a talking machine, the system can adjust the tone of conversation according to the user's mood. Fall detection is perhaps one of the most important monitoring systems that provides elderly users a peace of mind as they undertake physical activities. Connected technology comes in two ways, either active where counter-balancing mechanisms can help reduce the risk of a fall which we will not discuss the details as this involves a substantial amount of mechanical designs and their implementations are still not practical; or passive in detecting any loss of balance so that an alert can be generated. Passive fall detection alert can be provided with affordable solutions like accelerometers or video imaging.

Smart home assistive technologies also help with energy saving by regulating temperature and illumination; air conditioners and blinds can be adaptively controlled by sensors installed throughout the home. Likewise, lights can be automatically turned on when a user is in the room and the ambient light from the windows falls below a certain level. Further, medication dispensing can be connected to the system and automatically locks itself when not needed (Cheek et al. 2005). As with a refrigerator, for those who require long-term medication, each drug can also be automatically tracked and order new stock before they run out. Smart home technology offers a totally different flexibility and functionality than traditional home networks.

10.2 Smart Home Integration

Assistive care has become much more affordable as resource utilization becomes far more efficient with the help of telemedicine. Sensor-based consumer health devices make mass-produced healthcare products more available while supporting users on carrying out their daily activities (Garge et al. 2017). The fact that assistive care has become more readily accessible in recent years is not only due to lower manufacturing costs but also to the reduction in cost of mobile data access (Yang et al. 2017). We will take a look at how consumer electronics gadgets change our daily lives through digital health and assistive technologies.

10.2.1 Consumer Electronics in the Home Setting

In addition to the refrigerator, smart home technology can be implemented in virtually all kinds of home appliances for more automation and intelligence. When used in conjunction with a home network or the Internet, different devices can communicate with each other. They can even facilitate communication between different users, caregivers, and device manufacturers. Smart home technology has already been widely implemented in many kitchens (Kranz et al. 2007) and entertainment in living rooms (Palazzi et al. 2009) for various control functions. Integrating smart home technology with telemedicine in an older patient's home, a range of possibility can be offered in addition to the cooking and entertainment functions outlined above. Demiris et al. (2004) assessed the use of smart home technology for preventing or detecting falls, assisting with visual or hearing impairments, improving mobility, reducing isolation, managing medications, and monitoring physiological parameters, and reported that the main concerns users expressed were user-friendliness of the devices, lack of human response, and the need for training tailored to older learners. Rialle et al. (2002) also reported that the large diversity of needs in a home-based patient population requires complex technology. Such need demands data acquisition and wireless communication technology that older users with minimal training would feel comfortable with using at home.

Artificial intelligence (AI) plays an important role in providing assistive technology to older people. Looking back at the smart fridge example above, we can look at how AI works for therapeutic lifestyle changes (TLC). TLC includes management of diet, weight, and exercise, which are all effective for the control of nonalcoholic fatty liver disease (NAFLD), diabetes, and cholesterol (Bauer et al. 2016):

- *Diet*: based on an individual patient's health conditions, dietary recommendation such as that shown in Figure 10.1 is automatically generated for the purpose of recommending meals along with recipes with reduced saturated fat for cholesterol control (Sialvera et al. 2018). This is part of a smart home network that also links to the fridge so that dietary recommendation is optimized from what is currently in the fridge.

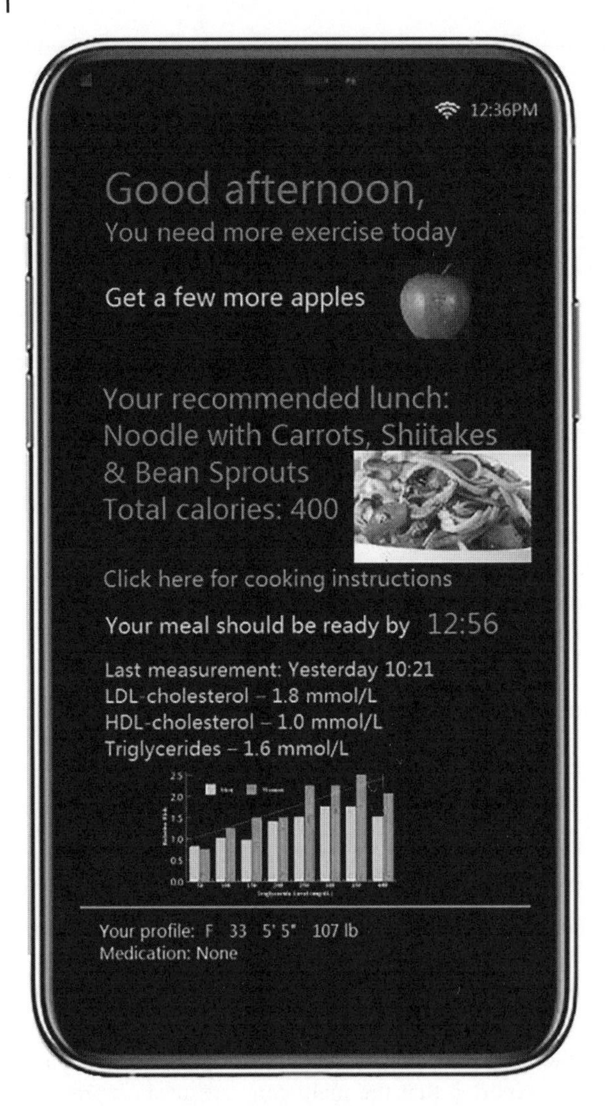

Figure 10.1 AI-generated diet for cholesterol control.

- *Exercise*: regular exercise plans can be computed according to the patient's cholesterol level and heart condition. Patients with an active lifestyle can significantly reduce the risk of coronary artery disease (CAD) by using smart technology to optimize their fitness regimes based upon their individual health condition (Olin et al. 2016).
- *Weight*: obesity is often associated with increase in low-density lipoprotein (LDL) cholesterol levels above 160 mg/dl (Giagulli et al. 2015). Weight watching that targets even as little as 5–10 pounds (2.3–4.5 kg) weight loss can significantly reduce cholesterol and triglyceride levels through TLC (Vilar-Gomez et al. 2015). Combining both diet and exercise management can provide recommendation for ways to reduce the intake of saturated fat below 7% of total calories (Sasdelli et al. 2016).

10.2.2 Integrating Healthcare and Lifestyle into the Home

Quality of life can be enhanced by monitoring the activity of the user and what is around them. Monitoring devices such as accelerometers, pressure sensors, motion detectors, and video cameras can be either discretely or collectively installed in the smart home to collect details about the status of an older person. Sensors are used in areas from logging when the door has been opened to tracking the movements of a user. As discussed throughout the text that there are different sensors for health monitoring. Computational intelligence can also collect user data to learn and analyze from long-term patterns of user behavior. This can serve many objectives, including the rehabilitation progress, warning of abnormality, and active prevention of a fall.

Sensors can also help those who have cognitive or visual impairments; users can be reminded of daily activities such as switching off gas stoves and taking medication and can be warned of any forthcoming hazards like walking toward a staircase or slippery surface. Smart home technology can provide contextual guidance and warnings in hazardous situations according to environmental conditions so that preventive measures can be taken.

Used in conjunction with a telemedicine network, a doctor can retrieve information about the user and view the up-to-date electronic patient record, and see whether the user has been eating or drinking properly as well as other behavioral variations.

10.3 Digital Health in Improving Treatment

Smart home assistive technologies also help with energy saving by regulating temperature and illumination. Air conditioning and window blinds can be adaptively controlled by sensors installed throughout the home. Likewise, lights can be automatically turned on when a user is in the room or the ambient light from the windows falls below a certain level. Further, medication dispensing can be connected to the system and designed to automatically lock itself when not needed (Cheek et al. 2005). As with a refrigerator, for those who require long-term medications each drug can also be automatically tracked and new stock ordered before it runs out. Smart home technology gives a totally different flexibility and functionality than traditional home networks. As we learned at the end of Section 4.5, telemedicine provides an important platform for medication management, but we will now look at how digital health and telemedicine revolutionize the way traditional pharmacology blends into smart technologies.

10.3.1 Treatment Innovations

Computer-aided drug design has been used extensively over the past two to three decades for the discovery and development of new drugs (Åqvist et al. 1994). Over the past decade, AI has been widely used throughout the process of structure-based drug design (Duch et al. 2007). While the biopharmaceutical aspects of ligand in drug design is not within the scope of this text and specific details can be found in references such as Schneider and Baringhaus (2008) and Zartler and Shapiro (2008), our objectives here is to look at how smart and assistive technologies contribute to drug innovations. This is a particularly timely topic as numerous health supplement products claim to have many benefits, and we generalize these along with medications as preventive and pharmacy medicine without going into the details of validating their benefit claims.

To illustrate the role of smart and assistive technology, we depart from the general thoughts of pills, which we elaborate on below by discussing the technologies behind smart pills. Instead, we look at a case study of hair dye, which may not sound relevant to our discussion on drugs. One example of a hair dye that uses natural ingredients as opposed to chemical color dyes (illustrated in Figure 10.2) is illustrated as a medication that induces hair color changes (Lademann et al. 2007) while delivering nutrients into porcine hair follicles (Lademann et al. 2006). The idea of using smart technology here is very simple: use a digital camera with appropriate color calibration as shown in Figure 10.3 to compare the hair color of the last time the dye was applied. Color matching is then

Figure 10.2 Hair dyes for medication binding: mixing to yield a natural transition and maintain consistency is possible through a color analyzer.

Figure 10.3 Color analyzer for tone matching and clinical trial reporting.

carried out so that the hair can maintain the exact same color and any fading can be detected by comparing the images on a regular basis. Automated image process can generate the desired and consistent effect. Integrated into a telemedicine network, participants included in clinical trials can be linked together so that effectiveness can be collectively measured while analyzed off site so that clinical trials can be conducted with participants from anywhere.

Smart and assistive technologies extend far beyond color matching and nutrient penetration through hair. Dynamic compression can be implemented for the relief of varicose veins, as shown in Figure 10.4. The idea is to optimize circulation through dynamically adjusting the pressure based on the state of the varicose veins, which is of course different in individual patients. The common problem with traditional compression socks is consistent pressure in the legs (Gaied et al. 2006) such that the uniformly exerted pressure does not consider the individual patient's needs in that compression socks generally increase blood flow through exerting more pressure near the ankles and feet (Fujii et al. 2017). This issue associated with one sock design for all patients is best solved by

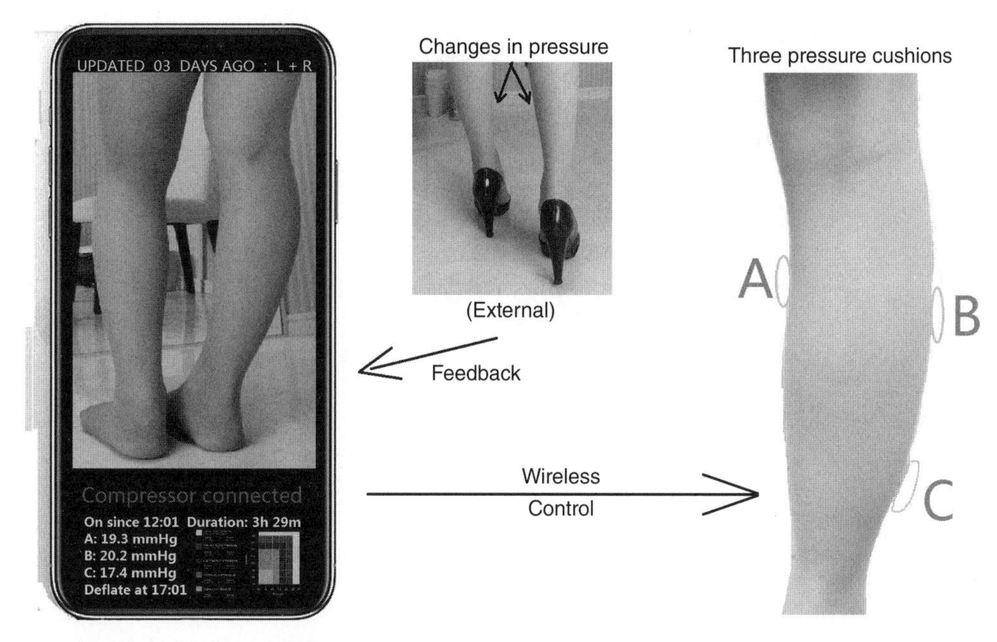

Figure 10.4 Support for varicose veins relief through dynamic inflation of air cushions instead of traditional compression socks.

exerting different amounts of pressure according to the condition of individual veins. Another feature of smart technology here is that appropriate adjustments can be made when oral medication, such as pycnogenol, is used in conjunction with the support system (Nafisi and Maibach 2017).

Using a reference image on a smartphone to determine patches that require relief, this support system dynamically adjusts the pressure between 17 and 22 mmHg at different locations mapped across the leg (Baxi et al. 2019). The pressure will be reduced as leg swelling improves. Since pressure is exerted by small padded actuators, unlike compression socks (which wrap the entire foot and leg), this system is designed so that most parts remain uncovered to encourage better ventilation, thus minimizing skin irritation and dents. It also reduces any impacts of peripheral neuropathy through the careful placement of actuators. Like prescription compression stockings, the system should be fitted by a specialist to ensure that the delicate electronics are not damaged and the sensors are correctly placed on the skin.

These are a couple of examples of how to develop new treatment methodologies through applying assistive and smart technologies in unconventional ways. As technologies make devices smaller and smarter, patients will certainly benefit from better pharmaceutical, preventive, as well as therapeutic solutions over the next few years. We shall continue our discussion by going back to the more traditional approach of using pills, but in a smart package.

10.3.2 Smart Pills

A smart pill is commonly a wireless, ingestible capsule that carries out various measurements within a patient's body. Some have an integrated camera that works in conjunction with a low-power light source to capture images along the digestive system for gastrointestinal diagnosis (McCaffrey et al. 2008). Virtually all smart pills have a built-in wireless transmitter so that

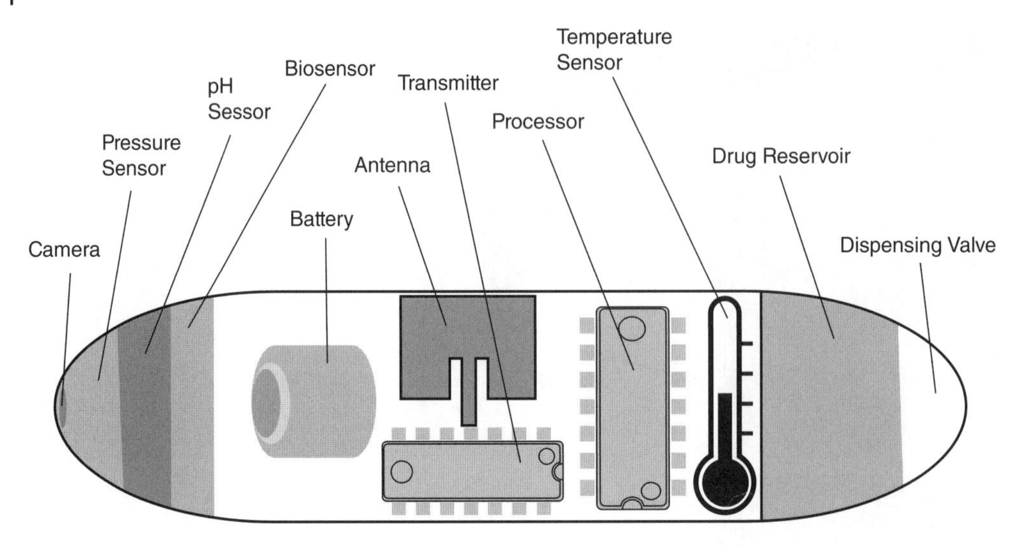

Figure 10.5 Dual function smart pill for health data acquisition and drug delivery.

the captured data can be sent for diagnosis and analysis. Note, incidentally, that smart *pills* have electronics to perform various functions and are completely different from smart *drugs* prescribed for the purpose of enhancing cognitive functions (Stoeber and Hotham 2016). Smart pills are connected devices that form an important part in telemedicine for both diagnosis and treatment.

As the outer casing of smart pill is ingestible, it is not normally intended to carry any drugs – although it is actually possible to deliver drugs at a specific location (Goffredo et al. 2016). One of the major advantages of local drug delivery is to enhance the effectiveness per dosage, yet the timing of releasing the drug requires careful control (Zarekar et al. 2017). A smart pill therefore acts as a tiny device for health data acquisition as well as drug delivery. Figure 10.5 shows the key components within a smart pill that carry out these two major functions. It contains pH, pressure, and temperature sensors as well as a complementary metal–oxide–semiconductor (CMOS) light sensor, wireless transmitter, a light-emitting diode (LED) light source and a power source. A receiver outside the patient's body is needed to pick up the transmitted data from the pill.

More recent developments in smart pills includes one that carries capecitabine, which is a cancer chemotherapy for regular intake (Staines 2019). In this example, a signal is triggered when the pH sensor detects that the pill has reached the patient's stomach. A receiver patch is affixed to the patient's skin for the tracking of drug adherence. The same mechanism can be used for tackling diseases other than cancer by using a different chip that is programmed for its intended purpose, such as wireless capsule endoscopy (WCE) in optometry (Chuquimia et al. 2018).

Smart and assistive technologies can also help identify suitable candidates for clinical trials, such that algorithms can be developed to optimize the distribution for trial subject groups through pattern recognition (Leyens et al. 2017). The use of prognostics, which we discuss in Section 10.4, will provide an advanced warning mechanism for a clinical trial that can facilitate early intervention so that more conclusive results can be derived for biomarker discovery.

10.4 Prognostics in Telemedicine

Telemedicine is certainly an important core technology for healthcare service delivery. Evolving technologies make data communication faster, safer, and more economical. Unlike in consumer electronics, where a malfunction will most likely cause no more than inconvenience, with perhaps losing unsaved work when our home computer freezes, in the case of telemedicine any failure can have far more serious consequences, such as delay in treatment or even avoidable death. Ensuring maximum reliability is therefore consideration of the utmost importance in telemedicine.

Reliability is the most important aspect of any system, as we learned in Section 5.4. Indeed, an unreliable system would be useless no matter what it is capable of when functioning, and so now we should look at how reliability can be optimized.

The word "prognostics" usually refers to forecasts of what may happen based on signs or symptoms. This implies prognostics can predict what may happen to a system so that reliability can be assured, for example when calibration or preventive maintenance has to be carried out before the system fails. Prognostics is defined in *wiki* as:

> "Prognostics" is an engineering discipline focused on predicting the time at which a component will no longer perform a particular function. Lack of performance is most often component failure. The predicted time becomes then the "remaining useful life" (RUL). The science of prognostics is based on the analysis of failure modes, detection of early signs of wear and aging, and fault conditions. These signs are then correlated with a damage propagation model. Potential uses for prognostics is in condition-based maintenance. The discipline that links studies of failure mechanisms to system lifecycle management is often referred to as "prognostics and health management" (PHM), sometimes also "system health management" (SHM).

Note, incidentally, that the word "health" here refers to the system's health status rather than human health, which this text is about. In essence, what we would like to accomplished is by deploying prognostics and health management (PHM) techniques to optimize the health of medical systems so that these systems can in turn optimize human health. Based on this definition, PHM can be used for condition-based maintenance for any system taking into consideration any degradation during its operational life. Indeed, PHM is a proven technology that is widely used in many consumer electronics products (Fong and Li 2011). Of course, medical devices are made up of electronic components, and the main difference between those for consumer electronics and medical systems are mainly reliability and precision, since the impact of a failure would be far less on the former than the latter. PHM ensures the reliability of electronic components and devices, electronics packaging, product reliability, and systems risk assessment (Lau and Fong 2011). Proper prognostic health management can ensure hardware reliability.

10.4.1 Smart Network Management in Telemedicine

"Network outage," usually the main cause of telemedicine system failure, refers to the problem where the wireless link is temporarily disrupted for any reason, including intentional activity such as system maintenance or upgrade. The weakest link of the entire telemedicine system lies with the network transport section, which, depending on the type of wireless network used, can span several kilometers within a city to thousands of kilometers across continents. As discussed earlier in Section 2.4, a number of factors can cause severe disruption along the signal propagating path.

Network breakdown is usually due to stochastic link failures (Fong et al. 2012), where statistical modeling can describe its occurrence due to certain events. Prognostics techniques will require information about network data traffic to be collected and analyzed in order to ensure maximum reliability and availability. It uses data transmission performance of the wireless network to detect potential and future problems. In wireless telemedicine systems, most problems are caused by either wireless link or hardware failure. Prognostics enables link outage prediction through statistical modeling as well as to maintain optimal balance between reliability and performance. With condition-based monitoring, the network health can be maintained by adjusting a number of parameters in response to any performance degradation. For example, adaptive power control and data throughput can be dynamically adjusted according to network condition. Prognostics may also entail the use of different modulation schemes. Although QPSK is very robust with a relatively long range offered when compared to higher order modulations. More spectrum may be necessary, particularly in drier areas where less rain is recorded.

Rain is usually the most influential factor in the reliability of outdoor wireless communications. As such, an adequate link margin must be allocated to combat the effects of rain-induced attenuation (Shogen et al. 2016). Selection of an appropriate carrier frequency, primarily determined by licensing, will provide a tradeoff between bandwidth and range. Generally, frequencies of no more than 10 GHz will be much less affected by rainfall while having a channel of narrower bandwidth. Hub placement is also an important consideration to ensure maximum network reliability, and infrastructure cost and coverage will decrease with increased hub spacing, thus there is an economic trade off. This also leads to the issue of selecting the optimal point-to-multipoint (PMP) antenna patterns (Chou and Su 2017). Condition-based network monitoring also allows control of sector-to-sector interference with frequency diversity, and spatial diversity enables high frequency reuse. This would eliminate the requirement for media access control (MAC), which would save overheads for improved bandwidth efficiency. The statistical information obtained can be used for computing the adequate margin to ensure network reliability irrespective of any change in the operating environment.

Having looked at how PHM can monitor the condition of various parts of a telemedicine system, let us take a look at how PHM can be implemented. PHM relies on computational modeling of a known data set (Fong and Li 2011). Relevant data can be collected during the normal operation of the telemedicine system. For example, information about the data in transit can be used to construct a statistical model that describes the network status. Any abnormally long packet delay or excessive data packet loss may indicate a network congestion or node failure. This kind of problems can be easily diagnosed using PHM techniques. In some systems, PHM can be implemented with diagnostic built-in test circuitry installed. Other implementation options include software/firmware systems for fault identification and isolation that incorporate error detection and correction circuits, self-checking, and self-verification circuits. These circuits can be small pre-calibrated cells that fit into small biosensors. The task that they all share in common is to collect operational data to monitor any performance degradation. In addition to operational reliability, PHM models and tools can also optimize maintenance planning and assess return on investment (ROI). Figure 10.6 summarizes the process of PHM implementation.

Statistics about a network's "health" (its condition) is usually collected by a network management system (NMS). The NMS is usually a piece of software installed on a computer that monitors the network condition and predicts a network outage when performance deteriorates. Figure 10.7

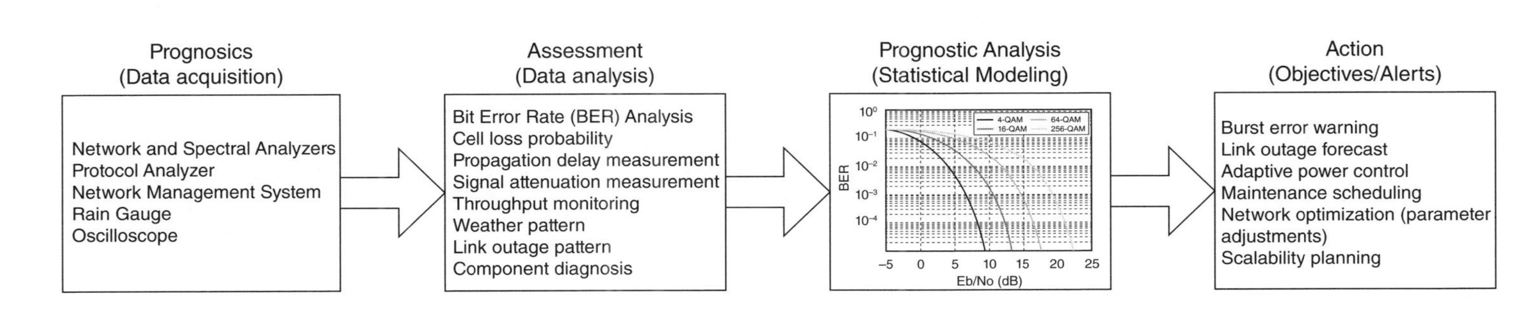

Figure 10.6 Prognostics framework.

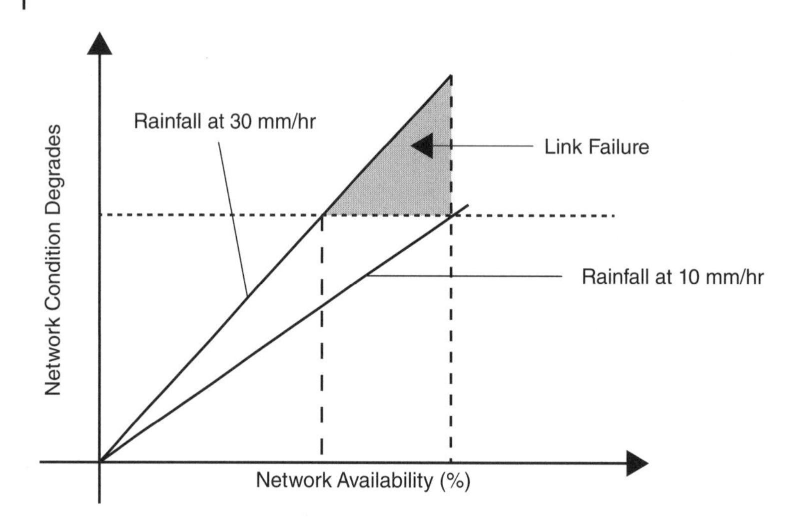

Figure 10.7 Network failure model.

shows a scenario where a link failure can be expected when the rain becomes heavier. Heavier rain causes more signal attenuation and hence reduces link availability. The link condition is continuously monitored based on information about data transfer so that certain network parameters can be adjusted in order to ensure data transmission reliability as the network condition degrades. Some networks do not have a direct link between the transmitter and receiver, and so data transmission must go through certain nodes or repeaters. When the network degrades, certain paths along the network may be temporarily disconnected from the overall network to avoid network disruption. When a node fails (as shown in Figure 10.8, which depicts part of a network with multiple nodes), each data packet can travel through any path along a combination of nodes across the network. When a link within the network fails, data packets can be re-routed based on known information about the network condition and the location of the failed node. Packets that experience abnormal delay or loss that have gone through a certain route would indicate that the route concerned is no longer reliable and hence no more packets should be routed through there. The lost packets may need to be re-transmitted via other routes.

Data-driven prognostics techniques monitor network health through an analysis of various network parameters. These include data loss, packet delay, latency, BER, and E_b/N_0 (the energy per bit to noise power spectral density ratio), which tell us how well the network is performing. NMS or protocol analyzer, usually a piece of software package that is installed on a network computer console, provides such information about the health of the entire network. Typically, an NMS or protocol analyzer will generate a list of information related to packets that travel across the network. An NMS also proactively detects abnormality, such as that shown in Figure 10.8 with a link outage somewhere across the network. Data packets can be automatically diverted to the bottom path that does not exhibit any known problem.

Fault detection is often carried out when certain network parameters fall below a certain predetermined threshold level. Further diagnostics can attempt to fix a problem, depending on its nature. For example, more system fade margin can be assigned to areas where heavy rain can severely affect a wireless link.

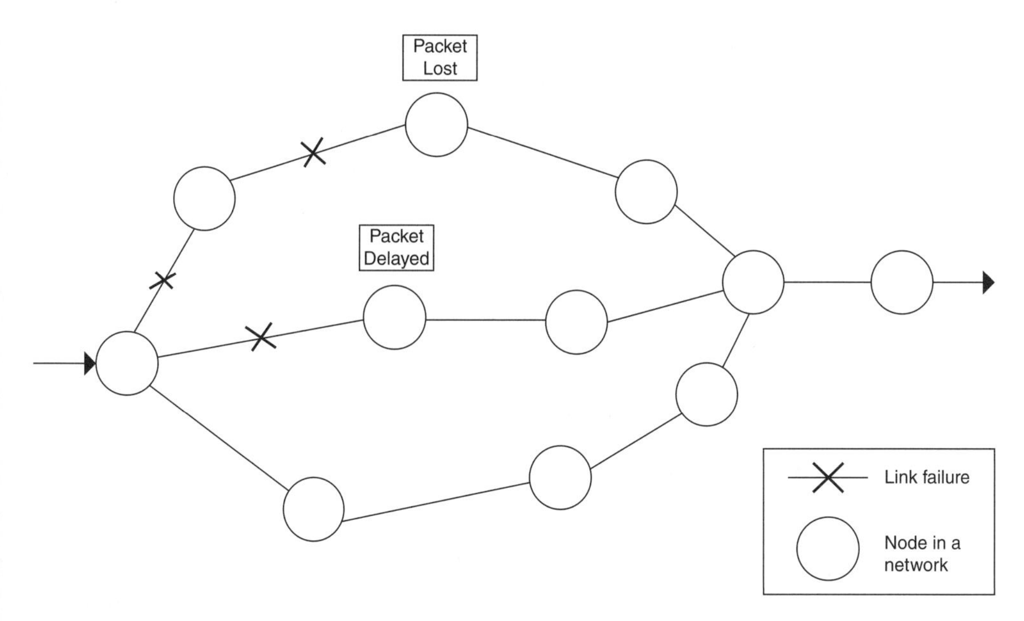

Figure 10.8 Network breakdown with re-routing.

10.4.2 Self-calibration

Many medical devices degrade over time that in turn require calibration, preventive maintenance, as well as scheduled repair (WHO 2017). Such processes are particularly problematic when implanted devices have to be extracted from a patient (Bayrak and Çopur 2017). To minimize the need for removing the implanted device from the patient, it can be programmed to continuously monitor its performance against certain preset baselines using algorithms that fuse sensor data, discriminate transient or intermittent failures, correlate faults with relevant system events and mode changes, and predict failures in advance based on actual operating conditions (Fong et al. 2018). Prior to installing an implantable device on a patient, fault identification and prognostics algorithms that utilize a mixture of advanced symbolic time-series analysis, optimal feature selection, and physics-of-failure analysis to predict system degradation can provide important information on how often the device needs to be calibrated (Vališ et al. 2016). One of the important steps for determining when a device should be calibrated is to discriminate transient and false alarms from actual failures, so that we can understand when the device no longer performs as it is designed to, while ensuring that unnecessary calibration is not carried out under situations like temporary loss of data due to transmission error or when the patient undertakes certain activities that cause abnormal readings. Calibration should therefore be carried out with algorithms that recognize degradation patterns to forecast failures and ensure that the device is properly calibrated prior to an actual failure.

Anomaly detection is an important part of determining when a device should be calibrated. While the classification task focuses on minimizing the misclassification of various discriminating failure states, the anomaly detection task focuses on minimizing the time-delay of detecting performance degradation that may suggest the need for calibration. Prognostic algorithms can be run within the device that take the time-dependent data and extract the key characteristics of parameters that could impact the accuracy of a data set being captured by the device. Such characteristics can be

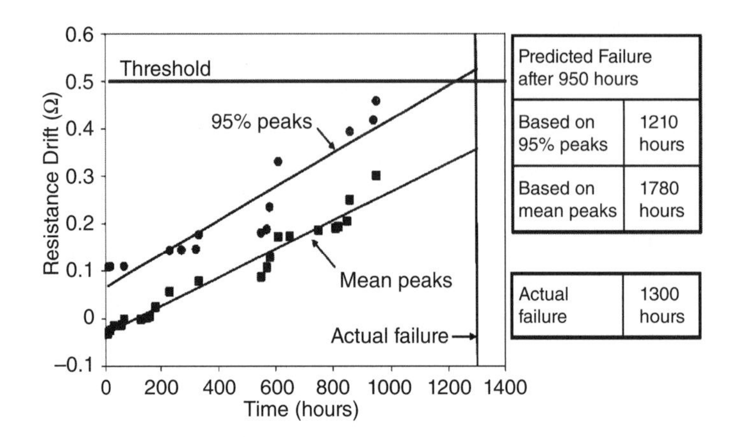

Figure 10.9 Performance degradation to initiate self-calibration prior to an actual failure.

fitted into a degradation model that is predetermined in a controlled laboratory experiment to estimate the performance degradation as the device is being operated over time. It is then possible to program the device so that it automatically executes self-calibration when the actual performance falls below a preset baseline, which indicates that readings taken are no longer trustworthy relative to design specification. This can be illustrated by the failure identification and prognostic-bounding through monitoring performance degradation, as shown in Figure 10.9, where an example of the drift of resistance value that affects the measurement is monitored as a baseline. It shows that, by considering different metrics (95% peaks or mean peaks in this particular example), the device can be programmed to monitor predicted failure at different confidence levels, which can be used to trigger a self-calibration prior to failure.

10.5 Clothing Technology in Telehealth

As with smart home technology, artificial and computational intelligence can also be embedded in clothing for various tasks, ranging from lost person tracking to professional sports training (Mann et al. 1999). Smart clothing involves far more than self-heating and glowing textiles (Grancarić et al. 2018). Traditionally, smart clothing has been deployed in specific areas. For example, space suits used by astronauts are dense with miniaturized electronics, which is also increasingly possible for telemedicine applications as electronics components becomes smaller, cheaper, and more structurally flexible. Smart clothing technology can trace a lost person by embedding a radio frequency identification (RFID) transponder chip (Foroughi et al. 2016). Similar to that used in airports and postal systems for item tracking, RFID tags are used for older people who are disoriented or having lapses of memory (Akmandor and Jha 2017). Such personal identification can also serve as an alternative to door locking so that keyless entry can be supported. Conversely, such technology can be used to restrict individual freedom. The locking of doors and similar measures can be used to control access in restricted areas or for child safety.

Smart clothing has been used for monitoring bodily fluids, in rehabilitation, and chronic disease monitoring for over a decade (Coyle and Diamond 2016). The underlying mechanism of smart textile functions as an active device. Many smart textiles are equipped with embedded electronics that can store and manipulate data, display information, input data, and communicate with the outside world. Some can offer active protection in much the same way as air bags in motor vehicles, for example detecting the presence of hazardous chemicals in the air, rapidly deploying a protective filtering mask, changing color according to the environment for camouflage, and projecting an image of the scene behind the wearer for perceived invisibility. As we discuss in Section 9.1, the type of battery used plays a substantial role in the device's size and weight, minimizing power consumption and making smart clothing thinner and lighter.

10.5.1 Self-powered Devices

The size and weight of a battery and how much charge it holds are very much dependent on its charge density, which falls within the field of material science (Berg et al. 2015). Most wearable devices currently on the market are powered by batteries just like consumer electronics, such as smartphones and watches. Some can actually generate power from the wearer's movement in much the same way as the winding mechanism of an automatic wristwatch, as shown in Figure 10.10. Its operating principle is quite simple. The eccentric weight of the rotor that turns on a pivot caused by movements of the user's wrist causes the rotor to pivot back and forth on its shaft, which is attached to a ratcheted winding mechanism. The motion of the wearer's arm is thereby translated into the circular motion of the rotor and hence, through a series of gears, the mainspring is wound automatically by the natural motion of the wearer's wrist. Embedded electronics in clothes can be powered by such mechanisms so that they can operate once worn.

Although the power supply may not be as bulky as a battery, electronic components are usually rigid and bulky, which contradicts the fact that clothes are made to be as soft and light as possible.

Figure 10.10 Automatic winding movement mechanism.

The wearer's comfort therefore becomes an important design issue. Another important design consideration is the clothes' washability, where designers aim to find something that can be washed just like ordinary clothes made of fabric and possibly with plastic buttons. With these basic requirements understood, let us take a look at a case study with an "intelligent wristband" that continually monitors the blood glucose level of a type 1 diabetes patient.

10.5.2 Noninvasive Glucose Monitoring Wristband: A Case Study

A fabric wristband that consists of a light source, a photo sensor, a timer, and a Bluetooth transmitter is shown in Figure 10.11. The electronic components are embedded in the fabric wristband. The controller illustrated in Figure 10.12 drives the infrared light beam and photo sensor pair that measures the blood sugar level, where a certain percentage of the infrared beam is absorbed by the blood depending on the sugar content. The amount of the beam that is reflected therefore represents the amount of sugar present. The controller is also responsible for capturing and forwarding the reading through a wireless link with a patch antenna. The captured data is stored and forwarded to a Java-enabled mobile phone via a Bluetooth link. The mobile phone serves as a console for storing and analyzing the measurement data. It can also be connected to the user's family doctor where appropriate. Any abnormality detected can therefore alert the physician for any necessary follow-up action.

Calibration prior to use is an essential step to ensure reliability. This involves testing on a subject with a known blood glucose level. A blood glucose laboratory test will be performed for the purpose of setting a reference value to calibrate the reading obtained from measuring the amount of infrared light absorption, since the total amount of light absorption also accounts for skin and

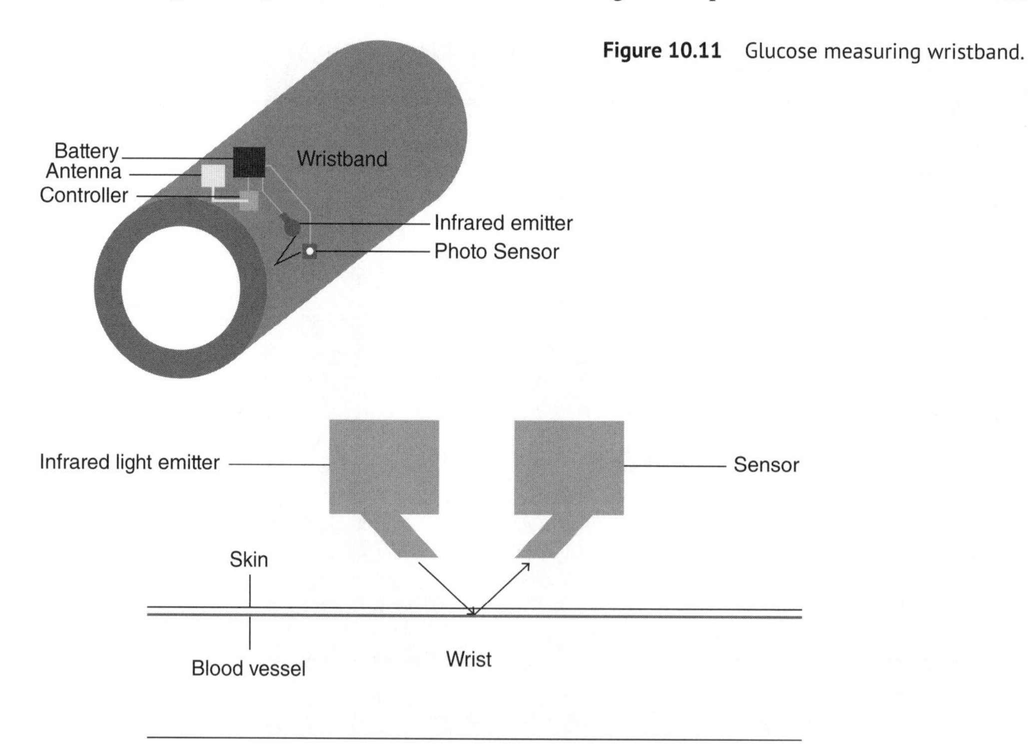

Figure 10.11 Glucose measuring wristband.

Figure 10.12 Noninvasive optical glucose measurement.

tissue absorption. This will normally be accomplished by fasting blood glucose (FBG) level measurement prior to the first use. While calibration guarantees measurement accuracy for a certain period thereafter, prognostics and system health management can be effective in deducing the deviation from expected precision and any impact on measurement due to changes in environmental parameters, such as ambient temperature, humidity, shock, and skin condition.

In this chapter, we have looked at a number of recent examples where smart and assistive technologies have changed the way preventive care as well as treatment can be supported through telemedicine integration that generates a number of novel solutions. Coupling miniaturization with faster and more reliable wireless communication technology will see more of these innovative solutions being applied in different sectors of the healthcare and medical industry in the foreseeable future as the possibilities of what can be done are seemingly endless.

References

Åqvist, J., Medina, C., and Samuelsson, J.E. (1994). A new method for predicting binding affinity in computer-aided drug design. *Protein Engineering, Design and Selection* 7 (3): 385–391.

Akmandor, A.O. and Jha, N.K. (2017). Smart health care: an edge-side computing perspective. *IEEE Consumer Electronics Magazine* 7 (1): 29–37.

Bauer, I.E., Gálvez, J.F., Hamilton, J.E. et al. (2016). Lifestyle interventions targeting dietary habits and exercise in bipolar disorder: a systematic review. *Journal of Psychiatric Research* 74: 1–7.

Baxi, O., Yeranosian, M., Lin, A. et al. (2019). Orthotic management of neuropathic and dysvascular feet. In: *Atlas of Orthoses and Assistive Devices* (eds. J.B. Webster and D. Murphy), 268–276. Elsevier Health Sciences.

Bayrak, T. and Çopur, F.Ö. (2017). Evaluation of the unique device identification system and an approach for medical device tracking. *Health Policy and Technology* 6 (2): 234–241.

Berg, E.J., Villevieille, C., Streich, D. et al. (2015). Rechargeable batteries: grasping for the limits of chemistry. *Journal of the Electrochemical Society* 162 (14): A2468–A2475.

Cheek, P., Nikpour, L., and Nowlin, H.D. (2005). Aging well with smart technology. *Nursing Administration Quarterly* 29 (4): 329–338.

Chen, D., Yang, J., Malkin, R., and Wactlar, H.D. (2007). Detecting social interactions of the elderly in a nursing home environment. *ACM Transactions on Multimedia Computing, Communications, and Applications* 3 (1): 1–22.

Chou, H.T. and Su, H.J. (2017). Dual-band hybrid antenna structure with spatial diversity for DTV and WLAN applications. *IEEE Transactions on Antennas and Propagation* 65 (9): 4850–4853.

Chuquimia, O., Pinna, A., Dray, X., and Granado, B. (2018). Smart vision chip for colon exploration. 13ème Colloque du GDR SoC/SiP 2018, Paris, June 2018. https://hal.archives-ouvertes.fr/hal-02089846/document (accessed 20 January 2020).

Coyle, S. and Diamond, D. (2016). Medical applications of smart textiles. In: *Advances in Smart Medical Textiles* (ed. L. van Langenhove), 215–237. Woodhead Publishing.

Demiris, G., Rantz, M.J., Aud, M.A. et al. (2004). Older adults' attitudes towards and perceptions of 'smart home' technologies: a pilot study. *Medical Informatics and the Internet in Medicine* 29 (2): 87–94.

Duch, W., Swaminathan, K., and Meller, J. (2007). Artificial intelligence approaches for rational drug design and discovery. *Current Pharmaceutical Design* 13 (14): 1497–1508.

Fong, B. and Li, C.K. (2011). Methods for assessing product reliability: looking for enhancements by adopting condition-based monitoring. *IEEE Consumer Electronics Magazine* 1 (1): 43–48.

Fong, B., Ansari, N., and Fong, A.C.M. (2012). Prognostics and health management for wireless telemedicine networks. *IEEE Wireless Communications* 19 (5): 83–89.

Fong, A.C.M., Fong, B., and Hong, G. (2018). Short-range tracking using smart clothing sensors: a case study of using low power wireless sensors for patients tracking in a nursing home setting. In: *2018 IEEE 3rd International Conference on Communication and Information Systems (ICCIS)*, 169–172. IEEE.

Foroughi, J., Mitew, T., Ogunbona, P. et al. (2016). Smart fabrics and networked clothing: recent developments in CNT-based fibers and their continual refinement. *IEEE Consumer Electronics Magazine* 5 (4): 105–111.

Fujii, N., Nikawa, T., Tsuji, B. et al. (2017). Wearing graduated compression stockings augments cutaneous vasodilation in heat-stressed resting humans. *European Journal of Applied Physiology* 117 (5): 921–929.

Gaied, I., Drapier, S., and Lun, B. (2006). Experimental assessment and analytical 2D predictions of the stocking pressures induced on a model leg by medical compressive stockings. *Journal of Biomechanics* 39 (16): 3017–3025.

Garge, G.K., Balakrishna, C., and Datta, S.K. (2017). Consumer health care: current trends in consumer health monitoring. *IEEE Consumer Electronics Magazine* 7 (1): 38–46.

Giagulli, V.A., Carbone, M.D., Ramunni, M.I. et al. (2015). Adding liraglutide to lifestyle changes, metformin and testosterone therapy boosts erectile function in diabetic obese men with overt hypogonadism. *Andrology* 3 (6): 1094–1103.

Goffredo, R., Pecora, A., Maiolo, L. et al. (2016). A swallowable smart pill for local drug delivery. *Journal of Microelectromechanical Systems* 25 (2): 362–370.

Grancarić, A.M., Jerković, I., Koncar, V. et al. (2018). Conductive polymers for smart textile applications. *Journal of Industrial Textiles* 48 (3): 612–642.

Grossi, F., Bianchi, V., Matrella, G. et al. (2008, January). An assistive home automation and monitoring system. In: *2008 Digest of Technical Papers-International Conference on Consumer Electronics*, 1–2. IEEE.

Kranz, M., Schmidt, A., Rusu, R.B. et al. (2007). Sensing technologies and the player-middleware for context-awareness in kitchen environments. In: *2007 Fourth International Conference on Networked Sensing Systems*, 179–186. IEEE.

Lademann, J., Richter, H., Schaefer, U.F. et al. (2006). Hair follicles: a long-term reservoir for drug delivery. *Skin Pharmacology and Physiology* 19 (4): 232–236.

Lademann, J., Richter, H., Teichmann, A. et al. (2007). Nanoparticles: an efficient carrier for drug delivery into the hair follicles. *European Journal of Pharmaceutics and Biopharmaceutics* 66 (2): 159–164.

Lau, D. and Fong, B. (2011). Prognostics and health management. *Microelectronics Reliability* 2 (51): 253–254.

Leyens, L., Reumann, M., Malats, N., and Brand, A. (2017). Use of big data for drug development and for public and personal health and care. *Genetic Epidemiology* 41 (1): 51–60.

Mann, W.C., Ottenbacher, K.J., Fraas, L. et al. (1999). Effectiveness of assistive technology and environmental interventions in maintaining independence and reducing home care costs for the frail elderly: a randomized controlled trial. *Archives of Family Medicine* 8: 210–217.

McCaffrey, C., Chevalerias, O., O'Mathuna, C., and Twomey, K. (2008). Swallowable-capsule technology. *IEEE Pervasive Computing* 7 (1): 23–29.

Nafisi, S. and Maibach, H.I. (2017). Nanotechnology in cosmetics. In: *Cosmetic Science and Technology: Theoretical Principles and Applications*, vol. 337 (eds. K. Sakamoto, R. Lochhead, H. Maibach and Y. Yamashita). Elsevier.

Olin, J.W., White, C.J., Armstrong, E.J. et al. (2016). Peripheral artery disease: evolving role of exercise, medical therapy, and endovascular options. *Journal of the American College of Cardiology* 67 (11): 1338–1357.

Palazzi, C.E., Stievano, N., and Roccetti, M. (2009). A smart access point solution for heterogeneous flows. In: *2009 International Conference on Ultra Modern Telecommunications & Workshops*, 1–7. IEEE.

Rialle, V., Duchene, F., Noury, N. et al. (2002). Health "smart" home: information technology for patients at home. *Telemedicine Journal and E-Health* 8 (4): 395–409.

Sasdelli, A.S., Barbanti, F.A., and Marchesini, G. (2016). How much fat does one need to eat to get a fatty liver? A dietary view of NAFLD. In: *Human Nutrition from the Gastroenterologist's Perspective* (eds. E. Grossi, F. Pace and R. Stockbrugger), 109–122. Springer.

Schneider, G. and Baringhaus, K.H. (2008). *Molecular Design: Concepts and Applications*. Wiley.

Shogen, K., Kamei, M., Nakazawa, S., and Tanaka, S. (2016). Impact of interference on 12GHz band broadcasting satellite services in terms of increase rate of outage time caused by rain attenuation. *IEICE Transactions on Communications* 99 (10): 2121–2127.

Sialvera, T.E., Papadopoulou, A., Efstathiou, S.P. et al. (2018). Structured advice provided by a dietitian increases adherence of consumers to diet and lifestyle changes and lowers blood low-density lipoprotein (LDL)-cholesterol: the increasing adherence of consumers to diet and lifestyle changes to lower (LDL) cholesterol (ACT) randomised controlled trial. *Journal of Human Nutrition and Dietetics* 31 (2): 197–208.

Staines, R. (2019). US patients receive Proteus' digital chemotherapy pill. *Pharmaphorum* (22 January). https://pharmaphorum.com/news/us-cancer-patients-trial-proteus-digital-chemotherapy-pill/ (accessed 20 January 2020).

Stoeber, J. and Hotham, S. (2016). Perfectionism and attitudes toward cognitive enhancers ("smart drugs"). *Personality and Individual Differences* 88: 170–174.

Vališ, D., Žák, L., Pokora, O., and Lánský, P. (2016). Perspective analysis outcomes of selected tribodiagnostic data used as input for condition based maintenance. *Reliability Engineering & System Safety* 145: 231–242.

Vilar-Gomez, E., Martinez-Perez, Y., Calzadilla-Bertot, L. et al. (2015). Weight loss through lifestyle modification significantly reduces features of nonalcoholic steatohepatitis. *Gastroenterology* 149 (2): 367–378.

Wachter R.M. and Chair M.D. (2016). Making IT work: harnessing the power of health information technology to improve care in England: report of the National Advisory Group on Health Information Technology in England. https://assets.publishing.service.gov.uk/government/uploads/system/uploads/attachment_data/file/550866/Wachter_Review_Accessible.pdf (accessed 20 January 2020).

Williams, J.B. (2018). Give someone a bell: telephones. In: *The Electric Century*, 126–135. Springer.

World Health Organization (2017). *Global Atlas of Medical Devices*. WHO.

Yang, G., He, S., Shi, Z., and Chen, J. (2017). Promoting cooperation by the social incentive mechanism in mobile crowdsensing. *IEEE Communications Magazine* 55 (3): 86–92.

Zarekar, N.S., Lingayat, V.J., and Pande, V.V. (2017). Nanogel as a novel platform for smart drug delivery system. *Nanoscience and Nanotechnology* 4 (1): 25–31.

Zartler, E.R. and Shapiro, M. (eds.) (2008). *Fragment-Based Drug Discovery: A Practical Approach*. Wiley.

Zhu, H., Chang, A.S., Kalawsky, R.S. et al. (2017). Review of state-of-the-art wireless technologies and applications in smart cities. In: *IECON 2017-43rd Annual Conference of the IEEE Industrial Electronics Society*, 6187–6192. IEEE.

11

Future Trends in Healthcare Technology

Over the past 10 chapters, we have discussed how telemedicine and related technologies assist various aspects of healthcare and medical practices. Most of the technologies have a long proven history. Data communications evolved from the first telephone by Graham Bell and Elisha Gray formed the basis of many modern telemedicine systems deployed throughout the world today (Bashshur and Shannon 2009). Technology advances and innovative breakthroughs are opening a wide range of possibilities in medical and healthcare services.

Telemedicine is certainly an important core technology for healthcare service delivery. Evolving technologies make data communication faster, safer, and more economical. More people can now benefit from digital health than ever before with a wide range of healthcare and medical services made readily available through telemedicine. In this final concluding chapter, we look at areas where telemedicine technology will gain more interest over the next decade. Unlike the previous chapters, where we look at proven digital health technologies in tackling real life challenges, the objective of this concluding chapter is to discover how emerging technologies can be reliably deployed in a number of selected scenarios in the foreseeable future in the telemedicine context, based on currently available technologies.

11.1 Haptic Sensing for Practitioners

Haptic is a tactile feedback technology. Haptic sensing reacts to the user's hand movement including forces, vibrations, and motions. This provides a user interface that utilizes the sense of touch. Note, incidentally, that tactile sensors that sense the amount of force exerted on the interface in a somatosensory system, of the peripheral nervous system (PNS) and the central nervous system (CNS), are not considered haptic sensors. Control based on haptic sensing would be limited by friction, precision, and lack of stimulus for the sense of touch (Smith 1997). To understand more about haptics, we look at a control glove in Figure 11.1, where a number of sensors are found around the palm. These sensors are driven by real-time algorithms that interpret the hand's movement and drive an actuator controller. Here, the haptic mechanism conveys forces from the user's hand to the remote actuators. On the remote side, there are actuators and control circuitry that act upon the user's hand movement. What needs to be considered includes actuator size, precision, resolution, frequency, latency requirements, power consumption, and cost of operation. The controller can be either closed-loop or open-loop. In closed-loop control, the controller reads sensor movement from the received signal and then computes and executes the haptic output forces in real time

Telemedicine Technologies: Information Technologies in Medicine and Digital Health, Second Edition.
Bernard Fong, A.C.M. Fong, and C.K. Li.
© 2020 John Wiley & Sons Ltd. Published 2020 by John Wiley & Sons Ltd.

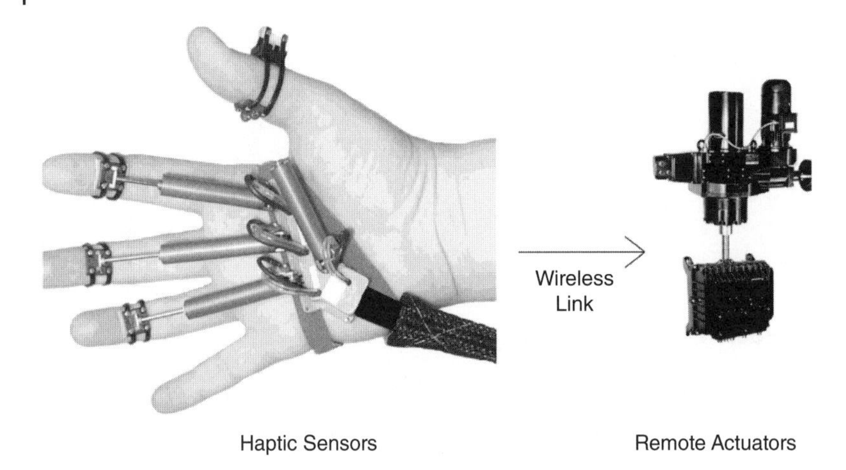

Figure 11.1 Haptic glove.

based on the sensor movement. In open-loop control, a triggering event will activate the controller to compute and relate the haptic output signal to the actuator in real time.

One obvious application of haptics in telemedicine is remote robotic surgery (Okamura 2004). One major advantage of using haptics for robotic surgery is for medical schools when students can practice on simulators so that there will be no risk of injuring a patient while learning to operate (Shen et al. 2008). Another important application is operations where visualization is not possible. The amount of force being exerted on an organ or tissue can be very delicately controlled and regulated by actuators. With robotic telesurgery set up, a patient can be prepared by local hospital staff and operated by expert surgeons who can perform from anywhere without traveling (Davies 2000).

Protecting veterinary surgeons is also one major advantage of haptics in surgery. Dog bite injuries are risks that can be eradicated if the surgeon does not make direct contact with the dog being operated on (Cameron et al. 2017). In fact, it is even possible to operate on a caged animal, by placing a robot inside the cage. Surgeons can easily perform the operation from outside the cage.

As shown in Figure 11.1, any system that supports robotic telesurgery requires a communication link that links the surgeon's hands to remote actuators, or a simulator in the case of surgical practicing. This link needs to deliver the information that replicates exactly the hand's movement in real-time. In addition to control information for the actuators, a camera that acts as the surgeon's "remote eye" also needs to transmit crystal clear images in real time back to the surgeon (English et al. 2005). The reliability and bandwidth requirements must be addressed. Here, we need to remember that video compression cannot be relied on because any loss in image detail can have extremely serious consequences during an operation. The challenges of optimizing bandwidth efficiency and reliability still remain.

11.2 Business Intelligence in Healthcare Prevention

Business intelligence (BI) supports healthcare service providers by providing insights that enhance operational efficiency and resource utilization thereby driving down costs, improve patient care and safety, as well as facilitate decision making (Kao et al. 2016). Common to BI applications in other industry sectors, healthcare service providers can identify highly profitable sectors like medical tourism and aged care while diverting resources away from underutilized services.

11.2.1 Medical Tourism

Medical tourism is a popular business for many to take advantage of cheaper surgical operations, particularly in developing countries where the cost can be only a fraction of that charged in the patient's home country (Kasemsap 2018). This is a substantial business since the cost saving is often much more than the total cost involved in traveling. The business incentives demand a comprehensive support plan for medical tourism.

To provide support for patients traveling across different countries, similar to that of cellular phone roaming, the service provider can utilize BI to link service providers and manufacturers of various healthcare devices and systems together, following through the process all the way from laboratory research to product launch and after-sales support. This is to ensure that interoperability, reliability assurance, and effectiveness are optimized for all parties involved (Baars and Ereth 2016).

Owing to the popularity of speedy and economical treatment, medical tourism has become increasingly popular in recent years. The ideal medical tourism conditions are those that allow the patient to electively travel for services or obtain medical devices and drugs from any country. As regulatory requirements vary across different countries, evaluation and assessments related to worldwide practices are all important steps. BI helps to gather the information necessary for providing solutions to traveling patients and to provide guidance to companies and organization on the following key topics (Laursen and Thorlund 2016):

- Risk assessment of providing and supporting medical tourism services.
- Logistics for medical tourism.
- Partnership and joint venture establishment.
- Provide a framework for automated computing of administrative and insurance cost.
- Maintaining a feedback scoring system for service rating.

11.2.2 Cyber Physical Systems

A cyber physical systems (CPS) are monitored and controlled by computer-based algorithms thereby reduces the need of electronic hardware while enhancing the utilization efficiency of resources (Mosterman and Zander 2016). Medical CPS combines telemedicine and sensing for low-cost patient monitoring through computational solutions that reduces the use of expensive medical devices (Gu et al. 2015). CPS uses software-based management tools for monitoring of various vital signs such as core body temperature, electroencephalogram (EEG), electrocardiogram (ECG), and heart and respiratory rates (Dey et al. 2018).

CPS is normally considered as a network of interacting computational algorithms using physical components instead of standalone devices. This yields a significant reduction in implementation cost when fewer devices are needed within the health monitoring system. A vital link of CPS is an Internet of things (IoT) platform that connects patients to physical objects such as actuators, sensors, smartphones, as well as embedded systems to computing elements that in turn provide data streams to different healthcare applications (Ochoa et al. 2017). The main feature of a medical CPS is to manage and coordinate communication between various components within a telemedicine network such as health monitors, smart home, and smart city infrastructure, which provide essential hardware connections for data exchange between a wide variety of healthcare and medical applications (Xia et al. 2015).

In practice, real-time remote monitoring and diagnosis entails the use of ambient sensors to collect biosignals from sources such as ECG, glucose level, etc. All these sensors are connected via the telemedicine network with an accurate location estimation via global positioning systems, as shown in Figure 11.2. The interference between wearable biosensors operating in the same network

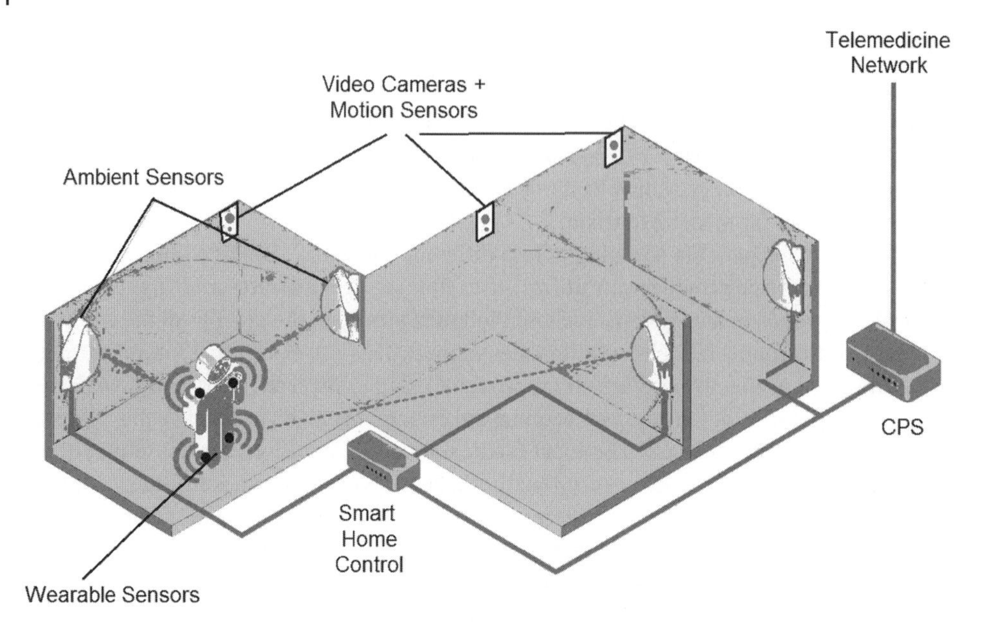

Figure 11.2 Remote health monitoring CPS within a patient's smart home environment.

within the frequency band can cause a number of issues, such as reduction in received signal strength and loss of data packets, which can lead to a significant degradation in signal detection performance and consequently affect health anomaly detection as well as diagnosis.

As we recall from Chapter 2 that additive noise has a substantial impact on any communication system, the impact of noise like Gaussian and salt-and-pepper noise is particularly problematic when analyzing medical images (Tu et al. 2017). The use of video cameras in CPS for assistive care for patients with developmental coordination disorder (DCD) will require noise. Removal of noise often requires the use of high-resolution (high bit depth) analog-to-digital converter (ADC) for restoration of sparse biosignals (Tsakalakis 2015). While we discuss more about telemedicine for tackling DCD in Section 11.4.2, we concentrate our current discussion on the CPS system itself. One of the challenges in implementing wearable sensors is to develop ultra-low-power transceivers that address the issues associated with sparsity in wearable biosensor networks (Wang et al. 2016). This is particularly problematic when using ADCs running high-speed sampling to process high-frequency signals in wireless networks like Wi-Fi and Bluetooth.

To grasp a better understanding about smart CPS based on machine learning, we continue our discussion by referring to medical CPS that supports remote patient monitoring, as shown in Figure 11.3, where health information about a patient is collected from various locations as the patient travels around within a particular city where the patient is covered by a smart city infrastructure. Health data of different types is collected from a selection of ambient and wearable sensors. In this CPS, there are two separate layers, namely the cyberworld on the top and the physical world at the bottom. The boundary between the two layers is a collection of IoT services that provide both physical and logical connection between different entities within the CPS so that data communication and analysis can take place.

BI is applied within the CPS as a means of promoting sales of medication and healthcare-related products based on individual patients' buying behavior, health condition, and any symptoms that the patient exhibits. The analysis of medication sales and health advice sought can provide important insights into disease outbreaks, as we discuss next.

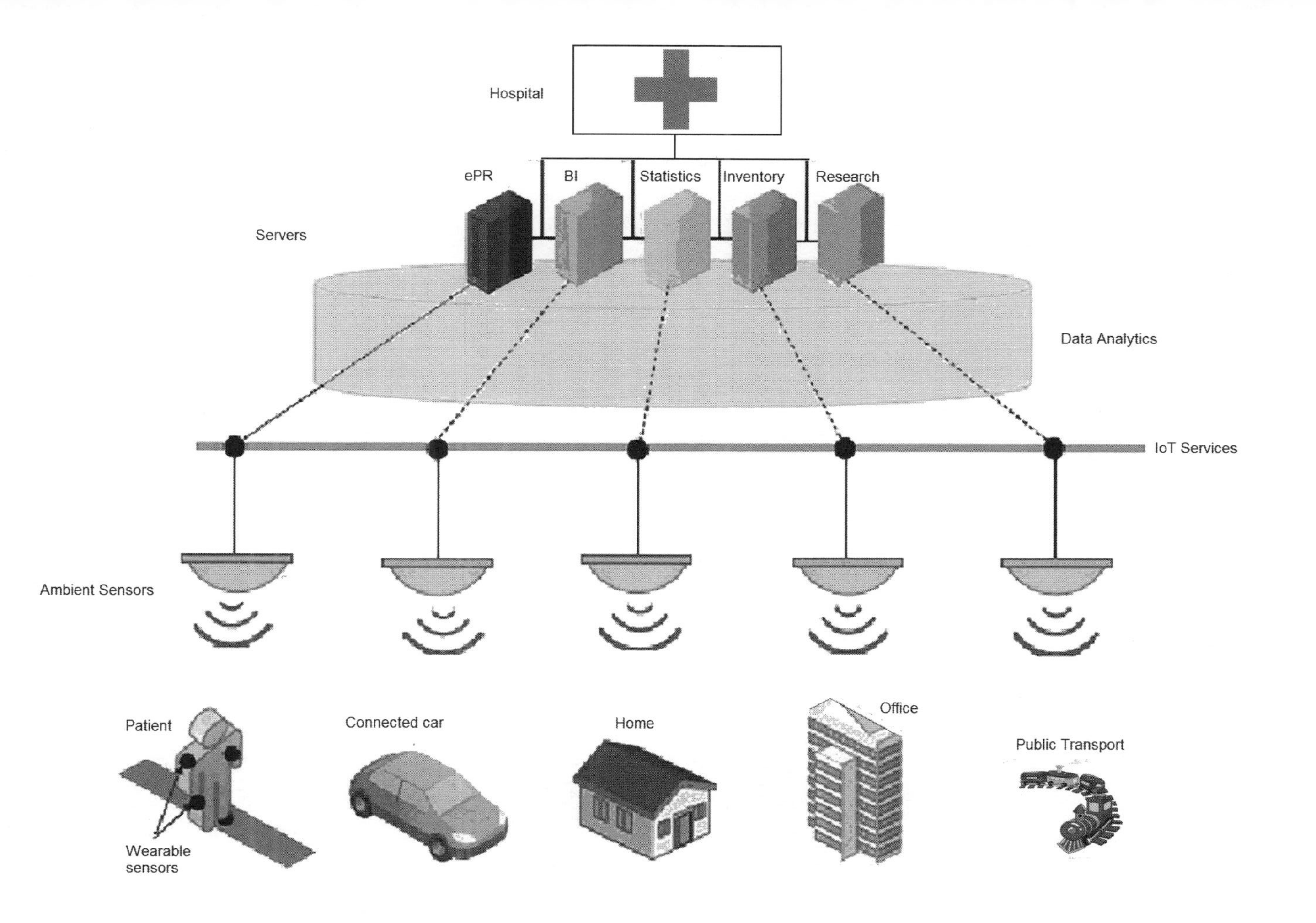

Figure 11.3 Remote patient monitoring CPS.

11.3 Cross-border Care: A Case Study of Syndromic Surveillance

The objective of syndromic surveillance is to detect disease outbreaks in a more efficient way than by using conventional reporting of confirmed cases (Lewis et al. 2016). Syndromic surveillance can be used for both natural as well as manmade intentional bio-attack (Roberts and Elbe 2017). Early detection can be accomplished by analyzing data from a variety of sources that are related to the outbreak, such as influenza-like illness (ILI) symptoms, over-the-counter (OTC) pharmacy sales, health advisory telephone calls, clinic visits, and various forms of Internet activity like forums and keyword search (Gardy and Loman 2018). By monitoring various disease-related indicators, an outbreak can be detected early so that countermeasures can be effectively and proactively implemented (Noufaily et al. 2019).

As observed in the outbreaks of SARS (severe acute respiratory syndrome), swine flu, and other infectious diseases over the past two decades, the vast volume of human traffic between international borders has posed immense challenges to the spread of commutable diseases (Chan et al. 2016), which also extends to the transportation of food and fresh produce (Garner and Kathariou 2016). In order to address the increasing risk of future pandemic, timely outbreak detection and effective disease-spread simulation analysis becomes vitally important to enable health resource management under pandemic outbreaks.

Syndromic surveillance implemented in telemedicine enables analysis of disease data through effective and statistically rigorous surveillance methods and will create a framework for information sharing between government agencies, healthcare service providers, and GPs. The main objective is to facilitate early outbreak detection methods, mitigation strategies, prognostics, and management (Tsui et al. 2008). One of the key features of telemedicine is the ability to connect devices and systems across a vast geographical area. This is particularly important in supporting real-time outbreak and disease surveillance (RODS) in collecting healthcare registration data in real time from hospitals and clinics (Hughes et al. 2019). As a practical example, Thomas et al. (2018) describes a public health bioterrorism surveillance framework for multiple data streams from clinic visits and emergency department visits and combines with OTC sales to detect and characterize outbreaks by disparate incident datasets as a true syndromic surveillance signal by analyzing respiratory syndrome cases during an event, such as the bioterrorism attack in Figure 11.4.

Automated time-series modeling and forecasting algorithms like regressions, autoregressive integrated moving average (ARIMA), and Holt–Winters (HW) exponential smoothing are syndromic surveillance algorithms commonly used for predicting the occurrence of future health events (Faverjon and Berezowski 2018). Additionally, spatiotemporal scan statistics have been used in a prospective setting for advanced detection of disease outbreaks (Mouly et al. 2018).

Spatial scan statistics have also become popular for the evaluation of geographical disease clusters in pediatrics (Burkom 2016), which is discussed further in Section 11.4.1 with a specific example on assisting DCD patients. In summary, telemedicine as shown in Figure 11.5 provides an analytic framework based on likelihood ratio statistics for both spatial surveillance and spatiotemporal surveillance under independent or correlated regions that can identify as well as bias toward areas with greater population, owing to parameter estimation errors by using spatial scan statistics (Lin et al. 2016).

Another concern in syndromic surveillance is that it encounters a multiple testing problem. There are multiple data sources in play, and within each data source there are usually multiple series; and many of these series are further broken down into subseries. Allowing the sharing of information across different locations facilitates faster detection by monitoring disease symptoms and correlated indicators in addition to monitoring confirmed incidents.

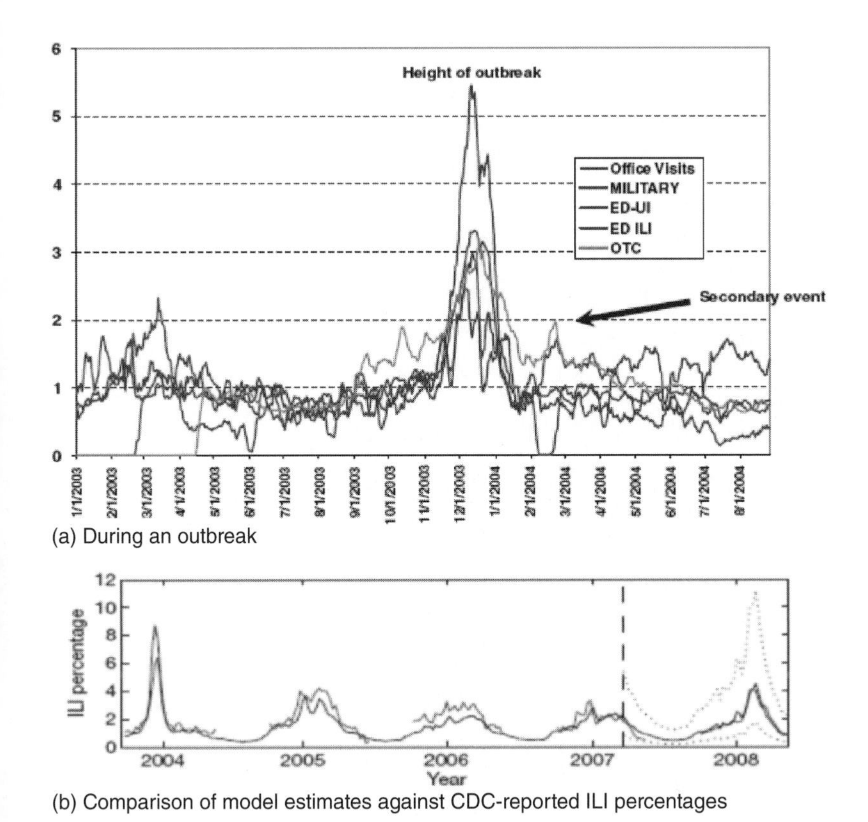

Figure 11.4 Respiratory syndrome data.

In addition to a number of areas already mentioned above, another major area is disease-spread simulation. It is important to quantify infection probabilities for microscale or society-scale models of infectious disease spread. Previous studies on pathogen-laden expiratory aerosols do not investigate infectious source strengths under the effects of particle size dependent dynamics on the removal and dispersion of expiratory aerosols in realistic hospital and clinical environments (Jones and Brosseau 2015). Environmental sensing networks can improve prediction models for infection probabilities in contained environments, such as hospitals and airports, to better understand the transmission mechanisms of infectious diseases. Simulation studies play a significant role in supporting pandemic scenario prediction and enabling the understanding of disease spread, which requires the acquisition and analysis of environmental data from different geographical locations (Charlton et al. 2018).

11.4 5G-based Wireless Telemedicine

Recent advances in wearable sensors make 5G an important infrastructure for remote health monitoring that can be used in different entities within a smart city like homes, buildings, and on the roads (whether for driving safety enhancement or as part of the transportation network). As discussed in Section 11.2.2, CPS enables a smart mobile healthcare service platform, which enables clinicians and caregivers to remotely monitor patients as well as provide advice or feedback to

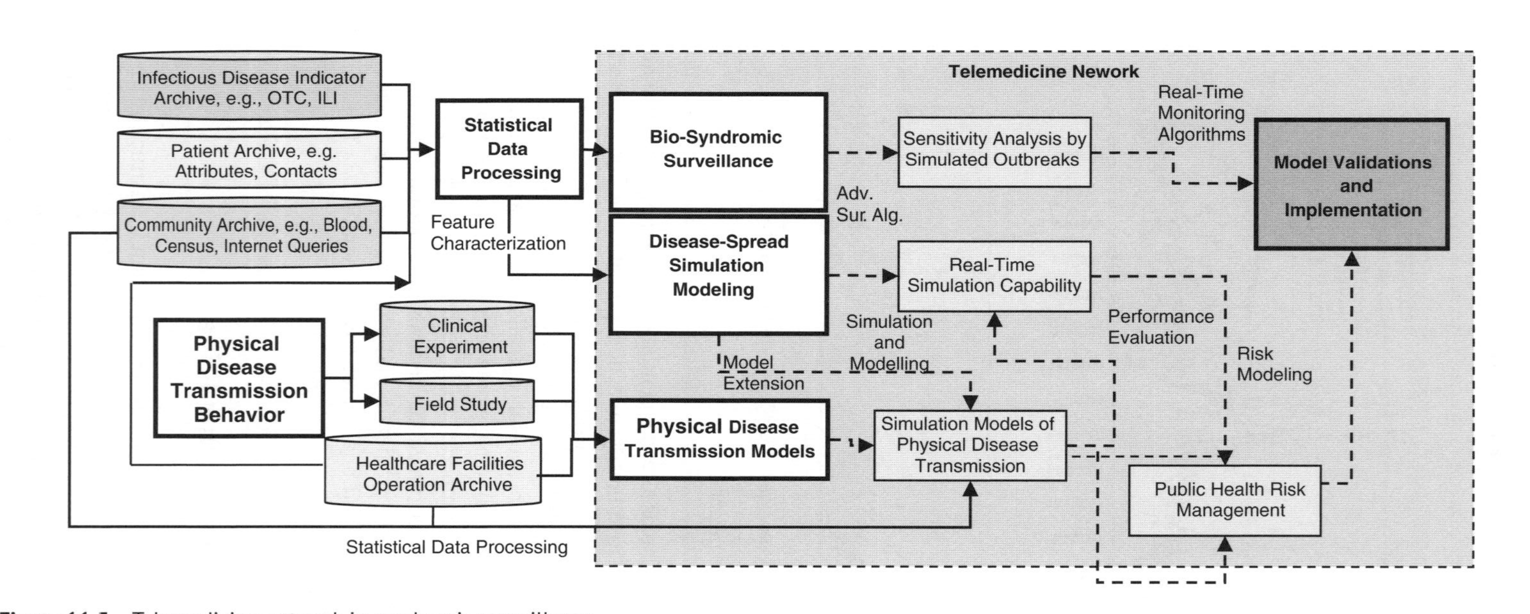

Figure 11.5 Telemedicine network in syndromic surveillance.

help patients maintain optimal health and support during surgical rehabilitation, irrespective of the patient's location. CPS consists of a large number of ambient sensors that collect a vast range of data from both the patient and the surroundings, all connected within a 5G network with adequate bandwidth to support virtually all real-time health applications. Connected sensors and devices allow remote patient monitoring to be carried out at all times where patient data is collected at various locations through wearable and ambient sensors and sent to different servers for analysis. The key point here is the data rate required to support the simultaneous connection of many devices and/or to enhance image clarity with high resolution and angle of view.

11.4.1 5G and IoT to Tackle DCD: A Case Study

Cerebral palsy (CP) and DCD have a substantial impact on the daily activities of children from a very young age (Acharya et al. 2016). These neurodevelopmental dysfunction affecting patients in early childhood can induce activity limitations that are attributed to nonprogressive disturbances (Wong et al. 2010). The motor disorders of these young patients are often associated with disturbances of sensation, cognition, communication, perception, as well as behavior that may require stimulation and assistive care. Analysis of gait using inertial sensors that facilitate DCD diagnosis can be used for testing (Mannini et al. 2017). In addition, video analysis using markers are also found to be effective in analyzing gait in DCD (Wilmut et al. 2017).

In addition to diagnosis, robotic-assistance is also reported to be effective in improving motor skills (Zhou et al. 2016), diagnosis, and assistive care for DCD, and can be supported through a range of sensors and actuators that assist the patient with making any adjustment in enhancing mobility (Lobo et al. 2016). Each of these components within an assistive framework requires data connection where 5G networks are capable of simultaneous support.

11.4.2 Faster Wireless Communications for Supporting Virtual Reality (VR) in Telemedicine

In chemotherapy treatment, the patient's body is constantly injected with a portion of drugs. These chemotherapeutic drugs are inhibiting and kill cancer cells that are entrenched in the body. These powerful drugs also erode normal cells while treating cancer, causing adverse side effects (Brown 2016).

For the long and painful process endured by the patient, the patient is put on VR with wonderful scenery to divert their attention from any discomfort. VR makes everything look realistic so that the patient sits in a closed, blue and white ward, and enjoys the virtual silence while receiving chemotherapy.

This is the most valuable use of VR in combating severe illness. Life is changing rapidly, and technological advances push our lives to new heights every day. The imagination of "the fantasy in the movie" has become a reality, or at least is perceived as such. VR has gradually become a medical tool surgeons and patients may not be aware of it. VR by itself will not cure any disease, but it can bring hope and comfort, to a greater or lesser extent, but they are all precious as far as a patient is concerned.

There is a palliative care unit in the municipal Ashiya Hospital in Hyogo Prefecture in Japan, where the hospital uses VR to let patients receiving palliative care to "go home" (Piazza et al. 2017). Piazza et al. report the case of a patient with pleural mesothelioma. This condition prevented the patient from resisting the bacteria of the outside world and so they could not leave the ward. Under such circumstances, he still issued a request for "One day, wanting to go home, and wanting to visit the former residence again." This was possible through equipping the patient with a VR device. The patient's wife imitated his usual way of life, using a camera to record the scene of the home, she sat on the sofa where her husband (the patient) used to sit, the channel tuned into his favorite golf program, and then came to the bedroom, living room, etc. The camera finally came to the car he had been driving for many years, and his wife struggled to imitate the way he usually drove. In addition to this male patient, more than a dozen other patients have received such VR therapy. Some of them put on their glasses and returned to their hometowns. Someone returned to the place where their wedding was held. Later, the hospital conducted a survey, which concluded that the anxiety level of these patients had been substantially reduced by using VR. Although this may not mean that their symptoms had been alleviated, the result was very significant in helping patients.

For patients who can be assisted through VR, a DCD sufferer can be shown scenes of successfully carrying out a range of difficult tasks. This is to give the patient a boost of confidence. This may have exceeded our imagination in the present way of treating children with DCD, but VR can bring to patients with severe illness and disorder hope, a psychological healing process of making a child with DCD seeing their own potential achievements.

DCD may make swimming extremely difficult, yet an underwater simulation program can let the patient see what it is like under the water, watching the jellyfish pass by and other marine life surrounding them. This is an example of where immersive VR can be perfected for the benefit of patients.

What many people may not be aware of is that VR with ultra-high definition (4K UHD) 360° images require a very high data throughput in the magnitude of tens or even hundreds of MB/s, so that we can bring patients into the illusory world. This puts a substantial demand on the bandwidth of any telemedicine network that supports VR.

In recent years, a project called Walk Again has used VR to help paraplegic patients recover the use of their lower limbs (Donati et al. 2016). According to the results of the publication, the eight patients who participated in the study received some noticeable changes. According to the researchers, they found that when these patients were imagining the walking process, they could not find any signals in their brains, which indicated that the concept of walking had been eliminated in the brains of these patients. In order to regain their athletic ability, the researchers put VR devices on these patients, to let them learn to use their brains to control their roles in the virtual world. By the time the brain regains the concept of walking, the patients can rely on the exoskeleton to try to restore the nerves. A 13-year-old patient has been able to move her legs slightly with this training. Although in the virtual world, they can only get "walking." This is a very bland experience for most people, but for those who have been immobilized for many years in wheelchairs, this is reported to be a very enjoyable experience by many patients.

11.5 The Future of Telemedicine and Information Technology for Everyone: From Newborn to Becoming a Medical Professional all the Way Through to Retirement

One of the major advantages of wireless telemedicine is being able to provide medical services with a high degree of mobility. Advances in wireless communications have made new services possible over recent years. Mobile monitoring yields significant cost savings. These include patients being discharged soon after surgery for home recovery so as to minimize the duration of hospital stay without any adverse effects, utilizing existing wireless home network for monitoring. Also, continual health monitoring can reduce demands for healthcare resources by maintaining optimal health. Other benefits include reduction in health insurance claims and loss in productivity. However, there are different levels of risks incurred when a patient is discharged from a hospital early, depending on the nature of their illness and physical state. Some may be taken care of by family members, while others may require medical attention. Since a vast range of possibilities exist in relation to different scenarios, e.g. the risks associated with a patient after a coronary artery bypass will be very different from that of an acute myocardial infarction patient even though both are cardiac operations.

Telemedicine technology offers healthcare professionals many different ways to monitor patients, for example as mentioned in Section 8.3, applications such as posture sensing for spinal injury or back pain can use accelerometers or video imaging. These technologies can be used in situations such as post-surgery rehabilitation, prevention of effects of backpacks on children, and design for consumer devices, such as massage chairs and calibration for optimal positioning of home audio-visual systems. Movement detection for knee and foot recovery, such as after an ACL (anterior cruciate ligament) operation or a monitoring mechanism for walking or jogging to record parameters such as pace, distance covered, heart rate, and calories burnt are possible using telemedicine applications. These parameters can be analyzed to build a profile on the recovery process. Wireless ECG measurement imposes far less movement restriction as well as reducing risk of infection due to pathogens on ECG telemetry lead wires.

Other services include alternative medicine properties of different herbal medicines, and support of acupressure treatments. Such scalable informatics frameworks can also provide a better understanding of the genetic bases of complex diseases by analyzing vast amounts of data collected in genetic computation and patterns of disease spreading.

Wireless technology is truly something for all ages. In the above example, we have seen telemedicine applications that potentially everyone can utilize. For the remainder of this chapter, we shall walk through telemedicine with a baby girl, Melody, to see how technologies will assist her throughout her life. We shall look at the possibilities that exist. Our aim is not to make wild forecasts on where technology is heading, but we would like to enlighten readers on how current technologies can be pushed forward to assist with various tasks for people of all ages by reminding ourselves of what the text has covered.

When Melody leaves her mother's womb, she will likely be given a radio frequency identification (RFID) tag to wear on her wrist. This is perhaps her first encounter with wireless communication technology. Since most babies look very similar to each other, RFID tags provide a safe and secure way to uniquely identify each newborn baby. Embedded in the tag is information including her mother's name, the date and time of Melody's birth, and any treatment provided as she is being monitored during the first few days of her life at hospital. Without understanding what is happening around her, Melody is already assisted by wireless communication technology that, although already in use elsewhere in very limited areas, was not available to her mother when born some three decades ago.

Melody's parents bring her home a few days after her birth. There is a good chance that her parents have bought her a baby monitor, one that is described in Section 7.3.2 and shown in Figure 11.6. With a video camera and sensors around Melody, her parents can leave her alone in the cot while enjoying some home entertainment in the adjacent room with the assurance that Melody is sleeping well. Pressure sensors ensure that the baby will not turn over while sleeping and alert her parents of any potential risk of rolling over. A microphone lets her parents hear what is happening., speech processing algorithms also analyze Melody's crying and suggest the likely actions needed, for example whether she wants attention or if she is hungry. Her parents can also look at how she is doing without going into the room. This also prevents disrupting Melody's sleep. Last but not least, the ambient environment is fully regulated with smart home technology.

When Melody becomes a little child, she can enjoy regular medical checkups to ensure normal grow with self-diagnosis and testing performed at home. All data will be automatically captured

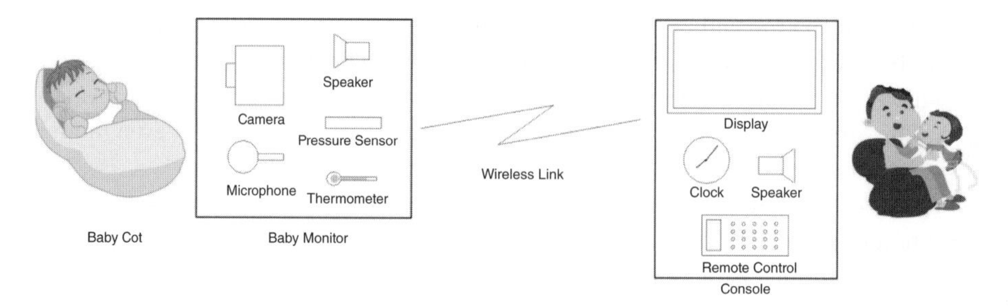

Figure 11.6 Wireless baby monitor.

and updated by linking to her electronic patient record. Parameters such as body mass index (BMI), blood glucose level, and ECG will all be recorded while she undertakes normal activities. She can even see the doctor remotely through video conferencing.

Melody grows and eventually becomes a medical school student. Mobile learning (m-learning) portals engage students to learn at any time, and anywhere, and to encourage truly active learning and teaching. "M-learning" refers to the use of mobile and handheld devices, such as personal digital assistants, mobile phones, laptops, and tablets to facilitate teaching and learning in a much more convenient and efficient way to learn anytime, anywhere. M-learning within other educational contexts often faces significant challenges in terms of technical support and infrastructure (Kukulska-Hulme and Traxler 2005). Problem-solving is recognized as the core competence by the statutory body and the stakeholders in local healthcare organizations (Lennox and Petersen 1998). New learners not exposed to a typical problem-solving environment may be neither aware of their roles required to address patients' problems nor ready to apply the knowledge or skills leading to a solution to meet patients' needs. However, the students may not realize that they are required to prove the mastery of intellectual and psychomotor skills within context-based practices. The process of problem-solving entails high-order thinking, in relation to skills in critical thinking and deriving an optimal solution. With the theoretical benefits learned in the classroom, physicians might adopt the cognitive procedures leading to possible solutions. The attributes and competence of problem-solving to provide safe practice are characterized with vigilance to individual and contextual issues, risk identification and management, error reduction, and research into practical solutions. In fact, clinical problem-solving is a mix of conceptual understanding and cognitive skills. Students may not be aware of the fact that their performance has to do with a range of integrated attributes and skills that involve the ability to integrate and synthesize factual information, theoretical concepts, and procedures. In response to the impetus of developing clinical problem-solving among medical students in an acute care setting, a simulated clinical problem-solving (SCPS) component could be developed in a mobile delivery format. Mobile learning platforms provide an SCPS component which is structured as a self-study element that requires initiative and active participation. Learners can be encouraged to drill their relevant skills in the SCPS with the integral pedagogy on creative thinking, self-directed learning, and experiential learning. So, Melody can get as many chances to practice her surgical skills as she wishes with the aid of haptic sensing surgical simulators, with the knowledge that even if she makes a serious mistake no real damage will be done. Telemedicine and related technology will certainly make learning a lot easier for future medical students.

As a physician, she can remotely track a patient's postsurgical recovery process with a health monitoring system based on existing home wireless networks for analyzing data captured by medical devices, such as an oximeter temporarily installed at home, for transmission to the hospital. The range of posthospitalization checks supported includes medication and nutritional administration, and body temperature and SpO_2 readings. Such information can be analyzed and appended to the patient's medical record. Such systems can help reduce demands on hospital resources as well as travel time for patients and caregivers, which is particularly helpful for older and disabled patients.

Utilizing consumer healthcare technology and network sensors for general health assessment for older and vulnerable patients is another major area that assists doctors. Telemedicine can integrate information technology and the bedside – a scalable informatics framework that will bridge clinical research data and the vast databases arising from basic science research in order to better monitor people's health. A home healthcare system is based on an existing home IEEE 802.11 WLAN (wireless local area network) that also facilitates simultaneous, independent connections between various networking devices such as computers and audiovisual systems. Biosensors can also collect a patient's physiological data, just like in the above example. The system will offer flexibility to support a wide range of healthcare monitoring services. Such patient monitoring systems are integrated into an existing home wireless LAN system that provides a number of network access and device control functions. This ensures minimal intervention is necessary within the patient's home throughout the monitoring process. This home network effectively provides an access point to an external telemedicine system that has a direct connection to the hospital. This means that, by using a public network, hospital staff can perform remote diagnosis and consultation without patients leaving home. Various instruments can be attached to the system, depending on the type of data sought to monitor the patient's progress and response to any sudden change. Its setup is simple and economical with all equipment at the patient's home installed on a temporary basis. Flexible monitoring can be offered to patients by making use of wearable computers for capturing data from the biosensors so that the patient can move freely when being monitored.

Advances in telemedicine technology help Melody through her career into retirement. Routine activities of senior citizens can be supported by a multisensory telecare system as an electronic guard. Older people with special needs such as memory loss and cognitive impairment sufferers can be greatly benefitted by technological advancements in HCI (human–computer interface) and wireless communications. A wearable therapeutic device provides general assistance. Health monitoring, calling for emergency assistance, alerts, and reminders can provide dementia sufferers with peace of mind. Mobility is also an important consideration as the current system is primarily designed for users remaining at home. User-friendliness is an important design consideration since most senior citizens are not familiar with technology. Another major function is to collect information about a user's health conditions, medication, nutritional intake, and fall history. Such clinical information will be analyzed on a regular basis for monitoring purposes. In addition, their clinical information can be connected to and shared with healthcare facilities, for example GPs or hospitals, using a wireless network. This feature is particularly suitable for older adults with cognitive impairments or users who are recovering at home after hospitalization (after hip fracture surgery, for example) who require close surveillance by hospital staff.

Alerts and reminders that assist routine activities such as medication intake, flushing the toilet after use, the safe use of gas stove and fire risk. The system can be design to help an older person with various tasks and daily routine activities with the attachment of appropriate instruments and biosensors. Medication reminders and instructions are automatically generated by embedding drug information on an RFID chip in the bag.

Here, we conclude the chapter by looking at how telemedicine and related technologies can assist a wide range of tasks for a person from birth to retirement. Technological advancements certainly bring countless exciting opportunities for medical science and healthcare that benefit both practitioners and patients.

References

Acharya, K., Pellerite, M., Lagatta, J. et al. (2016). Cerebral palsy, developmental coordination disorder, visual and hearing impairments in infants born preterm. *NeoReviews* 17 (6): e325–e333.

Baars, H. and Ereth, J. (2016). From data warehouses to analytical atoms: the Internet of things as a centrifugal force in business intelligence and analytics. In: *Twenty-Fourth European conference on Information Systems*, 1–18. ECIS.

Bashshur, R.L. and Shannon, G.W. (2009). *History of Telemedicine: Evolution, Context, and Transformation*. New Rochelle, NY: Mary Ann Liebert.

Bloom, D. E., & Canning, D. (2004). Global Demographic Change: Dimensions and Economic Significance. NBER Working Paper No. W10817 (19 October).

Brown, C. (2016). Targeted therapy: an elusive cancer target. *Nature* 537 (7620): S106.

Burkom, H.S. (2016). The role and functional components of statistical alerting methods for biosurveillance. In: *Disease Surveillance: Technological Contributions to Global Health Security* (eds. D.L. Blazes and S.H. Lewis), 55. CRC Press.

Cameron, O., Al-Himdani, S., and Oliver, D.W. (2017). Not a plastic surgeon's best friend: dog bites an increasing burden on UK plastic surgery services. *Journal of Plastic, Reconstructive & Aesthetic Surgery* 70 (4): 556–557.

Chan, J.F., Choi, G.K., Yip, C.C. et al. (2016). Zika fever and congenital Zika syndrome: an unexpected emerging arboviral disease. *Journal of Infection* 72 (5): 507–524.

Charlton, C.L., Babady, E., Ginocchio, C.C. et al. (2018). Practical guidance for clinical microbiology laboratories: viruses causing acute respiratory tract infections. *Clinical Microbiology Reviews* 32 (1): e00042–e00018.

Davies, B. (2000). A review of robotics in surgery. *Proceedings of the Institution of Mechanical Engineers, Part H: Journal of Engineering in Medicine* 214 (1): 129–140.

Dey, N., Ashour, A.S., Shi, F. et al. (2018). Medical cyber-physical systems: a survey. *Journal of Medical Systems* 42 (4): 74.

Donati, A.R., Shokur, S., Morya, E. et al. (2016). Long-term training with a brain-machine interface-based gait protocol induces partial neurological recovery in paraplegic patients. *Scientific Reports* 6: 30383.

English, J., Chang, C.Y., Tardella, N., and Hu, J. (2005). A vision-based surgical tool tracking approach for untethered surgery simulation and training. In: *Medicine Meets Virtual Reality 13: The Magical Next Becomes the Medical Now* (eds. J.D. Westwood, R.S. Haluck, H.M. Hoffman, et al.), 126–132. IOS Press.

Faverjon, C. and Berezowski, J. (2018). Choosing the best algorithm for event detection based on the intended application: a conceptual framework for syndromic surveillance. *Journal of Biomedical Informatics* 85: 126–135.

Gardy, J.L. and Loman, N.J. (2018). Towards a genomics-informed, real-time, global pathogen surveillance system. *Nature Reviews Genetics* 19 (1): 9.

Garner, D. and Kathariou, S. (2016). Fresh produce-associated listeriosis outbreaks, sources of concern, teachable moments, and insights. *Journal of Food Protection* 79 (2): 337–344.

Gu, L., Zeng, D., Guo, S. et al. (2015). Cost efficient resource management in fog computing supported medical cyber-physical system. *IEEE Transactions on Emerging Topics in Computing* 5 (1): 108–119.

Hughes, S.L., Morbey, R.A., Elliot, A.J. et al. (2019). Monitoring telehealth vomiting calls as a potential public health early warning system for seasonal norovirus activity in Ontario, Canada. *Epidemiology & Infection* 147: e112.

Jones, R.M. and Brosseau, L.M. (2015). Aerosol transmission of infectious disease. *Journal of Occupational and Environmental Medicine* 57 (5): 501–508.
Kao, H.Y., Yu, M.C., Masud, M. et al. (2016). Design and evaluation of hospital-based business intelligence system (HBIS): a foundation for design science research methodology. *Computers in Human Behavior* 62: 495–505.
Kasemsap, K. (2018). The role of medical tourism in emerging markets. In: *Medical Tourism: Breakthroughs in Research and Practice*, 211–231. IGI Global.
Kukulska-Hulme, A. and Traxler, J. (2005). *Mobile Learning: A Handbook for Educators and Trainers*. Routledge.
Laursen, G.H. and Thorlund, J. (2016). *Business Analytics for Managers: Taking Business Intelligence Beyond Reporting*. Wiley.
Lennox, A. and Petersen, S. (1998). Development and evaluation of a community based, multiagency course for medical students: descriptive survey. *British Medical Journal* 316 (7131): 596–599.
Lewis, S.H., Burkom, H.S., Babin, S., and Blazes, D.L. (2016). Promising advances in surveillance technology for global health security. In: *Disease Surveillance: Technological Contributions to Global Health Security* (eds. D.L. Blazes and S.H. Lewis), 179. Boca Raton, FL: CRC Press.
Lin, P.S., Kung, Y.H., and Clayton, M. (2016). Spatial scan statistics for detection of multiple clusters with arbitrary shapes. *Biometrics* 72 (4): 1226–1234.
Lobo, M.A., Koshy, J., Hall, M.L. et al. (2016). Playskin lift: development and initial testing of an exoskeletal garment to assist upper extremity mobility and function. *Physical Therapy* 96 (3): 390–399.
Mannini, A., Martinez-Manzanera, O., Lawerman, T.F. et al. (2017). Automatic classification of gait in children with early-onset ataxia or developmental coordination disorder and controls using inertial sensors. *Gait & Posture* 52: 287–292.
Mosterman, P.J. and Zander, J. (2016). Cyber-physical systems challenges: a needs analysis for collaborating embedded software systems. *Software & Systems Modeling* 15 (1): 5–16.
Mouly, D., Goria, S., Mounié, M. et al. (2018). Waterborne disease outbreak detection: a simulation-based study. *International Journal of Environmental Research and Public Health* 15 (7): 1505.
Noufaily, A., Morbey, R.A., Colón-González, F.J. et al. (2019). Comparison of statistical algorithms for daily Syndromic surveillance aberration detection. *Bioinformatics* 35 (17): 3110–3118.
Ochoa, S.F., Fortino, G., and Di Fatta, G. (2017). Cyber-physical systems, Internet of things and Big Data. *Future Generation Computer Systems* 75: 82–84.
Okamura, A.M. (2004). Methods for haptic feedback in teleoperated robot-assisted surgery. *Industrial Robot: An International Journal* 31 (6): 499–508.
Piazza, M., Casiraghi, L., Skok, M., and Rosa, D. (2017). Massage therapy and quality of life of cancer patient in palliative care: literature review. *Annals of Oncology* 28 (suppl 6): vi.
Roberts, S.L. and Elbe, S. (2017). Catching the flu: syndromic surveillance, algorithmic governmentality and global health security. *Security Dialogue* 48 (1): 46–62.
Shen, X., Zhou, J., Hamam, A. et al. (2008). Haptic-enabled telementoring surgery simulation. *IEEE Multimedia* 15 (1): 64–76.
Smith, C.M. (1997). Human factors in haptic interfaces. *XRDS: ACM Crossroads* 3 (3): 14–16.
Thomas, M.J., Yoon, P.W., Collins, J.M. et al. (2018). Evaluation of syndromic surveillance systems in 6 US state and local health departments. *Journal of Public Health Management and Practice* 24 (3): 235.
Tsakalakis, M. (2015). Design of a novel low-cost, portable, 3D ultrasound system with extended imaging capabilities for point-of-care applications. PhD dissertation. Wright State University.
Tsui, K.L., Chiu, W., Gierlich, P. et al. (2008). A review of healthcare, public health, and syndromic surveillance. *Quality Engineering* 20 (4): 435–450.

Tu, P., Bai, Y., Xu, W. et al. (2017). Digital volume correlation in an environment with intensive salt-and-pepper noise and strong monotonic nonlinear distortion of light intensity. *Optica Applicata* 47 (2): 209–223.

Wang, Y., Doleschel, S., Wunderlich, R., and Heinen, S. (2016). Evaluation of digital compressed sensing for real-time wireless ECG system with Bluetooth low energy. *Journal of Medical Systems* 40 (7): 170.

Wilmut, K., Gentle, J., and Barnett, A.L. (2017). Gait symmetry in individuals with and without developmental coordination disorder. *Research in Developmental Disabilities* 60: 107–114.

Wong, M., Fong, K., and Yiu, L. (2010). Profile of children with cerebral palsy at the child assessment service. *Child Assessment Service Epidemiology and Research Bulletin* 7: 53–56.

Xia, F., Wang, L., Zhang, D. et al. (2015). An adaptive MAC protocol for real-time and reliable communications in medical cyber-physical systems. *Telecommunication Systems* 58 (2): 125–138.

Zhou, S.H., Fong, J., Crocher, V. et al. (2016). Learning control in robot-assisted rehabilitation of motor skills: a review. *Journal of Control and Decision* 3 (1): 19–43.

Index

Telemedicine Technologies: Information Technologies in Medicine and Digital Health, Second Edition.
Bernard Fong, A.C.M. Fong, and C.K. Li.
© 2020 John Wiley & Sons Ltd. Published 2020 by John Wiley & Sons Ltd.